The Meaning Of Suffering

Recent titles in Contributions in Philosophy

The New Image of the Person: The Theory and Practice of Clinical Philosophy
Peter Koestenbaum

Panorama of Evil: Insights from the Behavioral Sciences
Leonard W. Doob

Alienation: From the Past to the Future
Ignace Feuerlicht

The Philosopher's World Model
Archie J. Bahm

Persons: A Comparative Account of the Six Possible Theories
F. F. Centore

Science, Animals, and Evolution: Reflections of Some Unrealized Potentials of Biology and Medicine
Catherine Roberts

The Philosophy of Human Rights: International Perspectives
Alan S. Rosenbaum, editor

Estrangement: Marx's Conception of Human Nature and the Division of Labor
Isidor Walliman

The Concept of Ideology and Political Analysis: A Critical Examination of Its Usage by Marx, Lenin, and Mannheim
Walter Carlsnaes

Soviet Marxism and Nuclear War: An International Debate
John Somerville, editor

Understanding: A Phenomenological-Pragmatic Analysis
G. B. Madison

Guilt and Gratitude: A Study of the Origins of Contemporary Conscience
Joseph Anthony Amato II

Nature, Human Nature, and Society
Paul Heyer

The Meaning Of Suffering

AN INTERPRETATION OF HUMAN EXISTENCE FROM THE VIEWPOINT OF TIME

Adrian C. Moulyn

CONTRIBUTIONS IN PHILOSOPHY, NUMBER 22

GREENWOOD PRESS

WESTPORT, CONNECTICUT • LONDON, ENGLAND

Library of Congress Cataloging in Publication Data

Moulyn, Adrian C.
 The meaning of suffering.

 (Contributions in philosophy, ISSN 0084-926X ; no. 22)
 Bibliography: p.
 Includes index.
 1. Suffering—Psychological aspects. 2. Time—Psychological aspects. 3. Space and time—Psychological aspects. I. Title. II. Series.
 BF789.S8M68   128   82-6171
 ISBN 0-313-22233-9 (lib. bdg.)   AACR2

Copyright © 1982 by Adrian C. Moulyn

All rights reserved. No portion of this book may be reproduced, by any process or technique, without the express written consent of the publisher.

Library of Congress Catalog Card Number: 82-6171
ISBN: 0-313-22233-9
ISSN: 0084-926X

First published in 1982

Greenwood Press
A division of Congressional Information Service, Inc.
88 Post Road West
Westport, Connecticut 06881

Printed in the United States of America

10 9 8 7 6 5 4 3 2 1

Copyright Acknowledgments

Grateful acknowledgment is given for permission to use the following:

Selections from THE PRICE by Arthur Miller. Copyright © 1968 by Arthur Miller and Ingeborg M. Miller, Trustee. Reprinted by permission of Viking Penguin, Inc., New York.

Excerpts from Papers Presented to the First Congress on the Meaning of Human Suffering, Notre Dame, Indiana. Reprinted by permission of Stauros International, West New York, New Jersey.

Figure 1 from Roy R. Grinker and Paul C. Bucy, *Neurology*, 4th revised edition, 1949, Figure 137, p. 358. Courtesy of Charles C. Thomas, Publisher, Springfield, Illinois.

Excerpts from Roy Schafer, "The Psychiatrist's Empathy," Lecture to the Westchester Psychoanalytic Association, Westchester Division of the New York Hospital, White Plains, New York, May 4, 1981.

Excerpts from Jurrit Bergsma, *Somatopsychologie: Op Zoek naar een Psychosociale Dimensie van de Geneeskunde* [Somatopsychology: Searching for Psychosocial Dimensions in Medicine]. Reprinted by permission of Uitgeversmaatschappij De Tijdstroom, Lochem, Kaatsheuvel, The Netherlands, 1975.

NOS DOULEURS NOUS RENDENT DIGNES DE NOTRE
ÉPANOUISSEMENT

(Our suffering makes us worthy of our development)
—Georges Duhamel

Contents

Figures	ix
Series Foreword	xi
Acknowledgments	xiii
1. Overview	3
2. Objective Time-Space and Subjective Time-Space	14
3. The Mental Triad and the Precious Present	33
4. Extrospection and Introspection	47
5. Causality and Goal-Striving Activities	63
6. Determinism versus Decision Making	89
7. The Mind-Body Relationship	145
8. Faults, Dichotomies, and Fractures	187
9. Suffering: The Core of Human Existence	222
10. Healing through Suffering in Four Modes	244
11. The Meaning of Suffering	289
Glossary	315
Bibliography	317
Index	327

Figures

1. Sensory-Motor Homunculus — 158
2. From Antithesis to Synthesis — 165
3. Communication from Concrete to Abstract — 176

Series Foreword

Inevitably all persons demonstrate to themselves the fact that they are human by seeking somehow to comprehend the two themes of this volume. They know that they have suffered and that they will continue to suffer as a result of physical and psychic pain, frustration, loneliness, and the anticipation of death. They are intrigued or dismayed by the experience of time: they look backwards and forwards, and they dwell in the present; they are convinced that on occasion time drags, or it may whip by. No one, consequently, is able or wishes to escape the problems raised by suffering and time. Each individual ponders over the eternal, cosmic questions associated with them. And every generation and every society offers its own replies.

Dr. Moulyn patiently, gently, and thoroughly explores two questions: What is the "intrinsic value" of suffering and "how do we human beings live in time?" His great achievement is to suggest again and again that the two questions are ontologically and psychologically related, inextricably so. Without confronting ourselves with suffering and time simultaneously, we are left with deep feelings of inadequacy which can produce philosophical and even psychiatric damage. The link between the two is not always easy to grasp, and therefore the exposition must move into more than a score of disciplines in order to produce the insight the author would provide.

Perforce his tapestry is large and diverse. He comes to the reader as a physician and psychiatrist, and hence he reports case histories out of his own experience. But he knows very well that suffering and time are not the exclusive property of the medical profession. He is a brave humanist in the great European, faustian tradition. He reaches into the realms of human endeavor related to suffering and time, beginning with architecture and art, with extended and pungent references to the great philosophers from Aristotle to Kant, and ending with both the natural and social sciences. He insists that we also peer at space, animal behavior, laughter, determinism, Freud, religion, and especially marriage. No, definitely no, this is not a random assortment of writers and topics: all, or almost all of the references are essential if our guide is to convince us and particularly himself that suffering and time are essential to our existence.

Just as playing a Beethoven piano sonata sometimes offers Dr. Moulyn profound relaxation and stimulation beyond the music itself, so each of us finds temporary or enduring salvation in his own way according to his nature, his interests, and his problems. Here in this book are suggestions that may possibly assist us, *provided* we look not for facile recipes but for provoking challenges. Still I do wonder whether suffering can ever be evaluated as other than "an undesirable evil," although I agree immediately that unavoidable suffering supplies the motivational incentive to achieve the values Dr. Moulyn and I treasure. What do you think? Good luck, I say to you, bon voyage through these pages, I believe you will find the destination deeply satisfying.

Leonard W. Doob

Acknowledgments

I am deeply obliged to Leonard Doob who suggested that I put my thoughts in the format of a book and to James Sabin for his confidence in my work right from the start of our contacts.

I have had helpful discussions with President Edmund Pellegrino, Catholic University of America, about religious topics, with J. T. Fraser about the time problem, and with Maurice Natanson in reference to causality.

Joseph Grassi has discussed René Descartes with me; James Long guided me to references about time in Saint Thomas Aquinas's *Summa Theologica*; King Dykman cleared up some controversial points in reference to Aristotle.

I have received encouragement from Stuart Spicker, Herbert Spirer and from the Reverend Joel Baehr by their sustained interest.

Bern Dibner has graciously allowed me to use his extensive library on the history of science and Chip Chillington of the Burndy Library has alerted me to recent sources on Kepler.

It has been a pleasure to work together with Margaret Brezicki and the staff of Greenwood Press.

The Meaning Of Suffering

1
Overview

It may seem rather strange for a physician to interpret suffering as a constructive element in human life. The goal of medicine is to alleviate suffering and if possible to eliminate it. Common sense too regards suffering as a blemish on the human existence and without intrinsic value. We ought not to suffer, instead we humans should strive for happiness as the highest good, the *summum bonum*. Aristotle says: "Happiness is that which everyone seeks. No one, if asked whether he wants happiness, would say: 'No, I want misery instead'."[1]

I shall develop an interpretation which regards suffering neither as an undesirable evil, nor as a state of mind which we should actively strive to attain. The intrinsic value of suffering lies in its propensity to clear a path toward mature, inalienable happiness. Going through the crucible of suffering, we reach a state of bliss.

I shall present a dualistic view of man as a rational as well as an emotional being. Following an interdisciplinary approach, one gets a panoramic view of man's dualistic nature, held together by the concepts of objective time and subjective time. In order to clarify the structure of these two contrasting concepts of time, I use the concepts of the Mental Triad, which describes the dynamics of human mentation, and the Precious Present, which is the appropriate form of time within which the mental triad operates. This view of human existence does not intend to uncover any new facts, but only to correlate well-known facts in an unconventional structuring.

Dualism as such is an unsatisfactory interpretation of human life, and, therefore, one strives to synthesize the contrasting aspects of our existence. These attempts to unify what is divided, generate inner tensions of an intellectual as well as of an emotional nature, insofar as we are thinking and feeling beings. Trying to synthesize the contrasting elements of our human nature as well as the discrepancies we notice in the world which surrounds us, creates feelings of tension, apprehension, anxiety, panic, dread, and suffering. Suffering has value and meaning because it is the inevitable concomitant within our search for the resolution of dualism.

It would seem as if we evade suffering at all cost. This much is certain, as Aristotle has said, that we do not go out of our way to make ourselves miserable by choice, at a time when we are happy. Few people evaluate suffering as a higher good than being happy. But this reasoning overlooks the fact that we have no choice in this matter. As Martin Heidegger points out, we have been 'thrown-into-the-world.' Although Heidegger does not mention suffering explicitly, his concept of *being-thrown* fits into the tenor of this book. He emphasizes the limitations of our actions, and he assigns a central role to emotionality which he conveys with the term *mood* (*Befindlichkeit*) as a fundamental constituent of our existence. Moreover he interrelates being-thrown with our shifting feeling tones. He describes the tensions between the several aspects of our living in the world in his complex language formations. "The expression, being-thrown, ought to indicate the facticity of surrender.... Being, of the character of being-there (*Dasein*), is its "There" in such a way that it, whether expressly or not, finds itself in its being-thrown."[2]

The tension between being-thrown-into-the-world, our feelings, and our activities of going-toward and going-away-from, constitute part of the background of our suffering, which I will examine from various angles. We have been thrown into a terrestrial existence which is rent by many blemishes, imperfections, contradictions, unresolved problems, dichotomies, and fractures. These blemishes and fractures are inherent in the human condition and they are the source of the many levels of suffering. Inasmuch as we have to accept our being thrown into the world, we also have to accept periods during which we suffer. But instead of devaluating suffering and rejecting it as something that should not befall us, we ought to make an effort to become aware of the meaning of suffering. Its value and its meaning are that it heals the blemishes and the fractures in our problem-ridden existence.

A physician is in an unusual position to observe and to evaluate the meaning of suffering. It is his task to soften, subdue, or if possible to eliminate suffering. He treats people who are suffering from illnesses that weaken or threaten or destroy their body or impede or distort mentation. He deals with destructive and meaningless suffering. But once the patients have recovered from their illnesses, doctors, especially surgeons, usually lose contact with them, and they lose the opportunity to observe how their former patients enter into a phase of constructive and meaningful suffering. On the other hand, health workers who treat terminally ill patients often take a different stance toward the suffering of their charges. If they are able not only to sympathize with their patients but also to empathize and identify with them, then they may be able to show them a way of meaningful suffering during the destruction of their psychophysical holism. If they can work together with the patient toward overcoming the dread of dying and of death, then the patient may be better able to maintain a heroic attitude toward the oncoming end. This difficult task requires that the health worker be consciously in touch with his own dread of his own inevitable death sometime in the future. But since most doctors treat patients during relatively brief periods, they ordinarily devaluate suffering. They view it as a blemish that they are called upon to eradicate, and they lose sight of the other mode of suffering which is constructive and meaningful. A psychoanalyst or a psychotherapist who does long-term intensive psychotherapy can experience directly how some patients are capable of releasing themselves from meaningless suffering, heroically choosing the path of meaningful, constructive suffering.

This outlook upon suffering takes issue with the teachings of many religions that sin is the cause of suffering. We are being punished for sins which we ourselves, our parents, and our progenitors have committed either recently or in a dim and distant past. Our sufferings as a sign of our being sinners make us unworthy of a close relationship with the gods, the godhead, or with God the Father of Christianity. Sin separates us from God, which separation is the most excruciating modality of suffering for the religious believer. We have to suffer in order to become worthy again of a close relationship with God. If we cleanse ourselves of sins, then we will move closer toward God and the yoke of our suffering will become lighter. Religions prescribe various belief systems, rituals, actions, and nonactions (asceticism) to which the sinner must submit in order to become re-ligated with God. If he travels this road, if he obediently performs these duties of repentance and contrition with sincere and fervent religious faith, then he will be reunited with God and he will enter into a state of bliss.

I will try to dissect the hidden meaning of suffering out of the conceptions which human beings have constructed around the painful subject of suffering. This alternative valuation of suffering agrees with the common-sense opinion that no one seeks out suffering for its own sake. But with an enlightened view of suffering one becomes aware that it is a healing agent, a power which brings together the warring strivings within our personality, that it prevents dichotomies from splitting us apart, and that it tames fractures in our existence from breaking us asunder. The meaning of suffering flows from its power to heal the imperfections in the human condition. As an ephemeral result, suffering's healing power bestows the gift of periods of firmly grounded happiness and bliss upon us.

Human existence fluctuates between periods of happiness and periods of suffering. We search for lasting happiness, but we are besieged by imperfections under which we suffer. We immerse ourselves in happy states of mind without questioning because we believe that this is how life ideally ought to be always, or at least most of the time. But when happiness is interrupted, or even more when it is being destroyed, we begin to ask why we should suffer. The question why we suffer has generated many answers of a cosmological, ethical, religious, logical, and philosophical nature. Some think that our psychological development may answer this question: we are not happy as an adult, but we were happy in childhood and we have lost, or we have spoiled, or we have been expelled from this presumably blissful childhood paradise. Whatever the answers, the question why we suffer always starts from the premise that it is against the natural state of affairs, which is happiness. This premise is grounded on the belief that suffering is a sign of evil, that our wrongdoings and our sins are the causes of our sufferings. One concludes that suffering is valueless because it destroys happiness.

In contrast to the pessimistic devaluation of suffering I shall develop the thesis that suffering has intrinsic value and that it is a positive, constructive power in life. Suffering heals fissures, dichotomies, and fractures. In short, suffering has meaning.

In order to develop an enlightened, optimistic thesis about suffering, we must construct a perspective about human existence through the framework of objective time-space and subjective time-space. This spatiotemporal structure must possess counterposed attributes similar to those experienced in daily living. The contrasts between happy and sad periods must fit into the two contrasting forms of time-space to provide a coherent background against which we can project the drama, the comedy, and the dull phases of our existence. We must evaluate and interpret the many imperfections and dichotomies in our human condition against the framework of our dual spatiotemporal structure. On the one hand, the forms of objective time-space and of subjective time-space have enough attributes in common to help build a harmonious image of human life. On the other hand, the two forms of time-space display essential contrasts which we can use as a theater in which the human tragedy of suffering and the happiness of human comedy can be made manifest. Because suffering can exert its healing power within the ambience of subjective time-space and in objective time-space, our dual spatiotemporal structure as the grounding constituents of man can demonstrate the value, power, and meaning of suffering.

Problems arise due to our complex temporal structure. We live in objective time as well as in subjective time, but for practical purposes we constantly use and measure objective time. Therefore, our thinking about time usually turns toward clock time and calendar time, which is objective time. We are drawn toward thinking about objective time because our practical life is driven by the clock and is ensconced in the calendar. We hardly consider the need for or even the possibility of another form of time, unless for various reasons we begin to see that there are problems hidden within the so-familiar clock and the ubiquitous calendar. It is mostly theologians, philosophers, physicists, psychologists, and psychiatrists who become seriously involved in contemplating the dualistic characteristic of the concept of time.

Alfred North Whitehead stresses the importance of the insight into our temporal structure from a social and historical viewpoint. He warns against our ingrained tendencies to allow past experience to dominate our 'foresight', our looking forward into the future:

> The note of recurrence dominates the wisdom of the past, and still persists in many forms even where explicitly the fallacy of its modern application is admitted.... We require such an understanding of the present conditions, as may give us some grasp of the novelty which is about to produce a measure of influence on the immediate future.[3]

In order to describe and to understand man's complex, mundane existence, we need not only the form of objective time but also the medium of subjective time. At this stage in the development of my thesis it is sufficient to point out that objective time flows forward into the unlimited future, while the future of our personal, subjective future is very short indeed in comparison with the objective future.

This particular difference between objective and subjective time raises a question which demands an answer: How do we live into the future? In the following pages I shall analyze and describe the manner in which we relate to the objective and to the subjective future, how we relate and interweave living into the objective and into the subjective future, and how we bring about a synthesis of the two forms of time within the dualistic temporal structure of our existence. I shall place the tensions, the dichotomies, and the fractures within the context of our creating this synthesis.

Our dualistic temporal structure brings to light two of the major fractures in our existence: the fear of loneliness and the dread of death. The deepest source of human suffering is the dread of death, of which we do have some foreknowledge. Loneliness is the second major source of suffering. This state of mind is engendered by our dualistic nature which encourages and cajoles us to live the life of an average man among others, versus the urge to be alone like an island onto ourselves. We are likely to be overcome by the fear of loneliness in either life-style. However, we may not feel lonely when we are with others or when we are alone. Yet when the feeling of loneliness overcomes us, it is as deep a source of suffering as when the dread of death takes hold of us.

How is loneliness related to our temporal structure? When alone, objective time takes on less importance; when we live as average man, subjective time assumes a secondary role. Put in a somewhat crude metaphor: When we live the island existence, we live, mostly, in subjective time, and we may hunger for those human contacts which we experience, mostly, in objective time. When we move around in the continent existence of human society, we exist, mostly, in objective time and we may long for being alone, mostly, in subjective time. The polarity of the contrasting temporal structures of these two modalities of existence is one of the fractures in our human condition, which makes us suffer. Does this suffering display the emotional physiognomy of dread, similar to the dread of death? A sensitive French saying expresses this similarity: *'Partir, c'est mourir un peu'* ("to part is to die a little"). Fear of loneliness and dread of death are fleeting experiences, and their similarity is the reason why we battle them with the strategies of forgetting, denying, and sublimation. We suppress and evade the fear of loneliness as much as we try to chase away the dread of death from our consciousness. For instance, people entering a theater, a concert hall, or a lecture hall anxiously look for a familiar face; feeling all alone in the crowd, the fear of loneliness suddenly bores into them.

Since we live into the future in objective time and into the future of subjective time, we are concerned with what we will do in the immediate, personal, subjective future and what will happen to us in this short span of subjective time. We also contemplate how the plans on which we are working right here and now, will work out in the distant, objective future and what may happen to us in times quite far ahead of the lived-in present. Foreknowledge of the future can be rather accurate in certain sectors of life, but the distant, objective future is mostly hidden from us. With one exception, we know with apodictic certainty that sometime in the objective future we will die. This foreknowledge of living versus dying engenders the dread of death. We may experience this dread more than once during our lifetime if our physical existence is threatened by accident or by serious illness and some people experience this dread when the end is near. Some people are convinced that they have never experienced this dread. Nevertheless, everyone has the foreknowledge of his eventual death, but they may deny that they connect this knowledge with any frightening emotion. This means simply, that the psychological defenses against the dread of death are more effective in some people than in others.

Even young children are concerned and worried about the fact that not only pets and other animals die, but that humans may die also, especially their parents and their own siblings, and they often worry about dying themselves.

Many ancient myths have conceptualized the dualism of human existence, for instance, the Chinese circular emblem of yin and yang. The yin side of the circle stands for the female, negative, evil principle, or the dark side of a hill, while the yang side represents the male, positive, beneficent principle, or the south and sunny side of the hill. One can "see" the dynamic interaction between the opposite elements in human nature and how the light side overcomes the dark side. The dark side, taken to represent the dread of death, initiates germinative powers in the light side which combats the dread of death with various cultural means. The elaborate burial procedures of the ancient Egyptians were designed to ensure the eternal life of the dead Pharaoh's *ba*, or soul. These immense cultural artifacts show to what lengths men do go to soothe the dread of death. In a metaphorical sense, the sophisticated efforts of the ancient artists, builders, and mathematicians are still alive in these admirable works of art and science.

One can interpret cultural achievements as countermeasures against the dread of death. The scientist, the philosopher, the artist, conceives and produces true and beautiful constructs during brief periods of subjective time, and then they are concretized into cultural artifacts in objective time, after painstaking effort. These long-lasting testimonies to man's creativity deny the fragility of our temporal existence. Hence, they are a counterbalance against the dread of death. Besides displaying communal values, cultural achievements also exemplify the value of suffering, since it is the binding agent which holds the two aspects of our existence together. They transilluminate the interaction between the yin and the yang principles by synthesizing objective time with subjective time.

The synthesis of these opposing aspects of our living in the world can be accomplished only if we adopt the heroic attitude toward life. I do not mean the life-style of the heroes of yore, who went out to destroy some monster, or the enemy of their people. The heroic attitude in day-to-day living sets free the search for the realization of a goal, an ideal, or values, either in the face of external hardship and disappointments or confronted by internal doubts, such as whether one will ever make an invention into a useful tool, or perform a surgical operation successfully, or discover what one is searching for. We meet life's problems head-on, we don't pretend that they do not exist or that they do not concern us personally. The heroic life-style tackles disturbing problems and is bent on finding solutions; even if the problems appear to be unsolvable, we accept this disappointment without flinching. Suffering is irrevocably included in this stance.

We do not remain locked into the immediate present when we face a problem or when danger threatens us, but we see it in a wider temporal perspective of objective time interwoven with subjective time. For this very reason we are tense and fearful, depending on what we intend to accomplish, or try to improve, or seek to eliminate. We do move ahead into our subjective future notwithstanding this inner tension or dread about what the objective future may have in store for us, provided we adopt the heroic life-style.

We try to overcome the blemishes, to solve the contradictions, to heal the fractures in our existence, and because this synthesis is never quite a completed process, we suffer. But our dual temporal structure is also the underpinning of our most treasured human attributes. We are conscious, we are rational and emotional beings, we communicate with our fellow humans and we increase our knowledge of our surrounding world by leaps and bounds. Because of our complex temporal structure we can objectively evaluate ourselves and others and we are guided by our conscience when we weigh and judge the results of our actions. The development of language, of different cultures, and especially of our personal and our communal history are possible, thanks to our dualistic temporal structure.

Paul Fraisse points out how important the way in which we live in time is for the development of thought: "Each new system of philosophy has its own conception of the opposing forces exerted on us by time. Philosophies are born from our attitudes to time and they rationalize and give value to these."[4]

The cultural aspects of our terrestrial existence are part and parcel of our drive toward the realization of values. This drive is a momentous source of suffering. But there is another side to human life: insofar as we are living in objective time we are tied down to being an anatomical structure with physiological functions. Values lie beyond this mode of existence and we can reach this region only when we live as total human beings in subjective time integrated with objective time. Man puts enormous effort into this synthesis and he seldom reaches his goal of giving concrete form and substance to the material that he is trying to mold after the mental, ideal image which is his lodestar.

Rockwell Kent's *Almost* shows the efforts of a man climbing almost to the top of a tree, reaching out for the light rays that come from up above.

Goethe makes us feel the painful discordance between opposite strivings within the human person in Faust's lament:

> Two souls, alas, dwell in my breast.
> One wants to separate itself from the other;
> One of them holds itself fast, in coarse lust of love,
> To the world with grasping organs;
> The other lifts itself forcibly from the dust
> Toward regions of high forebodings.[5]

Auguste Rodin's *La Centauresse* illustrates dramatically the struggle between the human part of the mythical creature, which tries to break loose from her equine hindpart that holds her down to a lower modality of life: "It is a frightening separation of the two natures of which the poor monster is composed. Image of the soul, whose ethereal flights remain the miserable captives of the corporeal mire."[6]

The artist, the musician, the poet, the architect, the philosopher, the theologian, the historian, the scientist, suffers because his creative efforts force him to go through the painful process of bringing the two worlds of reality and of ideality together. The truly creative person does not give up on his efforts to transpose his ideals, which have been hovering before his mind's eye, into concrete reality for everyone to see, hear, or read; regardless of the anguish of the struggle he embraces the heroic life-style.

We try to heal the fractures in our existence in many different ways and on different levels. We try to reach goals, to realize values for which we stand personally. We go through the valleys of suffering courageously and heroically in order to lead a worthwhile life. When we succeed in what we set out to accomplish, sometimes to a lesser, other times to a greater extent, we reach a state of bliss as the capstone of the painful period which now lies behind us. Man strives to do ethical deeds, to create beautiful constructs, to formulate true propositions. Suffering is the *conditio sine qua non* and it is the hallmark of our striving for these values.

We can make sense out of life's vicissitudes from the viewpoint of time. Progressing from minor blemishes to serious problems, to bothersome contradictions and dichotomies, and finally to the two ultimate fractures of our existence, we can notice a gradually increasing complexity and scope of the temporal panorama in which we deal with these imperfections. On one side of this spectrum, moving into the subjective future generates only mild tension and contemplation of the objective future reaches only a short time span ahead of the present. But if we are faced with fractures in our existence, and especially if we are faced with the possibility of our personal imminent death, then projection into the future of subjective time assumes the emotional coloring of dread and the outlook upon the objective future is truncàted. In between the two limiting points of the temporal spectrum we traverse a wide variety of emotional states in keeping with our balancing objective time and subjective time over against each other. Our existence is overshadowed by the ultimate destroyer and sometimes it is frozen by loneliness. But the pride in our achievements transilluminates our life.

Clearer insight into the way we live in time deepens our understanding of the meaning of suffering. But we have to go a step farther and we ought to tackle the question of how we live in space. I propose that we should include the concept of objective space and subjective space in the description of the temporal structure of man's being in the world.

However, this fusion of subjective time with subjective space has a quite different structure compared with Einstein's time-space continuum. First, Einstein, thinking as a physicist, operated within the forms of compartmentalized, objective space and time. Second, the theory of relativity was created to describe phenomena that proceed with the velocity of light, the universal constant c. Subjective time and subjective space cannot and should not be compartmentalized; they serve as the appropriate milieu in which we perform purposive movements with velocities of a few centimeters per minute and at the very fastest a few centimeters per fraction of a second. There are two universes of motion: one with the speed of light which lies far beyond our powers of direct sensory observation, another in an utterly different form of time-space—that of our low-speed, purposive movements.

Chapters 2, 3, and 4 analyze our spatiotemporal structure by contrasting objective time-space, which is a homogeneous, compartmentalized, and measurable milieu, with subjective time-space, which is a heterogeneous, interpenetrating, not measurable, but nevertheless meaningful milieu. The mental triad and the precious present represent the interpenetrating qualities and the emotional loading of our activities in subjective time-space. Extrospection is the indicated method of observation of phenomena occurring in objective time-space, which are causally structured and emotionally neutral, while introspection makes us aware of the subjective time-space structure, the purposive organization, the feeling tone, and the interpenetrating quality of our goal-striving actions.

Chapters 5 and 6 show that causation "nests" in goal-striving activities and that the latter do not suspend or eliminate causation. Decision making is a limited conquest of freedom over determinism. This process is impossible according to Freud's metapsychology since he uses only the form of objective time.

Chapter 7 depicts the multifaceted arrangement of time-space forms, structuring categories, compartmentalization and interpenetration, emotional neutrality and feeling tone, and the contrasting methods of observation that interweave and interact in the mind-body relationship. We suffer not only when we attempt to make mind overpower body, but also when the recalcitrant body overpowers mind.

Chapters 8, 9, 10, and 11 work further to secure the thesis that suffering is meaningful. The imperfections in our existence display an array of blemishes, fissures, contrasts, dichotomies, and fractures which range from the minor and merely bothersome, to the disturbing and frightening, to the outright threatening and dread inspiring. In keeping with the progression from minor blemishes to the two most severe fractures in our existence, suffering increases in depth from annoyance, to apprehension, to anxiety, and finally to existential dread. Fissures and minor dichotomies are of a more rational nature, such as the contraposition of objective versus subjective time-space, and we can heal them *in abstracto* by deriving objective time from the precious present and, similarly, by deriving causality from the mental triad. The fractures of loneliness and of dying are the most severe sources of our sufferings, characterized by heartrending emotional turmoils, which we try to heal *in concreto*. Suffering in the physical, emotional, and mental modes and the possibility that it can be healed by medicine and psychiatry, show the differences between destructive, meaningless suffering and constructive and meaningful suffering. We heal the dichotomies and the fractures *in concreto* with the aid of religion, philosophy, the arts, literature, music, and humor. Personal vignettes are concrete instances in which suffering did heal dichotomies and fractures and they show *how* suffering fulfills its mission. For suffering to be meaningful we have to adopt the heroic life-style. Suffering unifies determinism with decision making, optimism with pessimism, and we ascend while suffering toward securely grounded, indefeasible periods of bliss.

I shall rely in the analysis and the synthesis of the human condition on examples taken from many disciplines. Although one runs the risk of skimming over the surface of what others have studied in depth, this kaleidoscopic approach intends to unify what theoretical analysis has torn apart. Huizinga apologizes thus: "In treating the general problems of culture, one is constantly obliged to undertake predatory excursions into provinces not sufficiently explored by the raider himself."[7]

Frequent references to physical, emotional, and mental illnesses serve the purpose of comparing how healthy and how sick people live in time-space, since we can learn from these undesirable conditions how we synthesize the diverse elements of our mundane existence.

Jean Piaget's studies of child development help us to understand our own adult life-style. This interdisciplinary investigation is limited to man's terrestrial existence, and it leaves to others to contemplate what happens to us after we die.

To clarify abstract concepts, I shall use concrete, particular examples which cannot be treated statistically. Because this approach emphasizes the temporal structure of personal experiences and actions, it differs from the widely used statistical methods in biological and natural scientific research. However, Leonard Doob presents ample evidence that statistical methods are indeed important in research which covers the time problem. This research compares, mostly, how processes in objective time interrelate with experiences in subjective time. The former can be compartmentalized and they can be treated statistically; the latter interpenetrate and they cannot be so treated.[8] Statistical analysis does not explain how these contrasting constituent elements of our human nature interrelate.

Charles Sherover says about examples of a subjective nature: "The instrument of inquiry is, after all, ourselves."[9] I use not only the personal experiences of others, but also incidents in my own personal life to undergird theoretical discussion.

At this point one is obliged to focus on religious thinking and feeling for two reasons: religions give answers to the question why we humans suffer and religions try to come to grips with the fear of loneliness and the dread of death, contemplating what happens to us after we die. Religious metaphysics hold that man suffers because he is sinful and that the fractures in his existence can only be healed by a suprahuman Being or by suprahuman Beings. The fear of loneliness is dissolved in the union with the godhead and the dread of death is overcome by the belief in our immortal soul, which survives the death of the body. Religious interpretations of human life rest on the conviction that a personal God or that several personal gods urge and invite man to partake of ideals and values which emanate from the supreme Being or the supreme Beings.

The great Flemish poet, Guido Gezelle, a Roman Catholic priest, says that a poet is inspired by the benevolence of the supreme Being and therefore, to write poetry is not an art, it is a privilege, a gift from God.

The contrast between Rockwell Kent's *Almost*, Rodin's *La Centauresse*, and Faust's "Lament" on the one hand and Gezelle's little gem of a poem on the other, confronts us with opposing convictions concerning the human condition. One view describes how man struggles to ascend from the lower world of reality to partake of the sphere of the ideal, by dint of his own efforts. Gezelle rejoices when man in the nether regions of reality is open and willing to receive what God's beneficence and power bestow upon him from on high.

A dualistic humanistic interpretation of human life holds that we suffer because we strive to grasp and realize ideals and values within our temporal-spatial existence, this side of the grave. We have to live with the dread of death which makes us suffer, but which is also the source of our drive to create cultural artifacts. The battle against the fear of loneliness adds force to these cultural drives. Torn between living, mostly, in objective time-space and, in contrast, mostly in subjective time-space man suffers as he strives to heal the dichotomies and the fractures in his existence.

Human suffering is meaningful because it is a healing power.

NOTES

1. Mortimer Adler, *Aristotle for Everybody: Difficult Thought Made Easy* (New York: Macmillan, Co., 1978), p. 93.

2. Martin Heidegger, *Sein und Zeit* (Halle a.d.S: Max Niemeyer Verlag, 1929), p. 135 (Author's translation).

3. Alfred North Whitehead, *Adventures of Ideas* (New York: MacMillan Co., 1935), pp. 17-18.

4. Paul Fraisse, *The Psychology of Time*, translated by Jennifer Leith (New York: Harper & Row, 1963), pp. 292-93.

5. Johann Wolfgang von Goethe, *Faust: Eine Tragödie*, pt.1 (Leipzig; Utrecht: Pfeil Verlag G.m.b.H., 1924), p. 30 (Author's translation).

6. Auguste Rodin, *L'Art: Entretiens avec Paul Gsell* (Paris: Bernard Grasset, 1924), p. 213.

7. Johan Huizinga, *Homo Ludens: A Study of the Play Element in Culture* (Boston: Beacon Press, 1966), p. 1.

8. Leonard W. Doob, *The Patternings of Time* (New Haven: Yale University Press, 1971).

9. Charles Sherover, *The Human Experience of Time: The Development of Its Philosophical Meaning* (New York: New York University Press, 1975), p. 552.

2
Objective Time-Space and Subjective Time-Space

Conventional interpretations of human existence from the viewpoint of time separate time from space as if they were two independent, concrete elements in our existence and our experience. This separation is highly problematical; it stems from the ingrained conviction that we can perceive time and space, as if they were two thinglike objects. But one searches in vain for organs of time perception and of space perception, organs which might give us the sensory material upon which to construct concepts about time and space.

Instead of separating time from space, we ought to differentiate and contrapose the twofold forms of objective time-space versus subjective time-space. Pierre Teilhard de Chardin repeatedly emphasizes the need to understand our temporal structure and he does interconnect time with space.

> What makes and classified "modern man"....is [his] having become capable of seeing in terms not of space and time alone, but also of duration, or—it comes to the same thing—of biological space-time. Spacetime is this, that everything that up to then we regarded as points in our cosmological constructions became instantaneous sections of infinite temporal fibers.... the entire spatial immensity is "no more than a section at time 't' "....into an unfathomable past and somewhere of a future that, at first sight, has no limit.... After the walls of space, shaken by the Renaissance, it was the floor of time which, from Buffon onwards, became mobile. Yet in the first states...space and time, however vast, still remained homogeneous and independent from each other....And without the assurance that tomorrow exists, can we really go on living, we to whom was given—perhaps for the first time in the whole history of the universe—the terrible gift of foresight?[1]

Objective time-space is a homogeneous continuum, and it is the milieu in which natural scientists construct the world of physics, chemistry, astronomy. Subjective time-space is a heterogeneous discontinuum, and it is the ambiance in which purpose-striving, rational, emotional, communicating human beings live. Subjective time-space allows for meaning to unfold. Meaning is irrelevant in a world which occurs in objective time-space.

We perceive neither time nor space, but we do perceive changes and movement in time and in space. We have to go another step forward: not only are space and time inseparable, in addition, time, space, and movement interlock. Since we observe the two modalities of mechanical movement and of purpose-striving movement, we need two forms of time-space as the milieu in which these two kinds of perceived movement take place. Mechanical movement occurs in objective time-space; we perform purpose-striving movements in subjective time-space. Since we perceive mechanical movement separate from purpose-striving movement, we can separate objective time-space from subjective time-space. But we do not perceive and we do not perform purpose-striving movements separate from mechanical movement. Therefore, we cannot separate subjective time-space from objective time-space. When we perform purpose-striving movements, we superimpose them upon and we integrate them with mechanical movement. Even as we integrate the two modalities of movement, we synthesize the two forms of time-space.

We meet with such an integration of the two forms of time-space when we study the use of a calendar and the reading of clocks. These two ever-present aids with which we measure objective time reveal, surprisingly, activities which synthesize objective time-space with subjective time-space. The function of the calendar grid is to compartmentalize objective time. We mark an activity which, we trust, we will carry out in that distant future, in its appropriate compartment on the calendar, while we have either memorized where this activity will take place in objective space, or we have recorded it in some spatial slot other than the calendar. Now, if this appointment in the objective future has an exceptional, emotionally charged meaning for us, then we may conjure up in our imagination what we are going to do in that future, here and now, lived-in, subjective present. In steps, subjective time-space has attributes which differ vastly from objective time-space. Liberated from the spatialized time slot on the calendar, we partake momentarily of a fantasized present in subjective time-space, that is, we synthesize the two forms of time-space.

An analysis of the workings of a clock and of clock-reading ferrets out how we deal with mechanical movement of the clock in conjunction with purposive activities; here too, we are locked into the interrelations between objective time-space and subjective time-space.

OBJECTIVE TIME-SPACE

We represent and can measure objective time, clock time, calendar time, astronomical time, biological time, with spatial schemas. We compartmentalize time into intervals which are sharply demarcated from each other with punctiform instants, which represent the "now." We posit that the flow of time is created by a punctiform instant which moves from the past toward the future and, thus, the punctiform "now" creates a continuous line. We cut up this line with geometric points into equal stretches, which stand for equal intervals of objective time.

Instead of representing the flow of time and its intervals with a straight line, we can also represent it with movement around a circle, which movement represents the perceived circumvolution of the sun around the earth. De Solla Price has reconstructed how this imitation of the sun's apparent movement was first concretized by Andrikos, about 50 B.C. in Athens, with the aid of a clepsydra in the Tower of the Four Winds.[2] ("Clepsydra", with an illustration, in Webster is a water clock; a contrivance for measuring time by the gradual flow of a liquid, as of water, through a small aperture.")

We draw a circle on the face of a clock and divide the circle into twelve equal sectors, each of which represents one hour or five minutes. Inside the clock an intricate set of intermeshing gears produces the relatively slow movement of the hour hand and the relatively fast movement of the minute hand. These movements are initiated by the escapement wheel which performs a regular rotatory movement, interrupted by its teeth which interlock with two pallets on the balance which rocks up and down. The movement of the escapement wheel depicts the forward flow of time and the interruptions by the balance portray punctiform instants. The distance traversed between two punctiform instants depicts an interval of objective time. The Graham escapement mechanism and the pendulum make it possible to impress objective time into spatial schemas.

It is necessary to describe the punctiform instant and its genesis in greater detail. This concept has always dominated our thinking about time and it still exerts a powerful influence on theorizing about time. To understand the punctiform instant we have to give some thought to the geometric point. Euclid defined it as "that which has no parts, or which has no magnitude." In Euclidean geometry the line is created by the movement of a geometric point, the translation of a line generates a plane, and the movement of a plane into the third dimension forms a solid. I dare say that it is difficult for most adults to grasp the concept of the totally abstract geometric point and even of a line. But Henri Bergson claims that children accept these concepts readily and they seem to feel quite comfortable with them.[3]

However, for the reflective mind the geometric point remains an elusive concept. The geometrist-mathematician is used to thinking in abstract terms, but one can also start out from a position which is couched in the concrete terms of sensory impressions. Instead of the dimensionless geometric point, let us start out from a three-dimensional solid. Suppose you are asked to close your eyes and to touch a solid with your fingertips over which you move in the directions of three spatial coordinates; you will soon come to the conclusion that this felt thing is a solid. Now you are asked to palpate a plane surface over which you can move your fingers only in the direction of two coordinates, say to the right and the left and toward and away from your body; you will report that you are touching a plane surface. Next, you are asked to observe the very sharp side of an object, for instance of a knife blade, and because you can move your fingertips only back and forth in one spatial coordinate, you interpret your sensory perceptions as being a line. Finally, the point of a sharp needle is presented, over which you cannot move your finger at all and now you decide that this is a point. Over the solid you can move in three directions, over the plane in two directions, over the line in one direction, but movement is excluded from the point. From the viewpoint of concrete sensory impressions upon which we elaborate by apperception, we create the concept of the geometric point by totally abstracting from movement.

This simple experiment points up the kinship between the geometric point and the punctiform instant: nothing can move in a geometric point and time stands still in a punctiform instant. Hence, we can represent the punctiform instant, the "now", with a geometric point. We stop movement deliberately and regularly in geometric points-punctiform instants with a clock, in order to measure time. A clock delivers equal stretches of movement which are delimited by and sharply separated from each other by geometric points and so are the intervals of objective time, represented by these stretches of movement, separated from each other by punctiform instants. This manipulation is a piece of evidence of the close interrelation between objective space and objective time, both of which forms we can compartmentalize.

Another common attribute of these two forms makes the creation of the concept of the straight line possible. Archimedes has defined the straight line as "the shortest distance between two points". Shortest distance means: if a body moves with constant velocity V from a point A to a point B, it will require the smallest possible amount of time to complete its journey only if it moves along a straight line. Should A and B happen to lie on a solid sphere, then the circle sector A to B is a "straight line" because that represents the least possible amount of time for a body moving with velocity V to go from A to B. The concept of the straight line is not a purely spatially determined concept but it is a hybrid, temporal-spatial structure within objective time-space. The geometric point, the straight line, and the punctiform instant enable us to express objective time in terms of schemas in objective space.

We can measure objective time and objective space with these geometric constructs because we conceive of these two forms as homogeneous media which we can compartmentalize. Analytical geometry, which was developed by Pierre Fermat and René Descartes, has given a powerful impetus to this abstracting process and homogenization.

> Geometry, analytical, a branch of geometry in which straight lines, curves and geometrical figures are represented by numerical and algebraic expressions by the use of a set of axes or co-ordinates.... In analytical geometry any point in space can be located with respect to a pair of perpendicular axes by giving the distance of the point from these axes.... a straight line can always be represented by an equation containing X and Y.[4]

With the aid of three Cartesian coordinates one can describe the position of any point in space. No point in space has an intrinsic meaning compared to any other point in homogeneous objective space.

The intermeshing of the measurement of objective time and objective space is exemplified in quite a different context. Astronomers measure interstellar distances with the light year. When one first learns that a light year is not an exceedingly long span of time, but an enormous distance in space one feels a slight intellectual jolt. This is the measuring rod of the astrophysicist which is based on the constant velocity of light, so that he can be certain that during one year light always traverses the same distance. To call this highly sophisticated instrument a measuring rod may seem inappropriate. But any yardstick or other space-measuring device, concrete as it may appear in everyday use, is nevertheless a highly abstract means of space measurement: it represents a straight line, and its markings represent geometric points although the rod is not a straight line nor are the markings geometric points.

Compartmentalization of objective space and of objective time stands in the service of time and space measurement since it makes a homogeneous milieu out of objective time-space. For scientists of centuries ago space was not homogeneous. In the Ptolemeic cosmology, terrestrial events and terrestrial space were quite different compared to celestial events and celestial space.

Herder remarks about Copernicus, that it was for him the feeling of symmetry and harmony which was his lodestar, which allowed him to find the laws of the cosmos. Copernicus himself asks, "Should the world edifice be a draft with flaws, in which hand, foot, eye, head, heart, and all the limbs, however beautiful and lovely each taken by itself, but in which all of them together are a monster, not a whole? Who designs, what architect plans in this manner? And God should have designed our suns and earth in such a wise?"[5]

Kepler found certain flaws in the Copernican system, and he was determined to put astronomy on the firm basis of mathematics. His motivation for improving our image of the cosmos had a strong religious bias also. Kepler of course based his arguments on the fact that there are five and only five regular polyhedrons.

> This was the occasion and success of my labors and how intense was my pleasure from this discovery can never be expressed in words. I no longer regretted the time wasted. Day and night I was consumed with computing, to see whether this idea would agree with the Copernican orbits, or if my joy would be carried away by the wind. Within a few days everything worked and I watched as one body after another fit precisely into its place among the planets.[6]

One sees clearly in Kepler's writings that the creative genius is propelled emotionally toward his emergent discoveries and that rationality and emotionality work in tandem as if they operate as it were in a reverberating circuit. He was convinced that the velocities, time of rotation, and distances of the planets ought to be expressed by a mathematical formula. After twenty-two years of ceaseless struggle he discovered what came to be called the three laws of Kepler. He exclaims that he had the wonderful bliss "to discover the truth out of the wildest and quite absurd whims."[7]

Continuing the trends set by Copernicus and Kepler, analytical geometry refined the homogenization of objective space in which no point in space is preferred over any other point, since none has a higher value compared to any other point. This conception of space undermines the image of a teleologically organized universe and the belief that terrestrial space was structured around the point where Jesus was crucified, according to Christian tradition. The prayer rituals of Mohammedans exteriorize a similar belief five times a day, in the value and the meaning of the point in Mecca where the venerated holy Black Stone is housed in the Kaaba Mosque.

Thor Heyerdahl gives us a remarkable instance of the synthesis of objective space with subjective space during the voyage of Ra II. He says: "Praying toward Mecca, Mandani spreads his prayer rug on the cabin top. I give him the proper compass direction."[8] Heyerdahl thinks of Mecca as a geometric point in objective space whose location he can determine with sextant, chronometer, and within a two-dimensional coordinate system. Longitudes and latitudes compartmentalize the surface of the earth into a huge, geometric quilt work in homogeneous objective space. But for the Moslem this geometric point in objective space is meaningful, since Mecca is the center around which his life-style revolves in heterogeneous subjective space.

We should complete an analysis of the concept of objective time by a close look at the internal psychological processes which go on in the mind of someone who reads a clock. What happens when someone wants to measure an interval, say, of one minute with a clock that has a sweeping second hand? Let us assume that he wants to measure the interval from two o'clock to one minute after two. He watches the minute hand move toward twelve and he begins to count when the second hand goes through twelve at the same moment that the minute hand is exactly on twelve. Now he begins to count how many seconds it takes for the second hand to sweep again through the twelve mark and he is especially intent after fifty-nine seconds to register the moment when the second hand goes through twelve. He counts from one to sixty between the punctiform instants two o'clock and one minute after two; he compartmentalizes this interval between two geometric points-punctiform instants. This means that he stops the movement of the second hand in his imagination, in two infinitesimally small instants, which mathematical limit he tries to approach as close as possible. He performs this mental operation by anticipating the exact moment when the second hand goes through twelve at two o'clock and when it goes through this point at one after two; he lies in wait for these exceedingly short moments, that is, he anticipates an event to happen in his personal, private future.

While the second hand goes through sixty circumgyrations, the clock reader is counting all the while how many times it has completed its circle. While he receives concrete visual impressions which are accompanied by sensations from his intra- and extra-ocular muscle apparatus, he is also involved in a process of mental abstraction from these sensations. He observes the now-going-on present movement of the second hand, he remembers its immediately preceding past movement, and he anticipates its immediately following movement. However, he disregards the complex temporal structure of these sensory experiences and his concomitant and organizing intellectual processes of counting, and he is only bent upon counting from one to sixty seconds. Most important, he abstracts from the aspects of presentness, pastness, and futurity of his inner processes. He conceives of the movements of the second hand as proceeding in the bland present, shorn of pastness and futurity. For the measurer, the hand of the clock moves in homogeneous, objective time and space. He has stopped the movement of the second hand *in abstracto*, in two geometric points-punctiform instants, and therefore this compartmentalized movement can express compartmentalized objective time in terms of compartmentalized objective space.

Since clock reading is a matter of daily routine from childhood, we are apt to overlook the complex internal, psychological temporal-spatial processes which are going on in subjective time-space. Clock reading is a time-binding activity, which is based upon the interpenetration of our personal, private past, with our personal, private future, with our personal, private present.

It may seem as if the reading of a digital clock sidesteps these complex mental processes. This clock gives information about the appearance of two punctiform instants as the dial first shows 2 and next the digits 2:01 and it gives the numerical value of the interval. To be sure, the reader of a digital clock does not have to count 60 seconds and therefore he seems not to be doing anything at all. To the contrary, the reader has to take note of the punctiform instants when the dial changes from 1:59 to 2:00 and, again, when it shifts to 2:01. This fixation of two punctiform instants involves the reader's anticipation of the immediate future, awareness of the present here-and-now, and the memory of his recent past. Like the reading of a conventional clock, reading of a digital clock is also a time-synthesizing activity with the difference that the digital clock does the counting for us.

Merleau-Ponty's highly original insights into the problems of time and space are a welcome guidepost for progressing to the following section on subjective time-space, although he hardly thinks of time and space as belonging together: the section on time in his book is separated from the section on space by 125 pages. The following quote may give in brief the flavor of his trend of thought.

> Either I do not reflect, I live in the things and I consider space vaguely sometimes as the milieu of things, and then again, as their common attribute or, I reflect, I recapture space at its source. In the first instance, my body and things, their concrete relations according to high and low, right and left, nearby and far away can appear to me as an irreducible multiplicity, in the second place, I discover a unique and indivisible capacity to describe space.... When I say that I see an object in the distance, I mean to

say that I hold it already or that I am still holding it; it is in the future or in the past at the same time that it is in space.⁹

John Gay lets us see that the measurement of objective time-space is of little importance and therefore of little interest to the Kpelle's of Liberia.

> They measure the quality of the moment, not the quantity. Time is primarily qualitative, reflecting the character of the moment, not its numerical relation to other moments. Kpelle words show the character of time rather than the passage of a definite amount of time. The days of the week were counted or named with reference to market days. Reference is rarely made to more than two or three weeks. Months are rarely counted and then only in numbers no higher than about three. Most Kpelle are not aware of the Western calendar; they do not know their own age. A single expression refers to last week or next week; only by context can one tell whether the expression refers to past or future.
>
> The phenomenal ability of illiterate Kpelle adults to estimate the numbers of cups of rice depends on their need to buy and sell rice.¹⁰

I sense in the Kpelle's minimal use of space and time measurement that they live quite concretely in subjective time-space and that they find little practical or economic use in the quantification of temporal and spatial relationships. They live much closer to subjective time-space than we Westerners do. From these glances at two worlds widely apart from each other, the sophisticated philosopher and the primitive Kpelle, one can see how complex man's temporal-spatial structure actually is.

SUBJECTIVE TIME-SPACE

The forms of subjective time and of subjective space possess characteristics which differ markedly from those of objective time and objective space. Subjective time-space is heterogeneous and since its outstanding attribute is interpenetration, we cannot compartmentalize it. Therefore, we cannot represent subjective time-space with spatial schemas and hence we cannot measure it.

Let us pretend, temporarily, and for the sake of clarity of description, that we can separate time from space. Comparing objective space with subjective, personal space, one sees that subjective space displays many properties which are missing in objective space. The personal space of our lived-in world has contrasting aspects such as upward versus downward, right versus left, forward versus backward, which have specific meanings for a purpose-striving human being who moves about in this space. These attributes are utterly alien to objective space.

For instance, an outward-bound trip from home H to a destination D has a very different meaning and feeling tone for the traveller, compared to the homeward-bound trip from D to H. A careful introspective observer may become aware of and he can describe feelings of uncertainty, insecurity, tension or even apprehension when he is going from H to D. The explanation of these inner states is simple: he knows what he has left behind at H, but he does not know what he will face in D. Traversing the same trajectory in the opposite direction he probably experiences quite different feelings, dependent, in part, on what he expects to find on his return home. Usually there is less tension during the trip from D to H since there is less tension toward the future and besides, he returns to the well-structured area in space in which his home life, his social life, his business life, and other interests are centered. The localities D and H have quite different meanings for a human being who lives in his world structured in subjective space.

Instead of relying on introspective observation, one can choose to see the two trips as a mechanical movement, which goes from geometric point H to geometric point D and later on in the opposite direction. For a mechanistic interpretation, the movements from H to D and from D to H are essentially the same movement through objective space. The only difference between the two trips is that if one gives the first a positive sign, then the other lap must carry a negative sign, since objective space is a homogeneous milieu. If we could live exclusively in homogeneous objective space, we would be surd to value and meaning.

The above paragraph, which purports to describe a purposive activity, shows that objective space is an inadequate medium for such a description. This is so because one replaces purpose-striving movement with mechanical movement, and, in parallel, one uses the form of objective space and the category of causality. The human being who is more than a causally structured unity of structure and function is lost sight of, since man is also a teleologically structured whole. We ought to use the form of subjective time-space if we want to grasp the dignity of man.

Eugène Minkowski made some impromptu remarks during a lecture he gave in Vienna, about the limitations of a mechanistic outlook on human behavior in contrast to a holistic view.

> I have decided to go from here to the Danube to look at it in the moonlight. . . . I can measure the distance which I have to traverse, I can calculate the time it will take and which will be dependent on my speed of walking, and if I should happen to be a good physiologist I could determine as well the number of calories consumed. However, have I accounted for all the facts in this manner? Not at all. If I remain standing still halfway, have I reached half of my goal? No, because when I go to the Danube and while I am walking for several kilometers, during which activity minutes elapse, my consciousness of the act which I am completing is stretched out over the whole like an arch. This consciousness cannot be subdivided, it is continuously present as a whole and as such it contains in each moment that which has been accomplished, that is, all the past and the future phases in a specific temporal organization which is spread out without interruption

between the moment when we have made the decision until the fulfillment of the act.[11]

This holistic interpretation implies that time and space cannot be separated and that our purpose-striving behavior brings about a fusion of these two forms. Essentially, Minkowski shows that we cannot and must not try to compartmentalize an interpenetrating, purpose-striving activity because 'temps vécu', lived in time, cannot be interrupted with punctiform instants.

The same message comes from the psychological laboratory under the guise of the kappa and the tau effects which prove that our judgment of distances traversed in space is influenced by the lengths of the intervals of time consumed during these activities. Also, our judgment of intervals of time is influenced by spatial distances traversed.

These effects were established by the classical experiment in which a subject was exposed to three equidistant punctiform stimuli in succession on his forearm. If the time interval between stimulus *1* and stimulus *2* was equal to that between stimulus *2* and stimulus *3*, then the spatial distances between the three points were judged correctly. If the time interval between stimulus *1* and stimulus *2* was made shorter than that between stimuli *2* and *3*, then the subject judged the spatial distance between stimulus *1* and *2* to be shorter than the distance between points *2* and *3*. The inverse judgment was obtained when the first temporal interval was made longer because then the subject thought that the distance between points *1* and *2* was longer than the distance between points *2* and *3*.

One sees that the length of an interval in objective time influences our judgment of the length of a distance in objective space. Psychologists have invented numerous variations on the original experiment and I shall mention a few of these as described by Leonard Doob.

> In one part of an experiment, subjects walked and then ran the same distance (150 feet) without knowing that the two distances were equal; afterward they estimated those distances and the time taken to cover them. In the other part, the subjects first ran the distance of 150 feet and then were asked to walk the same distance; afterward they estimated the time consumed for each stretch. . . . there was a decided tendency for the distance walked to be considered longer than the distance run.
>
> Young children who were given the task of drawing lines for 15 or 20 seconds concluded that the work lasted longer when they were instructed to draw the lines as quickly as possible than when they were asked to draw them carefully.[12]

The tau and the kappa effects show that time, space, and the velocity of movement are interdependent. The separation of these three concepts by physics and mathematics is artificial from the viewpoint of immediately given data of consciousness. For the sake of understanding the world in which we live as total human beings we cannot use compartmentalized objective time-space since it lacks the attribute of interpenetration. Especially the punctiform instant, the 'now' of physical science, has no function in subjective time-space with its quality of interpenetration. I have proposed that we use the time concept of the 'precious present' which illustrates the process of interpenetration and which is the form of time that structures our personal world. In contrast to the punctiform instant, the precious present possesses temporal dimension.

We could not do a single thing, perform a single activity, think a single thought, nor could we experience any feeling tone in a punctiform 'now'. This is true especially of movement which, we have seen, is totally excluded from the punctiform instant. The punctiform instant not only lacks temporal dimension by definition, it is also devoid of pastness and futurity. The precious present is the lived-in 'now' in which we accomplish the triplicate synthesis of present with past with future. The concepts of subjective time, psychological time, personal time, *'temps vécu'* (Minkowski), *'temps duré'* (Bergson), and the "specious present" (William James) have been developed in recent times. There are historical and developmental reasons why men have long sustained a one-sided outlook upon the time problem.

A HISTORICAL BIRD'S-EYE VIEW

Aristotle (384-322 B.C.) developed the following definition of the punctiform 'now' which has dominated thinking about time ever since:

> For the 'now' may be regarded as the limit up to which the past has run, none of the future being this side of it, and also as the limit from which the future runs, none of the past being that side of it. If then we can make good that it is really one and the same thing, namely the authentic 'now', that limits the past from the future, it will be clear also that it is indivisible.[13]

He pulls his discourse on time and movement together in his famous definition that "Time is the number of movement." Whereby he compartmentalizes time and movement with the number series.

Saint Augustine (A.D. 354-430) takes up the struggle with the time concept from two different approaches. First he follows faithfully in Aristotle's footsteps; as he reduces the present time of one hundred years to twelve months, to one day, to one hour, and finally he reduces the present moment to a punctiform instant:

> If any instant of time can be conceived, which cannot be divided either into none, or at most into the smallest particles of moments; that is the only it, which may be called present; which little yet flies with such full speed from the future to the past, so that it is not lengthened out with the very least stay. For lengthened out if it be, then it is divided into the past and the future. As for the present, it takes not up any space.[14]

This is Aristotle's 'now' which creates the flow of time as it moves from the past to the future. Saint Thomas calls this moment the '*nunc fluens*'. Augustine concentrates on the technical problems as to how we actually measure time, but then in contrast with this concern his subsequent argument takes on an amazingly modern and psychologically oriented turn. He places the activity of time measurement into the human mind and next he proposes that time has a triplicate structure. Then he goes on to describe the three elements which make up our mental activities.

> ...it seems unto me, that time is nothing but a stretching out in length; but of what, I do not know, and I marvel, if it be not of the very mind. 'T is in thee, o my mind, that I measure time. The impression which things passing cause in thee, and remain even when the things are gone, that is it which still being present I do measure.... perchance it might be properly said there be three times: a present time of past things; a present time of present things; and a present time of future things. The present time of past things is our memory; the present time of present things is our sight; the present time of future things our expectation.... in the mind which acteth all this there be three things done. For it expects, it marks attentively, it remembers; that so the thing which it expecteth, through that which attentively it marketh, passes into that which it remembereth.... thus the life of this action of mine is extended both ways: into my memory, so far as concerns that part which I have repeated already, and into my expectation too, in respect of what I am about to repeat now; but all of this while is my marking faculty present at hand, through that which was future, is conveyed over, that it may become past.[15]

Saint Thomas Aquinas (1225-1274) followed Aristotle faithfully by accepting the '*nunc fluens*' as the instant that gives birth to time. He seemed to be less interested in time measurement than Aristotle and Augustine were, but he was more intent on putting the concept of eternity on a firm footing. He differentiated between three aspects of time: time (*tempus*), eviternity, and eternity. Time has a beginning and an ending; eviternity began on the day of creation but it has no ending; while eternity has neither a beginning nor an ending. The '*nunc stans*', or the still-standing now is the instant in which we humans can experience eternity.

> The 'now' that stands still, is said to make eternity according to our apprehension. As the apprehension of time is caused in us by the fact that we apprehend the flow of the 'now' (*nunc fluens*), so the apprehension of eternity is caused in us by our apprehending the 'now' standing still (*nunc stans*).[16]

The *nunc stans* was developed further by the medieval mystics as the temporal medium of their ecstatic experiences and to this day it is occasionally being used by Western devotees to Eastern religions and also in the language of psychiatric metapsychology.

Meister Eckhardt (c. 1260-1328) takes up the concept of the *nunc stans* and applies it to the process of fetal development in the womb in which moment God creates the soul on the fortieth day of conception.

> All this is contained within the present Now-moment. It is the real Now-moment, which for the soul is eternity's day....A child, conceived in a mother's womb, has the form, the color, and the nature of the parents; it is nature's work, which is done in the first forty days and nights. On the fortieth day God creates the soul in less than the twinkling of an eye and all that nature can do, comes to an end with the making of form, color and being. Thus the work of nature ceases, but as it does so it reappears again in the work of the intelligent soul.[17]

Immanuel Kant conceived of the problems of time and space within his monumental theory of knowledge. His conceptions were built upon Newton's ideas about time and space, but instead of projecting time and space "out there," independent of the human subject, Kant internalized them. But he did uphold Newton's separation of time from space.

> Time is nothing else than the form of the internal sense, that is, the intuition of self and of our internal state....We...represent the course of time by a line progressing to infinity, the content of which constitutes a series which is only of one dimension; and we conclude from the properties of the line as to all the properties of time, with this single exception, that the parts of the line are coexistent, while those of time are successive....Space is nothing else than the form of all phenomena of the external sense....By means of the external sense...we represent to ourselves objects without us, and these are all in space....All which relates to the inward determinations of the mind is represented in relation to time.[18]

Kant conceived of time and space as forms of perceptions and as constituents of man's mental organization. Space makes it possible for us to have sensory impressions coming from the outside world. Time makes it possible for us to have awareness of our mental processes. Kant held that time is the form of the internal intuition and that space is the form of the external intuition. It seems that the external intuition within the form of space is similar to extrospection and that the internal sense within the form of time is similar to introspection. It seems that he held to the separation of time from space because the internal and the external intuition are two contrasting and separable methods of observation. Notwithstanding his trend toward subjectivity Kant did not arrive at differentiating objective time-space from subjective time-space.

Kant's ideas about time and space as forms of observation are germinative for an understanding of human existence from the viewpoint of time and space. He has stemmed the inveterate trend toward reifying time and space, and he has therefore opened routes to a psychological understanding of these forms of our existence.

Shifting to the field of psychology proper we see that the battle between the concepts of the punctiform instant versus the durational present goes on unabated. E. N. Clay argues at length about the nonexistence of the 'now' which he gives the unfortunate title of 'the specious present'. Specious "implies a fair appearance assumed with intent to deceive."[19]

Clay thinks that our thinking about past, present, and future is muddled because we think of the specious present as being the matrix from which the three concepts about time are developed. I read into Clay's statements that we deceive ourselves if we believe that there *is* such a *thing* as the durational present. This deception is the result of our conviction that time exists as if it were a thing with an existence independent of ourselves. But this conviction has been overcome by Kant's doctrine that time is the form of our inner perception of ourselves and that space is the form of our perceptions of the world about us.

> Time then, considered relative to human experience, consists of four parts, viz., the obvious past, the specious present, the real present, and the future. Omitting the specious present, it consists of three non-entities—the past which does not exist, the future which does not exist, and their coterminus, the present; the faculty from which it proceeds lies to us in the fiction of the specious present.[20]

William James quotes Clay extensively and he adopts the term 'specious present' which, unfortunately, has been widely used in discussions about time. Without warning us that he is in fact contradicting Clay, James gives a vivid and incisive description of the durational present by means of his famous saddleback simile.

> In short, the practically cognized present is not of a knife edge, but a saddleback, with a certain breadth of its own on which we sit perched, and from which we look in two directions into time. The unit of composition of our perception of time is a *duration*, with a bow and a stern, as it were—a rearward- and a forward-looking end.... The original experience of both time and space is always of something already given as a unit, inside of which attention afterward discriminates parts in relation to each other.... The specious present has, in addition, a vaguely vanishing backward and forward fringe; but its nucleus is probably the dozen seconds or less that have just elapsed.[21]

The hitch in most treatments of the time problem is that one emphasizes the past and that one almost overlooks the fact that human mentation moves forward into the subjective and the objective future. James falls into this trap also and, very significantly, he tucks away in a footnote the vexatious problem of how we live into the future. "Again, I omit the future, merely for simplicity's sake."[22] After he has added a past fringe and a future fringe to the specious present, it is as if James amputates the future fringe!

The work of Henri Bergson is a genuine breakthrough toward an original approach to the time problem. He uses three examples to show the differences and the interrelations between time and space and that measuring of time demands that we compartmentalize phenomena with punctiform instants.

We see a shooting star during a brief period of time so short that we seem to perceive this movement as occurring in space; instead of seeing it as unrolling (*déroulement*) we interpret it as unrolled (*déroulé*) and thereby transpose its temporal aspect into space. We spatialize time (*temps espace*).

We experience the drawing of a line with eyes closed in a different manner compared with looking at the finished line. While we draw the line, our hand and arm movements are unrolling in durational time. When we look at it, we see the unrolled movement in space and in spatialized time.

A melody is an experience of pure duration since it coincides with our stream of consciousness. Memory binds the phrases of the melody together and makes a cohering whole out of the successive sensory impressions. A melody is more than a succession of pure instants, since we cannot divide the unrolling music with punctiform instants. We cannot represent durational time with spatial schemas.

To measure the temporal element in phenomena, we divide them into compartments with punctiform instants. We make these punctiform instants coincide with the punctiform instants which delimit the intervals of time represented by the clock mechanism. Then we compare the number of the compartments within the phenomenon with the number of time intervals which we have simultaneously counted by consulting the clock.

> It is therefore the simultaneity between two instants of two movements exterior to us, which makes it possible for us to measure time.... Measuring time consists of numbering simultaneities; they do not partake of the nature of real time; they have no duration.[23]

Edmund Husserl too establishes the continuity of mentation on the basis of memory.[24] He shows that there is continuity between the lived-in present and what we remember about this present. He postulates, and shows in an ingenious diagram of time, that initially memories are part and parcel of the past aspect of the present and that they are contiguous with it; these most recent memories have barely "fallen away." But as the present continues to move ahead into the future, these memories are displaced by even more recent memories which in their turn are contiguous with the now present. Thus, memories fall farther and farther away from the lived-in present and they seem to assume different characteristics.

Husserl does not describe how we use these contiguous memories and the memories that have fallen away in our present activities. Nor does he explain that memories which we use in our various activities are not continuous but discontinuous; in fact they can be widely separated from the present and from one another.

Although Husserl has enriched the concept of intentionality, he speaks very seldom in the section of his book on memory of the future-striving attribute of mentation. Hence his conception of time appears to be dominated by the past aspects of mentation.

Merleau-Ponty, deeply influenced by Husserl, places memory in the foreground when he describes our temporal structure. Since he brings the future-striving aspects of mentation to the fore, his time concept is more encompassing than those of his predecessors. He makes manifest how we synthesize in the present our near-at-hand private future with our recent private past and with our distant past through a process of interpenetration.

> ...Just as my living present opens up to a past which, nevertheless, I do not live any longer, and upon a future which I do not live as yet, and to have a social horizon, so that my world is enlarged in keeping with the collective history of my existence which my private existence recaptures and assumes.... It is by the relation between subjective time and objective time that one can comprehend those of the subject to the world.[25]

CONCLUSION

Several factors play tricks on our thinking about time and space. We live a good deal of time as average people doing our daily duties in the world of the average person, Heidegger's *Das Man*, whose temporal structure is dominated by objective time-space. This aspect of living goes on, mostly, in continuous uninterrupted time.

In contrast to this life-style, our inner life, our mentation, thinking, and feelings are quite variable, sometimes contradictory to one another and discontinuous. We are quite reluctant to accept this uncomfortable and disturbing fact because it leads eventually toward the ultimate discontinuity of our mundane existence, namely the fracture: living versus dying. Our thinking about time and space is distorted by our foreknowledge of our personal death. We prefer to conceive of time as an unending medium in which one constituent element of ourselves will continue to exist after our body has stopped functioning.

The idea of linear continuous time is for many people the consoling ambience in which our mind, or our soul, will continue to exist indefinitely. The dissolution of the body existing in three-dimensional space and in objective time thus becomes more acceptable.

This emotionally charged motivation to construct the concept of homogeneous limitless time is an important reason why our thinking about time and space is inconsistent. We find other types of causes for this lack of clarity by looking at the development of these two concepts during childhood. Jean Piaget has shown that children progress from confused ideas about time and space toward the more abstract concepts of adults during a period of ten to twelve years. He shows how the child struggles with the construction of an increasingly abstract conception of time and space. I glean from this research that we never totally outgrow the child's 'intuitive' conception of time and space; to wit, the spatialization of time seems to hark back to the original 'centered' conception of time and space. In brief, not only the child but the adult also has difficulty separating objective time-space from subjective time-space and synthesizing these contrasting aspects of our existence.

> The child's original conception of time involves a complete lack of differentiation between time and space, with the result that in the case of the two motions, he "decenters" his attention to the endpoints alone.... The 'before' and 'after' in time are originally conceived as functions of spatial order.... This objectivation of physical time goes hand in hand with the subjectivation of psychological time.... The moment when the child first succeeds in organizing a complete temporal system is so sudden that we can never actually put our finger on it; a total process the speed of which is far greater than that of any conscious process.[26]

One can clarify this contrast and its synthesis by comparing human behavior with the behavior of animals. Our dog is rigidly attached to my wife so that it is almost impossible for me to take the dog out for a walk when my wife is home H. On one of the rare occasions that I was able to do this, the dog kept looking backward toward our house, and I had to cajole her and give her a gentle pull as she continued slowly on our walk. But when we turned around, she began to lead and she almost pulled me back home. The dog's space-time has very definitely a heterogeneous structure in which the point in time-space of the house with her mistress in it has a specific meaning for her. She acts this out in organismic behavior which is so different when she moves away or toward point H.

For us humans too, going from H to D has quite a different meaning and feeling tone compared to going from D to H, but now the temporal rather than the spatial aspect of this difference comes to the fore.

In addition to being aware of the distance in three-dimensional space which separates point H from point D, we evaluate also the span of objective time in the distant future which separates these two punctiform instants. We may be apprehensive about what we will find in D in a point of objective space and an instant in objective time. Since we infuse geometric point and punctiform instant with meaning, or value, or problems and therefore with feeling tone, we change the homogeneous structure of point D in objective time-space as we superimpose the heterogeneous structure of subjective time-space upon this point. We experience this synthesizing process subjectively as tension, apprehension, anxiety, and sometimes with dread. That is, we suffer.

Natural science sidesteps this dilemma of being torn between two forms of time-space since it separates time from space in the service of measurement. This separation raises no questions since one measures time with a clock, which measurement flows from compartmentalization of movement, and one measures space with a measuring rod in which procedure-movement plays no role. One might object that Einstein's theory of relativity has indeed created an objective space-time continuum in which there is no longer a difference between the three spatial coordinates plus the time coordinate, which have been superseded by four equivalent coordinates. However, this condensation of coordinates was created because the theory of relativity describes and explains movements of objects which have nearly or actually the velocity of light. These velocities fall outside the purview of mechanical and purposive body movements. Therefore the separateness of and the interconnection between objective and subjective time-space are of an entirely different constitution, compared to Einstein's time-space continuum. Fraisse's remark is pertinent: "This effort...does not affect everyday psychological life, which is not at its source".[27]

I have proposed the concepts of the precious present and the mental triad to describe our temporally organized behavior. These concepts facilitate the description of the relations between purpose-striving and interpenetration and with this insight gained one can define causality and compartmentalization more accurately. We can use them as aids in the process of staking off causality versus purpose-striving.

NOTES

1. Pierre Teilhard de Chardin, *The Phenomenon of Man*, translated by Bernard Wall (New York: Harper & Row, 1961), pp. 219, 47, 217, 229.

2. Derek J. de Solla Price, paintings by Robert G. Magis; The Tower of the Four Winds; Piecing Together an Ancient Puzzle, *National Geographic* 149, no. 4 (April 1976): 586-97.

3. Henri Bergson, *Durée et Simultanéité*, 7th ed. (Paris: Presses Universitaires de France, 1968), p. 51 (Author's translation).

4. *Universal Standard Encyclopedia*, 1958, s.v. "Geometry, Analytical."

5. A. Diesterweg, *Diesterweg's Populäre Himmelskunde und Mathematische Geographie*, (Leipzig: Akademische Verlagsgesellschaft, Becker und Erler Kam, 1941) p. 548 (Author's translation).

6. Hans Freudenthal, "Johannes Kepler," *Dictionary of Scientific Biography* (New York: Charles Scribner's Sons, 1970), vol. 7, pp. 289-313.

7. Diesterweg, *Diesterweg's Populäre*, p. 551.

8. Thor Heyerdahl, "The Voyage of Ra II," *National Geographic* 139, no. 1 (January 1971): 44-71.

9. R. Merleau-Ponty, *Phénoménologie de la Perception* (Paris: Edition Galimard, 1945), pp. 282, 306 (Author's translation).

10. John Gay and Michael Cole, *The New Mathematics and an Old Culture: A Study of Learning among the Kpelle of Liberia* (New York: Holt, Rinehart and Winston, 1967), p. 71.

11. Eugène Minkowski, "Das Zeit und das Raum problem," in *der Psychopathologie*.

Wiener Klinische Wochenschrift, Dr. Kronfeld, ed. 1931, vol. 12, p. 348d (Author's translation).

12. Leonard W. Doob, *Patternings of Time* (New Haven: Yale University Press, 1971), pp. 122, 248.

13. Aristotle, *Physics*, with an English translation by Phillip Wicksteed, and Francis W. Conford (Cambridge, Mass.: Harvard University Press, 1935), bk. 6, chap. 3, p. 117.

14. Saint Augustine, *Confessions*, with an English translation by William Watts (London: William Heinemann, New York: G. P. Putnam's Sons, 1931), bk. 11, chap. 15, pp. 243-45.

15. Ibid., chap. 26, p. 269; chap. 20, p. 253; chap. 28, p. 277.

16. Saint Thomas Aquinas, *Summa Theologica*, trans. A. J. Pomerans (New York: Basic Books, 1969), pt. 1, art. 2, objection 1, p. 41.

17. Meister Eckhart, *Sermons*, trans. Raymond Bernard Blakney (New York: Harper & Brothers, 1941), Sermon 14; "Nothing above the Soul," p. 190.

18. Immanuel Kant, *Critique of Pure Reason*, trans. J. M. Meiklejohn (London: Dell & Daldy, 1870), pp. 30, 26.

19. *Webster's New International Dictionary*, 2d ed., s.v. "specious."

20. E. R. Clay, *The Alternative*, quoted in William James, *The Principles of Psychology* (New York: Henry Holt & Co., 1870), vol. 1, p. 609.

21. James, *Principles of Psychology*, pp. 609-10, p. 613.

22. Ibid., p. 641.

23. Bergson, *Durée et Simultanéité*, pp. 53, 60, 47.

24. Edmund Husserl, *The Phenomenology of Internal Time-Consciousness*, trans. James C. Churchill (Bloomington: University of Indiana Press, 1973), pp. 48, 49, 50.

25. Merleau-Ponty, *Phénoménologie*, pp. 476, 489, 495, 492.

26. Jean Piaget, *The Child's Conception of Time*, trans. A. J. Pomerans (New York: Basic Books, 1969), pp. 117, 119, 261.

27. Paul Fraisse, *The Psychology of Time*, trans. Jennifer Leith (New York: Harper & Row, 1963), p. 287.

3
The Mental Triad and the Precious Present

CONCEPTUALIZATION

I intend to describe the dynamic and the temporal structure of mentation using these two concepts. I prefer the term *mentation* over the word *mind*, because the latter carries a static intent, while the former term adumbrates the flexibility and variability of our mental life. Let us set the concepts of objective time-space and of subjective time-space aside for the time being, since the mental triad and the precious present lift conspicuous temporal aspects of our being-in-the-world out of the stream of consciousness. These two concepts show that we do not act in an Aristotelian 'now', nor in a Thomistic '*nunc stans*', nor in Clay's 'specious present', and least of all in a punctiform instant, but in durational time ('*temps duré*', Bergson), in lived-in time ('*temps vécu*', Minkowski), in a durational present.

The mental triad and the precious present are a united twosome which gathers several aspects of our being-in-the-world under one umbrella. The mental triad describes several interrelated processes; it performs three functions: the act, condensation of selective memories, and projection of this condensed material into the future, a process in which act, condensation, and projection interpenetrate. During the act we choose certain pertinent memories for condensation, which condensed memories derive their preferred status from the mental triad's preparing to project them into the future.

Although the term *condensation* is borrowed and adopted from Freud's concept of condensation in the dream work, it is used in a different context. Freud described condensation as an unconscious process, whereas condensation as an element in the activities of the mental triad is a preconscious and conscious process; it is an ego function.

Each of the three functions within the mental triad depends on the other two for its effective operation, since one cannot separate condensation from the act, nor the act from projection, nor projection from condensation. They are not compartmentalized functions which proceed in the before-and-after in objective time, since their outstanding and specific attribute is interpenetration in subjective time.

The interpenetrating characteristic of the mental triad and of purpose-striving activities can be described with the precious present. It too has a triplicate structure in that it consists of a central, durational present, a past fringe, and a future fringe. The durational present forms an organic whole with its two fringes with whom it shares aspects of pastness and futurity, since the three aspects of subjective time interpenetrate with one another.

The past fringe stretches for varying amounts of objective time into the past. In each case when we act, the extent and the structure of the past fringe are determined by the location and the physiognomies of memories in the past, which the mental triad is here-and-now condensing. The past fringe is discontinuous and heterogeneous since some memories from the recent past and some memories from long ago are being chosen for condensation. These sets of memories of vastly differing ages do not cohere in objective time, as they are lifted out of separate phases of our personal and our communal history.

These attributes of the past fringe tend to make us overlook its very different structure compared to the past in objective time, which is homogeneous and continuous. It is the process of condensation which is responsible for the discontinuity of the past fringe. In contrast to the more concrete and much more extensive past fringe, the brief future fringe carries an ephemeral flavor, as we shall see. We seem to grasp the past fringe more easily and, therefore, theorizing about time has been dominated by concern about how memory influences our temporal structure, to the detriment of pondering how we actually live into the future in subjective time.

The etymology of the word *mind* points up our past-tending thinking about time: *gemynd* (Anglo-Saxon); akin to *minna* (Old High German), memory; minde (Dan.) memory, remembrance; *mnasthai* (Greek), to remember.[1]

In contrast to the past fringe, the future fringe is short, covering, sometimes only a few seconds, or a few minutes; at other instances it may reach somewhat farther ahead into the objective future, perhaps half an hour or so. Moreover, the future-in-the-fringe into which we project our acts is unstructured and unpredictable during the precious present in which we perform this act. The very brevity and unpredictability of projection into the future fringe underlies the ephemeral nature of the mental triad and of the precious present.

We cannot separate the durational present, the future fringe, and the past fringe from each other with punctiform instants; hence, we cannot compartmentalize the precious present. As a result we cannot represent it with spatial schemas, nor can we measure it with clocks. The future-in-the-fringe is unpredictable on account of the attribute of interpenetration which characterizes the precious present and the mental triad.

The act, which is the central, dynamic aspect of the mental triad, holds condensation of past memories and projection into the future fringe together in a durational present.

We cannot compartmentalize the three functions of the mental triad nor the three constituents of the precious present. It is not as if, first, condensation does its work in the past fringe and that being finished, 'now', in a punctiform instant, the act takes over and performs its tasks in a durational present. Nor does projection into the future fringe take over when the act is finished in a punctiform instant and, next, projection stops abruptly in a punctiform instant. Although these three interrelated processes do not go on simultaneously, neither do they go on in succession of before-and-after in three compartments in objective time. The present, the past, and the future aspects of the precious present are not like three intervals of objective time which start and stop in punctiform instants. In brief, we compartmentalize objective time-space which is spatialized time and homogeneous space. Lived-in time-space interpenetrates and we cannot represent it with spatial schemas.

· The analysis of the temporal structure of human mentation uncovers one of its most important attributes, which is emotionality of variable coloring and qualities. Emotional tension arises between condensation of known past experiences and projection of them into the future-in-the-fringe which we cannot predict. On the one hand, we are bound to our past and, on the other, we strive into our personal future. Since the two fringes and the durational present interpenetrate, we are pulled back and forth in opposite directions in time, in a fluctuating equilibrium between the three operations of the mental triad. As a result we experience mental stress and emotional anguish in a precious present.

I have chosen the prefix "precious," replacing Clay's "specious," for good reasons. Rather than on our rationality, the mental triad and the precious present lay the emphasis on our intentionality and our emotionality. "Precious" indicates that something happens in a durational present which has specific meaning and personal value for us when the mental triad integrates our personal past with our personal future in our personal present. Value and meaning interweave with feeling tone. This ephemeral, mysterious act, projecting and condensing, is the ground swell on which we create new emergents, ethical deeds, grasp true propositions, and give form to beautiful things. We partake of the world of ideals and of lofty goals in a precious present.

The contrast between compartmentalized, measurable, objective time and interpenetrating, not measurable, subjective time undergirds our dichotomized temporal structure. The askew structure of the precious present with its long past fringe and its short future fringe, in contrast to objective time with its unlimited past and its unlimited future, which are symmetrical, has hampered our understanding of the differences and the interrelations between the two forms of time. While we project our acts into the future-in-the-fringe, we also contemplate and we try to calculate and visualize how these acts will fare in the near and the far distant future. In other words, calculation of the objective future seems to overshadow and soon takes over from the mental triad's function of projecting into the future fringe. These teeter-tottering, back and forth intentions between projecting into the subjective future and calculating and predicting the objective future, blur the differences between the two forms of time.

In daily life we speak glibly enough about what we are doing now, what we did a while ago or quite some time ago, and what we are going to do in the next half hour or several years from now. Present, future, and past seem to us expressions about time which we understand quite readily. But when one asks the question, "What am I doing right now?", then the question about temporal structure of this 'now' suddenly becomes quite difficult to answer, because the question lifts the discrepancies between the objective and the subjective past, present, and future out of the ongoing flow of objective time. Each of the three terms indicates two connotations or meanings, one indicating objective time, the other referring to subjective time.

Some so-called primitive languages express a far more subtle stance toward future actions than our European languages provide. Kirwan and Gore specify this feature as follows:

> There are two future and two past tenses in Luganda, called the Near Future, Far Future, Near Past and Far Past. The Near Future tense relates roughly to events up to to-morrow morning, the Near Past to those of the last twenty-four hours and the Far Future and Far Past to more remote periods. These tenses are formed by inserting a tense infix between the subject prefix and the stem or the modified stem of the verb. They are as follows:
>
> | | | |
> |---|---|---|
> | Near Future | *n-naa-kuba* | I will hit |
> | Far Future | *n-di-kuba* | I will hit |
> | Near Past | *n-a-kubaye* | I hit |
> | Far Past | *n-a-kuba* | I hit[2] |

Professor Doob told me that he underwent a dramatic personal experience in Uganda when he was learning these refinements of Luganda which suddenly forces one to think constantly whether one is talking about the near or the distant future and past. He wrote:

> European languages lack this grammatical nicety, but it would be difficult to contend that Europeans are less concerned with drawing fine temporal distinctions than traditional Africans.... In two African languages... there are three future tenses: (1) action in two to six months, (2) action that will occur immediately, and (3) action in the foreseeable future, after this or that event.[3]

We can use the precious present to refine the meaning of the three time words. The durational present is not a punctiform 'now', and it covers a variable interval of objective time which we cannot measure accurately and which we may tentatively approximate in terms of a few seconds, and in other instances perhaps half an hour. The distant past-in-the-fringe is quite unlike the past in objective time, since it contains our personal and our communal and our historical past history, but not the past of calendar time, of geological time, or of astronomical time. The future-in-the-fringe is totally different compared to the future in objective time, since our private, personal future is very short compared to the unlimited future of objective time, which we can measure and which we can predict with a certain degree of accuracy, while we cannot calculate nor can we predict the course and the outcome of the act in the future-in-the-fringe. In contrast to compartmentalizable objective time, the precious present is an interpenetrating, indivisible whole.

We can understand the differences between the two forms of time more clearly if we look at the contrasting activities during which we bring forth exceptional productions only once in a while, and the pedestrian pursuits going on during much of our waking life. On occasion, we perform a truly creative act when the mental triad unfolds and when we partake of a precious present. But when we execute our routine tasks, the mental triad and the precious present recede into the background, while we perform compartmentalized activities in the before-and-after in objective time. Objective time, as it were, covers subjective time. On the other hand, the two forms of time also influence each other actively, so that the precious present lends aspects of subjectivity to objective time, and objective time intrudes into subjective time. This intrusion creates the illusory conviction that subjective time must be continuous, although experience denies this. The wake-sleep-wake cycle, drifting of attention, living creatively versus living as average man, and, most glaringly and disturbingly, the fracture of living versus dying, make us aware of the discontinuity of subjective time. Our thoughts about time and space are muddled because we do not keep in mind that the forms of objective time-space and of subjective time-space differ from each other and yet are interrelated and intermixed.

From the viewpoint of objective time, the precious present is an illogical concept, rent by contradictions, since on the one hand it seems to honor the uninterrupted flow of time from earlier to later, and on the other hand it incorporates the reverse temporal order. It portrays projection into the future fringe as influencing the act in the durational present, while projection and the act codetermine condensation of memories in a retrograde manner. These flows and counterflows within the precious present are held together by the present-past-future aspects of mentation in the mode of interpenetration. The act has the Janusian structure of looking in both directions in time, since condensation depends on what will be projected into the future-in-the-fringe.

Sherover's remarks make one wonder if the mental triad and the precious present are concepts which describe mentation and its temporal structure, or if they are only metaphors.

In examining some of the "perplexities" which he [Collingwood] finds in most discussions of time, he summed them up in a conclusion about our usual use of temporal language which only rarely can be taken as anything but metaphorical truth: "All statements ordinarily made about time seem to imply that time is something we know not, and make assumptions about it which we know to be untrue."[4]

I maintain that the mental triad and the precious present are efficient concepts because they clarify some of the "perplexities" inherent in our thinking about time and space. At a minimum, they are useful terms which make it easier to handle the many conflicting attributes and constituents of the two forms of time-space in a linguistic sense. The precious present clears up theorizing about time because it dethrones the punctiform instant as *the* central time concept. Therefore we can develop a concept about lived-in time-space within a wider horizon.

The mental triad lifts us temporarily beyond objective time-space onto the level of subjective time-space and we experience in the synthesis of past with future within the durational present an uplifting state of mind which is similar to what the medieval mystics have reported experiencing in a '*nunc stans*', in which they claimed that they partook of eternity. An alternative, humanistic interpretation of such experiences holds that the mental triad creates things of lasting value, first, for the acting person himself and, second, these values are often transmitted to his followers and to his community, which values may be held in high esteem for long periods of time into the distant future. This immortality in a metaphorical sense is grounded in values which are meaningful in one's lifetime, during the lifetime of those who partake of these values, and remain "alive" long after the death of their creator. This outlook upon man's temporal-spatial existence concentrates on what we are doing during our life this side of death and on activities which make our lives and the lives of others valuable.

PRACTICAL APPLICATIONS

The practical usefulness of the concepts of the mental triad and the precious present comes to the fore when we compare a series of activities which have varying dimensions in objective time-space and in subjective time-space. This series progresses from miniature, to average, to considerable, and finally to the huge dimensions of a truly creative activity. Whether they happen in a constricted format or on a huge scale, all these activities create an interpenetrating whole in subjective time-space; they form a gestalt.

> [The] Gestalt school of psychology interprets phenomena as organized wholes rather than as aggregates of distinct parts and maintains that the whole is more than the sum of its parts.... The characteristics of phenomena as perceptual delusions and binocular depth perception cannot be derived from a summation of the constituents into which they might be analyzed, but instead depend on a given whole.[5]

Gestalt binds compartmentalization and interpenetration together. It points beyond phenomena occurring in objective time-space and which are causally structured, toward an emerging whole, interpenetrating in subjective time-space, and which is purposively structured. Gestalts are created by the mental triad which, operating in a precious present, forges compartmentalized phenomena into interpenetrating phenomena. This is accomplished by making the walls between compartments "porous," or "transparent" so that both compartments can interact and are transformed into a new, emergent whole. The mental triad performs this amazing and fascinating decompartmentalization in a precious present. A newly formed gestalt scintillates with feeling tone, while the summation of component parts into a causal unity does not display such mental tension and emotional coloring. Interpenetration means that one activity fuses with another activity, that one thought, one emotion partakes of another thought, of another emotion. Gestalt urges summation on toward meaning.

A simple example of how the mental triad works within a very small spatio-temporal dimension is a mistake typists often make: the transposition of characters. They want to type *and*, but on the paper appears *nad*. If one analyzes what has happened here, something mysterious jumps out of the underbrush. They have depressed the *n* key with the right index finger, then the *a* key with the left fifth finger, and lastly the *d* key with the left third finger, in succession. They should have moved first, the fifth left, then the right index, and thirdly the left third finger. This means that almost unbeknown to the typist, each of these three fingers was "waiting" on its appropriate key *before* they intended to type this conjunction. It is as if a miracle has happened in this simplistic situation since this mistake and its analysis shows that the mind-body as a whole was prepared to perform an interpenetrating act in the immediate future. The positions and the movements of the fingers project the act of writing *and* into the future-in-the-fringe.

To understand what goes on in an activity like typing, we must keep in mind that the typist has learned this trade several months, or years, or even decades ago. The beginner is taught where to place each finger on each key and learns in slow progression to type out short words, one character after another, so that each finger performs a separate operation. Gradually, the compartmentalized individual finger movements begin to combine into patterns of two, three, or more characters, forming interpenetrating wholes. Now the student condenses bits and pieces of what he has learned into the past fringe of a precious present and he begins to type smoothly because he integrates the act in the present, with condensation of past memories, and he projects his act into the future-in-the-fringe. During the first stages of his training his ten fingers work in the before-and-after in objective time, later they work together in unison, mostly, in subjective time. The mental triad superimposes interpenetration in subjective time upon compartmentalization in objective time.

This seemingly very simple activity reveals an amazing complexity. Typists are aware of mistakes they have made, even before they see the wrong characters or the wrong sequence on the paper. Although typists seem to work almost automatically, they develop a feel for the cohering patterns of words and they sense when the fingers have slipped outside the structure of these patterns; the critical activity of their mind warns them of the distortion of the gestalt.

We enter into a wider temporal-spatial perspective when we study the physical and mental activities which are going on when we drive out of our driveway. We can catch the evanescent act on the run, as it were, by observing introspectively what one does and experiences during this brief span of time. We make sure that the motor is running smoothly, we may have one foot on the accelerator and the other one on the brake pedal, as we ease our car slowly toward the road, ready to stop if there is no break in the string of cars. We look right and left and we join the traffic as soon as the road is clear. During these few minutes we integrate past knowledge of how to drive a car, with the movements of arms and feet which are going on in the durational present and we project these activities into the future immediately ahead.

Let us assume that we are going from home H to a destination D where we have never been before. When we come to unfamiliar territory we may stop from time to time to read the road signs and at last we arrive at D with a sigh of relief. The first lap of this trip goes on, mostly, in subjective time-space; during the second lap we compartmentalize the trajectory as we go from sign to sign in a quasi-interrupted manner. In contrast, if we go to D again, and if we have good road sense, we do not slow down or stop at the signs but we go from H to D in one fell swoop. The first time we drive *from* sign *to* sign, the next times we drive *through* the signs. For the first time the trip felt awkward, tense, saccadic because it was compartmentalized; later on it felt pleasant, assured, smooth, because the time-space synthesizing power of the mental triad is a mood-lifting experience.

An experience which comes out of a sculptor's studio is an instructive example of the progression from compartmentalization in objective time-space to interpenetration in subjective time-space. Zorach tells us quite casually how he works, and he reveals how an artist transforms a multitude of discrete, compartmented elements into a gestalt.

> When I am in the country and I want to carve a rabbit, I take my pad and pencil out to the pen where there is a family of rabbits a farmer gave me. I just sit down and watch them, observing how they move and how they are formed. Then I begin to make studies of them; I sit on the ground and put my head down on a level with the rabbits and I make drawings of what I see. Then I study them looking down from the top, and make more drawings. I study them from every angle until I know their form, the roundness of the back, the triangular shape of the head, their long ears which are so alive, and always alert to vibrations of sound; the position of the eyes, the shape of the mouth, the construction of the legs. I make notes of the design of each part and do drawings of the rabbits in endless positions. I save all

the drawings so that when I start to carve or model the rabbit, I can tack them on the wall of my studio and they will bring back to me all that I have observed.[6]

Zorach's detailed description of how he creates this sculpture is an enlightening example of how the human mind shapes compartmented elements into an interpenetrating whole. We are apt to think of sculpture as an exclusively spatial form of art and I have so far emphasized the temporal structure of interpenetration. Therefore, this view of how a sculptor actually works, is of special value for the understanding of the spatial aspects of interpenetration. The sketches of the rabbits are quasi-compartmented elements which testify to innumerable observations he has made. They facilitate his condensing these compartmented visual data and specific memories as he is working. The details of each of the sketches have something to say to him as he fuses them into the finished sculpture. The multiplicity of observed forms and movements are no longer discrete entities and individual memories, but they are transformed and they partake of the final gestalt.

The creative artist ascends from compartmentalization to interpenetration, from summation to gestalt.

Zorach's notes are an outstanding description of the process of decompartmentalization of individual memories, which are in storage in our so-called memory bank. Condensation fuses them together into an interpenetrating new entity, which the act, 'now', in a durational present projects into the future in the fringe of a precious present. One thinks that a sculptor works in the medium of space, but Zorach shows us that he works both in time and in space. We should enlarge the scope of the definition of sculpture as a temporal-*spatial* form of art, with the emphasis on space. Music, to the contrary, is described as the temporal form of art par excellence. But if we take a close look at what composers and what performing musicians actually do, we see that music is a spatio*temporal* art form, with the emphasis on time. These two art forms back up the thesis that time and space ought not to be separated from each other. Both the sculptor and the musician work in the dyadic spatial-temporal milieu of objective time-space and subjective time-space, interweaving.

Music, which structures auditory perceptions, occurs mostly in subjective time, in contrast to sculpture which organizes visual perceptions and which occurs mostly in subjective space. In practice, music straddles the regions of objective time-space and subjective time-space, with the emphasis on the precious present. Though primarily an auditory art, visual and spatial elements are essential factors in Western music. There are kinds of music which are transferred from one generation to another exclusively through the auditory channel, but these musics are foreign to our ears and it is a world of which most of us cannot partake. Moreover, many Far Eastern peoples admire, enjoy, and perform Western music. The score is a representation of the auditory processes going on in subjective time-space by means of notes, staff lines, bar lines, and so forth, in objective time-space. Music synthesizes the dichotomous forms of objective and subjective time-space.

The musical score looks to a nonmusician as if it is a spatial, scientific, quantifying representation of music. Two notes, one on a staff line, the following one between staff lines, can represent a halftone interval and on paper any halftone interval looks exactly like any other halftone interval; but not so for a singer or a string player. The seventh note in the diatonic scale is slightly less than a semitone below the top eighth note of the scale, which is the tonic. This semitone being slightly compressed gives the seventh note of the scale a distinct, protensive drive toward its resolution in the tonic. It is therefore called the leading tone in the scale, that is, it projects itself into the future-in-the-fringe. This very slight change in the dimension of the semitone between the leading tone and the tonic is not spatially represented in the score. This is one region where music goes beyond spatial representation as it synthesizes objective with subjective time-space.

The bar lines give the impression of dividing the staff into spaces that represent equal measures of time. But in actual performance this is not so. A musician does not hear, play, or sing music *from* bar line *to* bar line, but he hears, plays, or sings *through* the bar lines, which is one of the most effective methods of breathing life into the score. During a sensitive performance a slight distortion of every measure takes place, even of part of a measure which lengthens one group of notes and shortens another group ever so little. This temporal distortion or rubato, which robs and gives back, undergirds the dynamism of music. Dynamics overcome the compartmentalization in objective time-space that is visible in the written score. The performer keeps the listener on his toes and alert toward what is going to happen next, by making the bar line "porous" or "transparent." He communicates his own experience of the functioning of the mental triad in a precious present to the listener, who by reverberation partakes of a precious present himself. A musician accomplishes the miracle of creating an interpenetrating, auditory gestalt in subjective time-space out of that which seems to be compartmentalized in objective time-space.

B. H. Haggin presents an unusually concrete example of projection into the future fringe of a precious present.

> [Toscanini was able]...to show unmistakably, unfailingly...what is going to happen...when he was conducting, especially in a performance, in a medium tempo—say an Allegretto or an Andante—about half a bar before the occurrence of a detail in the music, you saw already in his face and in his gesture what he was coming to and would want. This was extraordinary: the parallel conducting of what was going on now and what was coming the next moment, so that the musician felt he was being guided through the complexity of the music, like Theseus led by Ariadne's thread through the labyrinth. It was an utterly unheard-of ability, almost like the clairvoyance of a seer.[7]

An example of the tension between these two regions of spatiotemporal structuring is revealed in an incident that happened during a rehearsal of a famous conductor, who was a stickler for exact tempos, with his orchestra. He criticized the players for using too much rubato and told them to play a certain passage in exact timing. He underscored his entreaties by making them play as they followed the clicks of a metronome, which indicated the exact tempo he wanted. When they played the passage according to the directions of this clocklike instrument, the conductor got mad with the metronome, in his usual irascible manner, and admitted this was not what he wanted either. The dynamic distortions which make the tension of music toward the future-in-the-fringe perceivable, are so subtle that they cannot be compartmentalized into intervals of objective time by a clock mechanism.

Roald Amundsen's description of how and when he made the decision to become a polar explorer is an example of a synthesis of objective time-space with subjective time-space on a grand scale. His life is a glorious example of the awesome power of some human beings to integrate subjective time-space with huge periods of objective time-space.

In his teens he was living with his widowed mother who urged him to prepare for medical school. At this time Roald had no idea what he wanted to do as an adult and he had no interest in becoming a doctor. But then, "When I was fifteen years old, the works of Sir John Franklin, the great British explorer fell into my hands. I read them with fervent fascination which has shaped the whole course of my life. Strangely enough, the thing in Sir John's narrative that appealed to me most strongly was the suffering he and his men endured. A strong ambition burned within me to endure those same sufferings."[8]

But instead of lapsing into daydreaming about his hero, as so many perfectly normal teenagers do, young Roald immediately began to prepare himself toward withstanding the rigors of the arctic climate. He forced himself to sleep during the long, freezing, cold winter nights of Oslo with the windows wide open. Not much of a sports fan, he became an enthusiastic soccer player and, of course, he skied. Also, he read everything about arctic explorations he could collect.

After a few years he felt that he was ready for a rugged test. He decided to cross the plateau east of Oslo in winter with a school companion. This feat had never been attempted by anyone, and the local farmers with whom the two stayed before their ascent onto the plateau warned them that it was impossible to survive this long crossing in winter. The two daredevils went anyhow and the dire predictions of the farmers nearly came true. Roald thought that it would be more comfortable to dig out a big hole in the snow and sleep thus protected from wind and snow. In the morning his companion did not see him and started looking for Roald frantically. One little telltale sign saved his life: a slip of Roald's blanket protruded a bit from the snow. Inside, Roald was entombed in a block of ice and he could not move arms or legs. It took his companion three hours to dig him out of his cave.

Roald did comply with his mother's wishes and he went for two years to medical school, but when she died he felt free to pursue his chosen career in earnest. His first goal was to obtain a captain's license. He had learned from his voluminous readings that many of the troubles and the failings of previous polar expeditions were due to the captain of the ship and the leader of the expedition having different and incompatible opinions about the execution of their plans. Therefore he took all the necessary steps to get a skipper's license so that he would always be the unquestioned authority on his ship. His autobiography is replete with similar instances of meticulous preparations for his expeditions. He seems particularly proud of figuring out during his preparations for his dash to the South Pole and back how many kilos of food one should carry on the sleds. He realized that since these loads would diminish from day to day as men and dogs consumed their rations fewer dogs would be needed to pull the loads. Therefore, if he sacrificed one dog on precalculated days, the need for victuals for the remaining dogs would be less. He wrote, with no little pride, "This schedule worked out almost to the day and the dog."[9]

He sums up his life as an explorer in this closing sentence: "Man's triumph over nature is not a victory of brute force, but it is the triumph of the mind."[10]

This summing-up statement means in terms of our spatiotemporal structure that a mental triad broke forth in a precious present during Amundsen's fifteenth year, which set him going on a course of dogged preparation for exploits in a distant, objective future. Reading about the courageous exploits of Sir John Franklin, he was swept up in an emotionally charged vision which became the sustaining center which energized his life's work, culminating in his locating the exact position of the South Pole.

His life is an inspiring testimony to the meaning of suffering. This extraordinary fifteen-year-old saw clearly that he too had to dare to face the challenge of death, if he wanted to follow in the footsteps of Sir John Franklin. His first encounter with his possible, personal death in his icy cave was repeated time and again in the dangerous profession he had chosen. His life is an inspiring saga of man's struggle with the prospect of death and dying and of a heroic struggle against the ultimate destroyer. He perished during an ill-fated rescue operation to save Nobile, the Italian explorer who attempted to fly over the North Pole.

Roald said that Sir John's suffering attracted him most. But he did not want to make himself unhappy for the sake of suffering. He realized that he had to impose periods of physical and mental suffering upon himself in order to reach the goal which he had set for himself. He had to go through initial phases of doubt, of being ridiculed, of facing financial hardships, while he prepared for the goal he was determined to reach. He knew that if he should be successful he would experience gratification, omnipotence, pride, and bliss. But happiness was not his goal, no more than suffering. Amundsen felt attracted to suffering because he knew intuitively that suffering would generate the energy which would drive him onward, notwithstanding the misery, disappointments, and failures that were sure to block his way.

It was not as if first he chose to suffer, then he chose his goal, and then he looked forward to the ultimate goal of happiness. Striving toward a goal, organizing the means to reach it, he was fully prepared that he would have to suffer and if, finally, he should be successful, he would reach a state of bliss. But one phase of this process does not necessarily follow the other, as if they were compartmentalized elements in the before-and-after in objective time-space. They are the inseparable elements of one and the same interpenetrating whole. The mental triad and the precious present forge the constituent elements of goal-striving activities in an interpenetrating gestalt.

Roger Bannister has given us the rare opportunity to partake with him of the early experience when the decision to become a runner first began to take shape in his mind during a precious present. He communicates the intense feeling tone of this germinative experience in very poetic language.

> I remember a moment when I stood barefoot on firm sand by the sea. The air had a special quality as if it had a life of its own. The sound of the breakers on the shore shut out all others. I looked up at the clouds, like great white-sailed galleons, chasing proudly inland. I looked down at the regular ripples on the sand, and could not absorb so much beauty. I was taken aback, each of the myriad particles of sand was perfect in its way. I looked more closely, hoping that perhaps my eye might detect some flaw. But for once there was nothing to detract from all this beauty.
>
> In this moment I leaped for sheer joy. I was startled and frightened, by the tremendous excitement that so few steps could create. I glanced around uneasily to see if anyone was watching—a few more steps—self-consciously now and firmly gripping the original excitement. The earth seemed almost to move with me. I was running now, and a fresh rhythm entered my body. No longer conscious of my movement I discovered a new unity with nature. I had found a new source of power and beauty, a source I never dreamed existed.
>
> From intense moments like this, love of running can grow. This attempt at explanation is of course inadequate, just like an analysis of the things we enjoy, like the description of a rose to someone who has never seen one.[11]

Let us contemplate this inspiring moment in his childhood against the progression of his career as a runner, which culminated decades later in his stunning success. The initial uplifting precious present carried him along through decades of boring, interminable, and miserable periods of training, in order to reach the goal he had staked out for himself. The blissful moment of his success had the power to lift him out of the fainting spell which was the result of driving his recalcitrant body to the limits of endurance.

Suffering is the tunnel through which we have to pass in order to reach the wider horizons of the goals for which we are striving.

NOTES

1. *Webster's New International Dictionary*, 2d ed., s. v. "mind."
2. B.E.R. Kirwan and P.A. Gore, *Elementary Luganda* (Kampala, Uganda: Bookshops, 1951), pp. 62-63.
3. Leonard W. Doob, *Patternings of Time* (New Haven: Yale University Press, 1971), p. 122.
4. Charles M. Sherover, *The Human Experience of Time* (New York: New York University Press, 1975), p. 555.
5. *New Columbia Encyclopedia*, s. v. "Gestalt."
6. William Zorach, *Zorach Explains Sculpture: What It Means and How It Is Made* (New York: American Artists Group, 1947), p. 19.
7. B.H. Haggin, *The Toscanini Musicians Knew* (New York, Atheneum, 1975), p. 157.
8. Roald Amundsen, *My Life as an Explorer* (Garden City, N.Y.: Doubleday, Pace, 1927), p. 2.
9. Ibid., p. 70.
10. Ibid., p. 269.
11. Roger G. Bannister, *The Four Minute Mile.* (New York: Dodd, Mead & Co., 1960), p. 11.

4

Extrospection and Introspection

GENERAL REMARKS

Having separated phenomena which happen in objective time-space from phenomena which occur in subjective time-space, we are obliged to follow two contrasting routes of observation. We have to assume an outward-bound approach toward phenomena occurring in objective time-space and we must assume an inward-tending attitude toward phenomena which happen in subjective time-space. What goes on in objective time-space must be studied by extrospection and what happens in subjective time-space must be studied by introspection. Methods of observation and spatiotemporal structures of phenomena must correlate. Extrospection is the method of observation which the natural scientist follows, while students of human mentation and human behavior use the method of introspection.

Having neatly separated these two contrasting methods of observation one will see that this separation is not that sharp in the actual work habits of these two groups of scientists. Rather, one often sees transitions and collaborations between extrospection and introspection. The either-or of theory becomes more-or-less in scientific practice.

An essential difference exists between phenomena which occur in objective time-space and those which happen in subjective time-space. The former are emotionally neutral while the latter are emotionally charged. Extrospection is adequate for the first group; introspection is the indicated method of observation for the second group. Here too, the theoretical differentiation between emotionally charged and emotionally neutral phenomena is challenged by the fact that many mixtures and transitions exist between these two types of data and experiences. Hence, extrospection as a factual activity is always emotionally charged to a lesser or greater degree, while introspective inquiry can be performed with a minimum of emotionality. Living in the world brings about an intermingling of extrospection, with its tendency toward emotional neutrality, and introspection which tends to bring emotionally charged, private experiences into view.

Merleau-Ponty supports a similar dualistic view: "We are concerned with understanding the relations between consciousness and nature, of the interior with the exterior. It is a matter of knowing how the world and man are accessible to two kinds of research, one kind explicatory, the other reflective."[1]

His explicatory research is similar to the extrospective approach which seeks to quantify phenomena, versus reflective research which suggests that introspection foregoes the goal of quantification since it endeavors to gain insight into a human being as an interpenetrating whole.

There is a further complication which hampers our placing extrospection and introspection in stark contraposition to each other. One observes causally structured events which are emotionally neutral exclusively with the extrospective method, which can stand alone in this area of inquiry. But to observe purpose-striving events which are emotionally charged solely by introspection is a one-sided and faulty approach, since purposive movements always proceed in the company of mechanical processes. Therefore, introspection calls upon extrospection as its coworker. The search for reflective insight into purpose-striving phenomena requires not only that we be aware of the differences between the two methods of observation, but also that we follow both routes at the same time and in conjunction with each other. When we enter the inner, subjective world by introspection, we ought not to lose sight of the outer, objective world about which we know through extrospection. The need to follow two opposite methods of observation is one of the major considerations that plagues the work of students of purposive behavior (biologists, psychologists, psychiatrists, historians) who are interested in the inner states of the mind of man and the mentation of animals. On the other hand, if a physiologist should travel this path, he might introduce mentalistic attributes into cell and organ functions which are causally structured.

The other extreme approach must be avoided also: some biologists who study the behavior of animals, some psychologists, and some psychiatrists rely exclusively on extrospection in their research and they may prefer not to travel the introspective route. They construct one-sided and inadequate interpretations of the behavior and the inner states of man and of animals as whole living organisms and they fail to describe mentation.

Since our world is beset by many contrasts and dichotomies, our methods of observation must be geared to these blemishes, opposing tendencies, and fractures. In order to grasp man in his dignity as man, we have to go beyond the clear-cut extrospective approach and the monistic theory of natural science. Although extrospection does not provide a view of man as a whole, rational, emoting, communicating, suffering being, it does not follow that extrospection prevents our being emotionally involved in our extrospective work and its results. Extrospection has helped us create the tremendous edifice of natural science, whose coherence, power, and usefulness affects us with great emotional impact. Moreover, no natural scientist would have the drive and the persistence to follow the route of extrospection if he were not emotionally involved in his work. He will regard this emotional satisfaction, which he can observe by introspection, as an additional and incidental reward for his work, but he does not make his internal state part and parcel of the quantitative results of his extrospectively conducted research.

The contrasts between the two methods of observation come to the fore in particular when we observe ourselves as a whole of mind and body. By focusing our attention on what is going on in this or that part of our body we bring these events into view by extrospection, but introspective awareness of the meaning of these body sensations may be unbeknown to us. In some instances we may actively suppress introspection, especially in case we suffer from painful sensations. I am thinking of the physician who was awakened by pain in his chest, which gradually began to irradiate to the left shoulder and arm regions. He knew that this spelled out coronary occlusion. While he made the necessary arrangements to admit himself to the intensive care unit of the local hospital, he kept his emotions effectively under control, which means that he had to restrain introspective awareness of his state of apprehension and dread. Undoubtedly he was torn between emotionally neutral extrospection and emotionally charged introspection.

Physicians are caught in between the objective approach and the intuitive attitudes when they are called upon to treat seriously ill patients. In these situations introspection has to be subdued because extrospection has to take the leading role.

Emotionality accompanies both approaches; it is more in the foreground when we follow the introspective approach, but it is also essential to extrospection. Our feelings of tension, apprehension, anxiety, panic, and even dread, unify this dichotomy, as we shift back and forth between the outward- and the inward-bound attitudes. Concrete living and suffering within the contraposed dichotomies synthesize the two approaches toward ourselves, the others, and the world which surrounds us.

The two methods of observation do not develop in parallel. Extrospection is already active in a primitive form during the ninth month in utero, when the unborn infant reacts to various sensory stimuli which reach it in its aqueous environment. In contrast, introspection develops much later. As a child moves into his teens, introspection and extrospection begin to differentiate out of the primordial, undifferentiated fluidum. The preoccupations with self of the adolescent sets him apart from his former self-enclosure when he was a young child. The teenager becomes excessively and notoriously introspective.

At a later stage of maturation this excessively inward-turned tendency loses its urgency. As the teenager turns into an adult, he begins to observe his world and the people in it more and more by extrospection. He gradually develops an image of the external world which is predominantly causally organized, although some teleological admixtures are held over from earlier stages that lie now behind him.

It is more difficult to develop introspection into a reliable method of observation of ourselves and of others, than it is to learn about causality and the measurement of physical phenomena which occur in objective time-space and which we observe by extrospection. Perhaps this is one of the reasons why it has taken the humanities and the arts a long time to develop introspection into a systematic method of observation. Jacob Burckhardt thinks that Dante was the first poet to use this tool. "When one reads... the extraordinary fragments of his youth, then it seems as if all the poets throughout the Middle Ages avoided themselves, while *he* was the first to search within himself.... The spirit and the soul take suddenly a tremendous step toward knowing their most secret life."[2]

It comes naturally to us to look out upon the world that surrounds us. This is one of the first actions of a newborn baby. It is difficult to look into ourselves, and extrospection is likely to dominate the inward-turned attitude. Writers and poets have used and developed the skill of introspection for centuries since Dante. Similarly, historians, sociologists, anthropologists working in the field, and especially psychiatrists listen to, identify, and empathize with the human beings whom they study. They bring the rational processes, the feeling tones, the strivings, and the sufferings of their subjects into the open and into the descriptions of the human material which they have uncovered. However, these scientists do not follow the introspective route to the exclusion of extrospection. On the contrary, they observe and carefully describe the external appearance and behaviors of their subjects, using the extrospective method. The inward-turned attitude does not in the least prevent us from being also active extrospectively. The two, combined methods complete each other and give us an image of goal-striving human beings who are also causally determined, who live in subjective time-space and in objective time-space, who are interpenetrating wholes as well as compartmentalized units, whose structures and functions can be measured.

EXTROSPECTION AND SENSORY PERCEPTION

Extrospection is the final stage of a complex series of perceptual processes. It begins with sensory impressions which bombard the receptors in our organism from without and from within. Sherrington has brought some order to this multiplicity of sensory impressions. He has divided them into those that arise inside the body versus the distance receptors (vision, hearing, smell) which receive stimuli coming from the ambient world. He divided the first category into exteroceptive sensory impressions which arise in the skin and the mucous membranes; the proprioceptive variety which start in the muscles, tendons, and the labyrinth; and the interoceptive impressions which are generated in the heart, the pleura, the lungs, and the guts. The vast majority of these stimuli do not reach the cerebral cortex due to a filtering out process that starts in the periphery. Gunther Stent has described this process in the case of visual impressions which are being abstracted, starting in the retina and ending in the cortex.

> There [is] a two-dimensional array of about a hundred million primary light receptor cells—the rods and the cones. The retina contains not only the input part of the visual system, however, but also the first stages of the internuncial part. The first internuncial stages include another two-dimensional array of nerve cells, namely, the million or so ganglion cells. The ganglion

cells receive the electrical signals generated by the hundred million light receptor cells and subject them to information processing. The point-by-point fine-grained light intensity information is boiled down to a somewhat coarse field-by-field light contrast representation. Here (in the visual area of the occipital cortex) the signals converge on a set of cortical nerve cells. A given cortical cell becomes active if a straight line of a particular orientation...is present in its receptive field. The stimulus requirements of complex cells [respond] upon parallel displacements of the straight edge. Thus the process of abstraction of the visual input begun in the retina is carried to higher levels in the cerebral cortex. In the case of man, with his vast semantic capacities...the notion of a single cerebral nerve cell as the ultimate element of meaning seems worse than a gross oversimplification; it seems qualitatively wrong.[3]

Stent shows that a purely physiological model of perception is inadequate no matter how refined it becomes with advancing research. We add more and more mentalistic attributes to the physiological processes which develop through the stages of sensory perception to apperception and finally to extrospection. Wilhelm Wundt has further developed the concept of apperception conceived by Leibnitz.

If we say that representations which are given in a present moment...reside in the field of vision of consciousness, then one can designate these parts of the latter which are turned toward attention as the inner point of the field of vision. We wish to call perception the entry of a representation in the inner field of vision of consciousness and apperception its entry in the point of the inner field of consciousness.[4]

Furthermore, and I paraphrase Wundt, apperception is not only influenced by present impressions but also by earlier experiences. It depends on the total condition of consciousness. Since we experience apperception as an internal activity with subjective meaning, we begin to understand apperception's feeling tone. Going one step beyond Wundt's description, one can instill feeling tone, pastness, presentness, and futurity as constituent attributes of apperception. The original sensory impressions, filtered out to manageable dimensions, and with intentionality and the time-binding faculty added, become meaningful, useful, and valuable data of consciousness for an acting, rational, emoting, communicating human being. In the mode of apperception, sensory perception reaches the status of extrospection and on this level it can cooperate with introspection.

A problem which plagues the differentiation of extrospection and introspection is that the former method of observation is publicly verifiable and the latter is not. One challenges the usefulness of introspection because one doubts that it is accessible to truth-value tests. This critique of introspection is derived from a similar situation between interoceptors and distance receptors. When two persons see the same object, one presumes that they can compare their visual impressions directly and that they can subject them to a truth-value test. But if one person complains of pain in his right calf muscle, the other cannot observe the complainer's subjective sensations directly. He can only imagine and reconstruct indirectly what the other feels in his calf, by comparing the reported pain with remembered pain which he may have felt in his own calf some time in his past.

However, when these two people compare the visual image they receive and construct it through apperception, they cannot publicly verify this construct either; first, they see the object from different angles, and second, the psychological and mental elaboration of the causally organized original stimuli creates a different visual image in each of the two viewers. As Wundt points out, we bring our personal history into the process of apperception.

If one anchors the truth-value of perception on public verifiability then one is on the road toward solipsism and the denial of the possibility of interpersonal communication. Since we know for certain from our own and our fellow humans' experience that man is a social being, one who maintains that interpersonal communication is a figment of the imagination suffers from blind spots in his outlook upon man, as if he were an island unto himself. Introspection and its ancillaries correct this fallacious interpretation of human existence.

INTROSPECTION AND ITS ANCILLARIES
William James has not the slightest doubt that introspection is a valuable and scientific method of observation, *sui generis*.

> Introspective observation is that we have to rely on first and foremost and always. The word introspection need hardly be defined—it means, of course, looking into our own minds and reporting what we discover.... All people unhesitatingly believe that they feel themselves thinking and that they distinguish the mental state as an inward activity or passion, from all the objects with which it may cognitively deal. I regard this belief as the most fundamental of all postulates in Psychology.[5]

However, we should define the term *introspection* more clearly than James did, not only by contraposing it to extrospection but also by challenging the very idea that we "look" into ourselves. Introspection is of a different ilk than extrospection since it does not depend on stimuli coming into the organism from the environment, nor is it similar to the interoceptive types of perceptions which originate inside our body. Both terms are defective since they indicate two different ways of looking. Extrospection encompasses all sensory modalities, not just vision. The term *introspection* is even more misleading, since we do not look into ourselves— we listen to ourselves. Therefore I prefer the term *intra-audition*.

If intra-audition is not a form of perception, if it is not dependent on auditory stimulation, then what is it that one is listening to during intra-audition? One listens to oneself talking with, to, about, oneself. It is an internal dialogue which goes on some of the time and which is drowned out much of the time by the bombardment of external and internal stimuli to which we are exposed during our waking life. This is usually a silent dialogue but many people talk out loud to themselves. In either case, intra-audition operates within the medium of language as one speaks to oneself and listens to what one has to say. It emphasizes the linguistic and auditory, rather than the visual characteristics of introspection.

We humans are primarily visually organized beings, but it seems as if the dominance of the eye over the ear is turned around by the verbal exchange with oneself. Language, silent speech, and the inner ear create auditory dominance over the visual. The inner language activity goes on not only when we are puzzling over some intellectual problem but even more so when emotionally charged difficulties are harassing us. Intra-audition makes us aware of the feeling tone of our internal states. Thus far intra-audition completes the more rationally attuned functions of extrospection. We know ourselves as rational as well as emotional beings by the route of intra-audition. Over and above this inward-turned function it plays an important role in interpersonal communication. In order to fulfill this role introspection and intra-audition need the cooperation of their three ancillaries, which are the subtle attitudes of empathy, identification, and projection. Together this foursome composes the grounding functions which make interpersonal communication possible; they give us insight into the mental and the emotional experiences of our fellow humans.

Empathy with the other person, feeling into the others (*Einfühlung*), means that we absorb the meanings of the words which the other speaks and also that we are aware of the nonverbal messages which he sends by his manner of speech and its delivery, the expression of his eyes, his facial expression, his body positions, and his body movements. One sees at this point that introspection and extrospection collaborate during interpersonal communication.

Roy Schafer elucidates, in a thought-provoking paper on empathy, the functions of introspection and its ancillaries.

> We deal, so to say, with the analysand's organization and presentation of a second self. The second self is, of course a version of what that person presents in nonclinical relationships.... It does not follow from this proposition that the analytic self is in some absolute sense a better self.... I propose that he or she [the analyst], too, operates through what may be called the organization and presentation of a *second self*.... I wish to emphasize next that this second self presented by the analyst is analogous to the second self presented by creative writers.... Like empathizing, creative writing is in this respect no innocent activity; it is thoroughly transformed by existing systems of thought as these have been embodied in literature and other writings, past and present, including the author's own previous work.... As yet, however, we understand imperfectly the mediating actions by which analysts and writers organize and present these second selves, selves that can be different from, even if not thoroughly discontinuous with, their everyday selves.[6]

Elaborating on Schafer's presentation, I see the first self as one's everyday self which fits into one's superficial social roles; it is the self of the average human being, *das Man* of Heidegger. This self exists primarily in objective time-space, and it observes the world and the others and itself predominantly by extrospection. It is visually oriented, and it compartmentalizes its surrounding world and the human being who possesses this first self.

The second self is one's creative self which exists mostly, in subjective time-space, which observes primarily by intra-audition and by empathizing. It is auditorily inclined, and it conducts a dialogue with its first self. The second self communicates with one or more second selves of the other or of the others. They communicate with each other so that they are in touch with each other's inner states and feelings. That is, they empathize.

Identification rests on our conviction that the others possess mental attributes and have developed social attitudes and behavioral habits which are similar to our own. We rely on what we know about ourselves through introspection, and against this background of self knowledge we judge how far we can go in identifying with the others and whether we can or cannot trust that the verbal information we get from them is useful and valuable to us.

Projection is similar to identification since it invests the other person with characteristics which we find also within ourselves. The projective attitude has to be monitored carefully, and it must be critically evaluated since we may hastily assume that the other person possesses certain characteristics, an assumption which may prove to be totally erroneous in the long run. Faulty or too optimistic projection can interfere with personal relationships, and it can even ruin a promising relationship as one has to revise one's appraisal of the other's personality assets and deficits.

Harmonious, meaningful, interpersonal communication is built on the interrelated functioning of these four activities, and, if they are distorted or if one or more should be absent, interpersonal relationships degenerate into blandly sending and receiving sensory stimuli to each other. Jonathan Swift conjures up images of such distortions in Gulliver's travel to the Laputians on their flying island. His satirical fantasy isolates these astronomer-mathematicians on their island and they are also mentally isolated from each other because they are locked into fear-ridden preoccupations about future cosmic disasters.

> I observed here and there many in the habits of servants, with a blown bladder fastened like a flail to the end of a short stick, which they carried in their hands. In each bladder was a small quantity of dried peas or pebbles (as I was afterwards informed). With these bladders they now and then flapped the mouths and ears of those who stood near them, of which practice I could not then conceive the meaning; it seems the minds of these people are so taken up with intensive speculations, that they neither can speak, nor attend to the discourse of others, without being roused by some external taction upon the organs of speech and hearing; for which reason

those persons who are able to afford it always keep a flapper (the original is *climenole*) in their family as one of their domestics, not even walk abroad or make visits without them. And the business of this officer is, when two or more persons are in company, gently to strike with his bladder the mouth of him who is to speak, and the right ear of him or them to whom the speaker addresses himself.[7]

The verbal exchanges between these self-contained people is no more than sending and receiving of sensory stimuli within a mechanically contrived interpersonal contact. These compartmented activities are staked off by punctiform instants which the Flappers provide. They are imprisoned in a world of applied mathematics which shuts them off from their fellow citizens with whom they are unable to empathize. Theirs is a life without interpersonal communication.

In real life, the complex temporal structure of communication interweaves extrospection with introspection and its ancillaries. No doubt, communication is based mostly on the alternating auditory activities of speaking and listening, but visual contact adds a powerful dimension to the relationship. Eye contact seems to happen, mostly, in objective time-space because it can be nearly instantaneous, and it usually goes on simultaneously between two or more people. When we look intently at someone, our eyes and the other person's eyes seem to stand still in space and time. But this is not so. Our eyes move constantly over the outlines of an object, and even when the regard seems to be fixed upon one point in space, the eyes move over minute distances. Hans Syz has proved this with an ingenious instrument using corneal photography.[8] Eye contact gives us a great deal of subtle information about the inner mental processes of the other person by the combined methods of extrospection and introspection. Two lovers, looking at each other, share a highly charged emotional experience as they partake of a deep mystery. A third person who happens to be there, can be moved powerfully by this visual fusing of two human beings.

Compared to quasi-instantaneous eye contact, communication by speaking and listening takes a lot of objective time, since these alternating, more active and more passive sets of opposite behaviors go on in a rapid back and forth movement. The person who is speaking listens inwardly to himself, and he tries to fathom the other person's reactions at the same time. Then he waits for an answer or a retort. In the meantime, the listener is synthesizing what he has heard the speaker say, and these words are falling back into the immediate past. Now the listener takes his turn to speak, and the previous speaker who is now the listener judges if his message which he sent in the immediate past has come across. Now he integrates what he is hearing with what he did say and is planning what to say next. The two conversants are constantly shifting the emphasis from projecting into the future-in-the-fringe of a precious present to evaluating memories which they have condensed in the past fringe. The structuring of a conversation in subjective time-space stands out more clearly in the auditory sphere then in the visual realm. Also, introspection and its ancillaries are more in the foreground during verbal contact compared to visual contact when extrospection seems to dominate communication. The temporal structure of almost instantaneous visual contact, seems less complex compared to the back and forth flux of speaking and listening in alternating succession.

The combined activities of sensory perception, apperception, extrospection, and introspection, are completed by the time-binding functions of the mental triad in a precious present. At this point, emotionality enters into the picture. In this progression from physiological phenomena which go on in objective time-space to the creation of new emergents in subjective time-space, we go forward from man as a rational agent who compartmentalizes sensory impressions and who lives in isolation, to man as a complete, rational as well as emotional being, who synthesizes interpenetrating wholes and who communicates with his fellow humans.

MAN AS AN OBJECT VERSUS MAN AS A SUBJECT

We can study man as an object by extrospection, and we can understand man as a subject by introspection and its ancillaries. The work of a physician who examines, diagnoses, and treats sick people, is an instructive example of the separation and the reconciliation of these two opposite approaches to one's subject matter.

While a physician examines a patient, he relies almost exclusively on extrospection and isolates this approach from the introspective attitude. He regards this human being temporarily as a quasi object which exists in objective time-space, which is causally structured, and which he studies with the methods of natural science. Taking a history he listens to and he speaks to the patient for the purpose of getting clues to what organ or organ systems may be causing the patient's complaints. He looks at, listens to, touches, and manipulates the patient's body during the physical examination, and next he may order laboratory tests, X-ray pictures, EKGs, EEGs, and numerous chemical tests. He compartmentalizes the patient's body into organs and organ systems, and, when he makes a diagnosis, he puts these compartments together again to see how the diseased parts fit into and interfere with the normal structure and function of the body. Then he prescribes therapeutic measures that will, hopefully, rectify the abnormal physical processes that are going on in the patient's body. Up to this point the doctor's work is similar to that of a natural scientist who gathers data by extrospection.

However, most physicians do not break off their concern with the patient at this point. Another aspect of the doctor-patient relationship comes to the fore which so far has remained in the background, playing a subordinate role. Right from the start, most doctors see their patients not only as compartmentalized physical organisms, but also as whole, purpose-striving, rational, emoting human beings who, besides being ill in the physical realm, are also caught up in a network of interpersonal relations. It is important to be aware of one's patient's subjective attitude toward his illness, toward the doctor, the prescribed treatment, and the many subtle, nonquantifiable factors which have a decided influence on the course of physical illness. We cannot grasp these subtleties by extrospection, but we must be sensitive to them by introspection and its ancillaries. In short, the physician shifts from proceeding mostly by the extrospective route toward a combined extrospective and introspective approach toward the patient as a whole human being, thus building toward a person to person, doctor-patient relationship.

Physicians are trained to observe man as if he were a physical object, for the express purpose of suppressing and suspending their natural inclination to see their fellow humans as subjects. Within the limited medical-surgical context, one has to suppress introspection, identification, empathy, and projection so that one is able to be objective about the patient's plight. After his work as a natural scientist is completed the doctor can free himself from these restraints, and he may allow introspection and its ancillaries to join with extrospection.

As is the case in all areas of human existence, this is not an either-or dichotomy, but a more-or-less attitude, where one is shifting from a mostly extrospective toward a mostly introspective approach and then again toward extrospection. Since introspection does and extrospection does not bring emotionality into view, it is obligatory for the physician to forego introspection during certain phases of his work, so that he is minimally involved emotionally—doing his work objectively and therefore effectively.

The practice of medicine and surgery is an example of our complex spatiotemporal structure which is living, mostly, in objective time-space and shifting toward the other side of the spectrum which is living, mostly, in subjective time-space. *Mostly* is the key word in this statement since we never live at either extreme of the temporal spectrum during normal, waking life. Sometimes we live more as if we were an object which we can observe by extrospection, and at other times we live as if we were exclusively a subject, with whom we are in touch introspectively.

It is a rare physician who can separate extrospection and rationality sharply from introspection and feelings. But the intermingling of rationality with emotionality is also a fact of life for the astronomer. Astronomy is the extrospective science *par excellence* and it has been a predictive science from time immemorial. This fact is exemplified by the Maya calendar. Mathematics, ritual, religion, astrology, must have carried an enormous emotional impact for these people, as evidenced by their considering the five remaining days of a year's end as evil.

> The wheels of time ground exceedingly fine for the Maya. To be able to predict the seasons for farmers and astronomical events for religious rites, they utilized a calendar of two meshing repeating cycles. Maya mathematicians could project this calendar millions of years into the past and the future; time had no beginning, no end.[9]

Clearly, mechanistic techniques and animistic views of the cosmos intermingled, as is the case in all forms of astrology. It has taken a long time to cleanse astronomy proper from the emotionally motivated, astrological functions. Even Tycho Brahe and Johannes Kepler believed that astrology influenced human life.

> Even strong men of clear vision like Tycho Brahe and Kepler, could not quite liberate themselves from the influence of astrology; indeed, they were of the opinion that if one only cleansed it from the crude foolishness which was attached to it, then a valuable nucleus would remain.... Astronomy was practiced in the old monasteries because without it religious service could not be celebrated on fixed and definite days.[10]

The discovery of Neptune by Le Verrier is a fascinating example of the predictability of celestial movement within a mechanistic cosmic image. Astronomers had been puzzled for some time by unexplained deviations in the elliptical path of Uranus around the sun. Le Verrier hypothesized that if there were an as yet unknown and unobserved planet with a wider orbit than Uranus, the irregularities in its path might be explained. He calculated the size of the planet he envisioned and its orbit, and then he advised Galle of the astronomical observatory in Berlin to scan the evening sky on a specific day of the year 1846. The existence of the second planet was verified in the exact location and time which Le Verrier had predicted.

I am convinced that Le Verrier must have felt ebullient emotions of omnipotence for being able to make and verify this prediction, and de Galle and his colleagues undoubtedly had feelings of admiration for Le Verrier.

> Authoritarian and easily offended... Le Verrier conducted his projects like battles in which, however, he displayed more emotion than strategy. As a result, in the course of his career he acquired the friendship and often the admiration of the greatest foreign scientists while quarreling with most of his French colleagues who were in more direct contact with him.[11]

Kepler intermingled emotionally charged personal reactions to his disappointments and his discoveries with the coolly rational data of his observations. Le Verrier seems to give no direct evidence in his scientific writings of such emotional involvement in his work. We get only a brief glimpse of this side of his personality in Lévy's description of his personality. The excitement which Le Verrier's discovery caused is described by A. Pannekoek. "This course of events made a deep impression on the world of scientists, but no less on the world of educated laymen. From all countries honours were showered upon Le Verrier, and the discovery at a desk, of a body never seen, was the ruling topic for a long time."[12]

Modern astronomers have learned to separate their private feelings from those which flow from the results of their work as scientists. They do not project their own future-attuned actions and their emotions onto the heavenly bodies like the astronomers-astrologists of ancient times used to do. Natural scientists keep identification, empathy, and projection in check insofar as they are natural scientists. Personally, they are usually quite well aware introspectively of their emotional involvement in their ongoing studies.

Aren't we all overwhelmed on occasion by an oceanic feeling when we contemplate the magnificent sight of the sky during a clear night? But these feelings have neither rationale nor function within the mechanistic, causally structured events occurring in objective time-space which astronomers study extrospectively.

One can interpret this oceanic feeling from two opposite directions. This feeling can stem from our intuitively identifying with the natural phenomena which we observe, or we can make ourselves introspectively aware of the feelings of omnipotence over these phenomena on the rational level. Look at a fair-sized tree, and you can feel a tremendous power locked up within this static structure that is able to climb up so far away from the surface of the earth and in which innumerable and largely not-yet-understood physical, chemical, and electrical processes are going on. One is overwhelmed by the power and the complexity of the tree. One certainly did not make the tree; it is given to one through extrospective observation, which has been refined by the methods of natural science.

But there is another side to the feeling of power and the oceanic feeling: natural science *is* of our own making. Le Verrier was the one who calculated and who directed Galle where to look for Neptune. The primitive oceanic feeling belongs in the animistic world image, the rationally evolved oceanic feeling fits into the cosmic image of natural science.

We have no power over our being 'thrown-into-the-world', nor over the termination of our terrestrial existence. But in between these two states of powerlessness, we do partake legitimately of oceanic feelings and feelings of omnipotence. Thus we seesaw to and fro between impotence and omnipotence as we experience one of the many dichotomies in our spatiotemporal existence.

Mathematicians are acutely aware of the emotional impact which they undergo during their work in this most exact and abstract of all sciences. Saul Kripke was deeply interested in our emotionality. "What I think is important is whether emotions have introspective qualities. I would think that fear is not just a belief that something is dangerous and a corresponding tendency to avoid it. There is something more, and what is it?"[13] His offhanded mentioning of introspection shows that he is deeply interested in his own inner processes of mentation. This points up that scientists are not cold-blooded collectors of facts and that mathematicians are not feelingless constructors of intricate mathematical formulae. Another contemporary mathematician compares his battles with problems of formal logic with the exciting and frightening experiences during parachuting.

Kripke senses that there is more to fear than just a physical avoidance reaction; that "something more" is greater than present imminent danger—it is that existential dread which Kierkegaard derived from our dread of our future, distant death. Dread derives from our temporal-spatial structure, from the tension between our projecting our activities into the future fringe of a precious present versus our calculating and predicting of the objective future which reveals to us the foreknowledge of our eventual, personal death.

Psychiatrists, psychologists, and historians develop different attitudes toward the material which they study as compared with the approach of natural scientists. The former rely on introspection and they do not suppress or evade empathy, identification, and projection; on the contrary, this is part of their methodology. Professor Wallace Ferguson, a Renaissance specialist, told me that he feels he knows some of the Renaissance personalities, whose history he studied in great detail, better and more intimately than many of the people with whom he has daily contact in his professional life. The historian has to be emotionally involved with his human subject matter in order to make history really come to life.

Professionals who work in the humanities are scientists, not too different from natural scientists, since they too strive for objectivity, they sift critically what they see and what they hear, they organize accumulated data within specific frameworks. But unlike natural scientists, they do not separate extrospection from introspection. They strive to reach a balanced intermingling of these methods of observation since their goal is not to eliminate feelings, and purpose-striving, and values from the results of their work. If they left emotionality, interpersonal communication, and intentionality out of their scientific data, they would fail to reach the goal they endeavor to attain.

If one compares the methodologies of the natural scientist, who relies exclusively on extrospection, and those of the historian who adds empathy to his work with human subject matter, with those of the psychiatrist, who develops systematic cooperation between extrospection and introspection and its ancillaries, we see a progression from extrospection proper toward gradually increasing, complementary reliance on introspection. The historian adds empathy, as it were, intuitively, and he usually does not lean toward introspection, indentification, and projection. But the psychiatrist intentionally organizes the outward-bound and the inward-turned attitudes into a systematic cooperation between extrospection, introspection, and its ancillaries.

Psychiatrists are trained not only to be aware of the patient's reactions within the therapeutic hour, but also to their own reactions to the patient's verbal messages and behavioral signals. The psychiatrist empathizes with the patient in order to register as clearly as possible what the patient is feeling and to understand the meaning of what he is saying. He introspects so that he is aware of his own feelings which the patient arouses within him and of how he feels toward the patient. He checks whether he identifies unduly with the patient insofar as he may see some of his own neuroticisms in the patient, and he is careful not to let identification distort the transference relationship. He checks to see if he projects his own personal attributes onto the patient, which might blur the image which he is building of the patient's character.

The psychiatrist keeps the foursome of introspection, and its ancillaries in the foreground of his self-critical attitude, which processes he examines not only in the patient but also within himself. All the while he also observes the patient by extrospection, which intermingles with and completes the inward-turned attitude and processes. Harrison and Guiora stress that overvaluation of studying the world about us unfortunately can stifle the inward-turned attitude which opens up insight into ourselves.

> Over the ages an ever-growing discrepancy has developed between people's knowledge of themselves and the knowledge of the external world... infant maturation and development are marked by the baby's increasing capacity to focus attention externally—away from themselves.... children are encouraged not to explore themselves—either their physical self or their inner feelings and motivations. Is it conceivable that [the idea that self-analysis] is considered so daring stems from the thought that self-knowledge is potentially dangerous and upsetting to the point of dislocating humans' self-concept?... It is as if we are trapped in an epistemological cul-de-sac, so that those who place their faith in human intelligence to assure human survival may make the same mistake as those who rely on advances in technology.[14]

I want to use this theoretical scaffolding as a take-off point for examining the blemishes, the dichotomies, and the fractures in our existence. It consists of two contrasting and cooperative sets of concepts and methods. On the one hand, objective time-space, causality, and extrospection are adequate to analyze compartmentalized phenomena. On the other hand, subjective time-space, purpose-striving, and introspection study interpenetrating experiences. To understand these activities we need the mutual cooperation of both interlocking series of concepts and methods of observation. To reach our different goals of quantitative knowledge about compartmented phenomena versus gaining insight into our own interpenetrating holism, we need two forms of time-space, two categories, and two methods of observation.

NOTES

1. R. Merleau-Ponty, *Phénoménologie de la Perception* (Paris: Edition Galimard, 1945), pp. 489-90.
2. Jacob Burckhardt, *Die Kultur der Renaissance in Italien*, 14th ed. (Leipzig: Alfred Kroner Verlag, 1925), pp. 289-90 (Author's translation).
3. Gunther S. Stent, "Limits to the Scientific Understanding of Man: Human Sciences Face an Impasse since Their Central Concept of Self Is Transcendental," *Science 187* (March 21, 1975), 1052-57.
4. Wilhelm Wundt, *Grundlage der Physiologischen Psychologie*, revised ed. (Leipzig: Verlag von Wilhelm Engelmann, 1893), p. 267.
5. William James, *The Principles of Psychology* (New York: Henry Holt & Co., 1870), vol. 1, p. 185.
6. Roy Schafer, "The Psychoanalyst's Empathy," Lecture to the Westchester Psychoanalytic Association, Westchester Division of the New York Hospital, White Plains, N.Y., May 4, 1981, pp. 12, 29, 14, 16.

7. Jonathan Swift, *Gulliver's Travels* (New York: Dodd, Mead, and Co., 1950), pp. 157-58.

8. Trigant Burrow and Hans Syz, "Two Modes of Social Adaptation and Their Concomitant Ocular Movements," *Journal of Social Psychology* 44, no. 2 (April 1949).

9. George E. Stuart and Imboden Otis, "The Maya Riddle of the Glyphs," *National Geographic* 148, no. 6 (December 1975): 783.

10. A. Diesterweg, "Johannes Kepler," *Diesterweg's Populäre Himmels Kunde und Mathematische Geographie* (Leipzig: Akademische Verlags gesellschafteft, Becker und Erler Kom, 1941), pp. 545, 546.

11. Jaques R. Lévy, "Urban Jean Joseph Le Verrier," *Dictionary of Scientific Biography* (New York: Charles Scribner's Sons, 1970), pp. 276-79.

12. A. Pannekoek, *A History of Astronomy* (London: Allen & Unwin, 1961; New York: Barnes and Noble, 1969), p. 360.

13. Saul Kripke, "New Frontiers in American Philosophy," *New York Times Magazine*, August 14, 1977, pp. 12-67.

14. Saul I. Harrison and Alexander Z. Guiora, "The Intelligent Use of Human Nature," *Journal of Operational Psychiatry* 11, no. 1 (1980): 41-47.

5
Causality and Goal-Striving Activities

The differentiation between causality and goal-striving is not just an intellectual exercise. Besides separating these two categories, we also synthesize them constantly in practical daily living and also in science. We can describe and define this separation and this synthesis from the viewpoint of time-space.

We can separate causality from purposiveness because we can observe causally structured phenomena apart from purpose-striving happenings. But since we observe and perform purpose-striving activities always in conjunction with causally structured events, we cannot separate intentionality from causality. We have encountered a similar and related separability of objective time-space from subjective time-space, while we cannot separate subjective time-space from objective time-space. Causally structured phenomena occur in objective time-space, while purposive activities occur in the combined and interweaving forms of time-space. Causality versus goal-striving is one of the basic dichotomies in our existence, and this generates the divisions of mechanical movement versus purpose-striving movement and determinism versus indeterminism.

I do not pretend to develop the concept of causality along some new lines of thought; on a more modest level, I intend to free from remaining animistic contaminations the mechanistic image of the world which natural science has developed. On the other hand, I will also show how we intermingle causality and intentionality in our goal-striving pursuits.

Discussed through centuries of philosophizing, purpose-striving tendencies have been mixed with causal explanations. One of the reasons for this mixing of the two categories is that objective time-space and subjective time-space have not been clearly differentiated. Cause-effect sequences occur in the before-and-after in dyadic, objective time-space, while means-toward-a-goal relationships proceed in the present-past-future of triadic, subjective time-space. However, one can show a close kinship between causality and intentionality by deriving causality from the mental triad by a process of abstraction. (See chapter 10).

In daily life we think about causal interconnections between events so casually that this concept seems to need hardly any explanation or definition, all the more since language covers over the ambiguity between the categories of causality and intentionality. One might say, "I went to see my friend because we wanted to discuss something in which we are both interested, but I was half an hour late because I had a flat tire." Unless one exerts special effort to analyze this sentence, one is hardly aware of the different meanings of the word *because* in the first part and in the second part of this sentence. In the first instance, *because* refers to a purposeful activity, in the second instance to a causally structured event. The ambiguity of the word *because* exemplifies our confusing the causally structured world in objective time-space with the purposively structured world in subjective time-space—a confusion we make much of the time.

The two concepts about time-space can help to clear up this confusion. Causally related phenomena can be compartmentalized with geometric points-punctiform instants and lines, and we can enclose them in spatial schemas since they occur in objective time-space which we can treat in the same manner. We can apply the number series to cause-effect sequences, and we can express this relationship in the causal equation as a most precise refinement of measurement. Hence the classical statement, '*causa aequat effectum*'. The causal equation is the cornerstone that carries the natural-scientific, mechanistic image of the world. As one example out of many, the law of conservation of energy is based on the causal equation.

Means-toward-a-goal relationships cannot be expressed in terms of a causal equation, since subjective time-space does not admit compartmentalization. Therefore, the sciences which study the behavior and the inner states of man use concepts and follow methods of observation which differ from those used by natural scientists. But the two meanings of the conjunction *because* which are used almost interchangeably in daily life, point out that we mingle causality with intentionality subtly and surreptitiously.

Although the causal equation is a very old concept indeed, it has taken modern science a long time to cleanse the cosmic image of the ancient Greeks and other classical peoples of animistic and indeterministic imagery. One sees the progression from an animistic, teleological view of the world, toward a causally determined, mechanistic interpretation of natural phenomena in two areas of development. Aristotle's cosmic imagery went through various stages of development in the Middle Ages and the Renaissance, until Galileo constructed a clear-cut mechanistic conception of our surrounding world which he described in mathematical terms. The other line of development occurs in all of us: as children we live in an animistic world, and it takes a decade or more before we become aware of the causally determined, mechanistic structure of our surroundings.

Aristotle's system of four causes is the prime example of mixing purposiveness with causality.

> In one sense, then, (1) that out of which a thing comes to be and which persists, is called "cause," e.g., the bronze of the statue. In another sense (2) the form of the archetype, i.e., the statement of the essence, and its

genera, are called "causes" (e.g., of the octave the relation 2:1, and generally number). Again (3) the primary source of the change or coming to rest, e.g., the man who gave advice is a cause. Again (4) in the sense of end or "that for the sake of which" a thing is done, e.g., health is the cause of walking about.... It is plain then that nature is a cause, a cause that operates for a purpose.[1]

Aristotle posited the formal, the material, the efficient, and the final causes and applied this series to an activity such as building a house. The formal cause is the plan which the architect makes of the structure. This done, the material cause enters the picture; that is, one collects the necessary building materials. Thirdly, the efficient cause comes into action as the workmen begin the actual construction of the house. The final cause is represented by the owner taking possession of the house and living in it. In this and in other examples Aristotle hides the concept of purpose in his four causes system since he does not describe a series of cause-effect sequences, but he gives an impressive and coherent description of the purpose-striving activities of *homo faber*, man the maker.

G. B. Kerferd takes issue with Aristotle's theory of causality by saying that the formal cause and the final cause cooperate in the artist's or the artisan's mind which infers that they do not happen one after the other in objective time-space but that they interpenetrate in subjective time-space.

> For Aristotle, to know is to know by means of causes and it is clear that the four Aristotelian causes are necessary elements of things, which must be known or understood if full understanding is reached, rather than causes in the modern sense. The final cause is the form as known by the artist or manufacturer. In this last sense the formal cause may be the efficient cause as well, in that the form, as present in the artist's mind and desired by him in the object is the true source of change that results.[2]

Even in modern times, roughly since the Renaissance, we are still struggling to release ourselves from Aristotle's grip on our thinking about causality. We can fall back on the two time-space concepts, the mental triad and the precious present, as models from which we can fashion a concept about purpose-striving activities. When correlated with the separation of causality from purposiveness we can follow the extrospective route to study causal interconnections and the introspective method to become aware of our goal-striving.

Hume approaches the problem of causality from a new and primarily psychological angle. According to Hume, we unite simple ideas which resemble each other, or which are contiguous, or which stand in relation of cause and effect to each other, into complex ideas.[3] Hume the skeptic thinks that complete skepticism cannot be entertained because of the strength of our natural belief in the necessary connection between cause and effect. To salvage and shore up this subjective feeling of necessity he introduces the power of our natural makeup,

66   The Meaning of Suffering

and thus he creates a hybrid causal-teleological entity with human attributes out of the causal relationship. This deus ex machina creates a mixed causal-purposive cosmic image. Hume the skeptic and the Hume who lives in the world present themselves like two different persons within the same human being. This view of a dichotomized human being fits into the trend of this book, particularly since Hume says that reason ought to be the slave of the passions. This is patently an interpretation of man as a combined rational as well as emotional being.

A. J. Ayers summarizes Hume's definition of causality in his typical incisive, logical manner.

> First, that the relation of cause and effect was not logical in character since any proposition asserting a causal connection could be denied without self-contradiction, secondly, that causal laws were not analytically derived from experience, since they were not deducible from any finite number of experiential propositions, and, thirdly, that it was a mistake to analyze propositions asserting causal connections in terms of a relation of necessitation which held between particular events, since it was impossible to conceive of any observation which would have the slightest tendency to establish the existence of such relations.[4]

Von Wright develops the concept of causality further along psychological lines in that he brings causation and purpose-striving closer together.

> The idea that causal connections are necessary connections in nature is rooted in the idea that there are agents which interfere with the natural course of events. The concept of causation under investigation is therefore secondary to the concept of human action. . . . The act is to interfere with the course of the world, thereby making true something which would not otherwise (i.e., had it not been for this interference) come true of the world at this stage of its history.[5]

Von Wright's thoughts about the acting human being making changes in the future are compatible with the mental triad projecting an act into the future-in-the-fringe of a precious present. This extension of von Wright's proposition is in harmony with his vision of past and future as asymmetrically structured, which is similar to the askew appearance of the precious present with its short and its long future and past fringes.

Jean Piaget adduces evidence along strict psychological lines that causality and purpose-striving are interrelated. The philosopher struggles with the concept of causality on the intellectual level. During early childhood every one of us struggles on a concrete, personal level to get a handle on the concept of causality. Piaget has studied the development of this concept in young children.[6] He shows that the young child structures the world which surrounds him in a poorly organized nonsystem. Piaget concludes that children from ages four to eight years old go through four stages of developing the concept of causality. From age four to five, the pieces of a bicycle are examined in detail, but the children do not analyze the "how" of movement, and they have no idea of how the pieces act upon each other. During a second stage, ages five to six, they give irrelevant answers as to how the mechanism works, but during the third stage, age seven more or less, they at least search for an explanation of the interaction. It is not until age eight or older that little boys give a mechanistic explanation.

A four-year-old thinks that the bike would go even if it had no chain, and he sees no connection between the movement of the pedals and the wheels. A five-year-old says the pedals do not make the wheels go and adds, "It is hard to say." Some say that the bike does not want to go, or that it must go, which Piaget calls "explanation by moral determination." I would prefer to interpret this explanation by the fact that the child projects his own purpose-striving tendencies onto the mechanism. This is the nub of an animistic conception of the inanimate world. The four- and five-year-olds explain the mechanism by the movement of parts which have no spatial contact with each other. During the second stage (ages five to six) they draw a bicycle by juxtaposing parts "without correct insertions, often without even any connection between them." A five-and-one-half-year-old understood that the chain turns the cogwheel, but he thinks also that the cogwheel makes the chain move. Some of them say that the air in the tires makes the bicycle go, and some think once the cyclist has pushed the handle bars and the pedals it goes by itself. Toward ages seven and eight they begin to see that all the parts are necessary to make the bicycle run.

Piaget shows us how the child lives originally in an animistic world in which inanimate and animate objects are propelled by purposive strivings because the young child identifies with everything that he observes and handles. Piaget also shows how the child gradually begins to differentiate between himself and inanimate objects and learns to compartmentalize the world. When he begins to see the interactions between the parts of the bicycle when he becomes eight years old or older, he begins to see the interactions between the parts of the machine and how each compartmented detail assumes a specific function. The bicycle is no longer a telologically structured, purpose-striving whole, but it has become a mechanistically structured unity. Before age seven or eight the child lives predominantly in subjective time-space in the narrow confines of a precious present, and he relates to his surrounding world as if it were a noncompartmentalized, interpenetrating whole. Later on he realizes more and more that the external world is a compartmentalized, causally structured unity which we have gradually stripped of purpose-striving tendencies.

The formation of an abstract concept of causality without teleological admixtures is a slow process which is never quite completed. Kunstadter describes in detail the animistic, teleologically structured thinking of the Lua with whom he lived for almost two years.

Ai Dam began to complain of weakness, fever and "shaking heart." Vitamin pills, aspirins and antimalarial choloroquinine did not help. Now the villagers set about curing him.... The Karens [outside consultants]... had to summon one of Ai Dam's 32 souls that had wandered away. They went to the forest and called the soul, coaxing it into a hard boiled egg... the men stuck the bamboo pipe stem into the ground; if the egg balanced on the stem, it would prove that the errant soul was indeed inside. The Karens... after repeated trials... managed to balance the battered egg and carried it, and the soul up the steps [of Ai Dam's house]. Ai Dam ate the egg—and, sure enough, he was well for the harvests.... The Lua were chased into the mountains by a huge rolling stone—[it] asked a bird whether it had seen the Lua.... All the Lua knew that if they speak the language within the hearing of the stone, it will recognize the tongue and start chasing them again. The people hurried by the boulder in genuine fear.[7]

Piaget shows us how we begin to outgrow involvement in an animistically experienced world during our childhood. Kunstadter lets us see that the Lua remain almost totally submerged in animism. The shift from living in an animistic world, which occurs in subjective time-space and which is teleologically structured, toward becoming aware that the external world is a causally structured mechanism going forward in objective time-space is never quite accomplished. It seems that we never free ourselves entirely from teleological admixtures in our causally structured conception of the world around us. Hence, we tenaciously hold on to the belief that cause brings about effect *of necessity*. Piaget has observed that the young child is dimly aware of some kind of necessity that is operative within the movements of the bicycle. This is a psychological fact of great importance since it relates to our conviction as adults that cause-effect relationships are necessary relationships.

In contrast to Piaget's painstaking research, let us take a look at a bit of common, spontaneous behavior that anyone may observe. This instance of combined causally as well as purposively structured behavior shows that time, space, and movement can be separated only by rational analysis. As you are walking somewhere, you kick a tennis ball that is lying around, and you watch it roll down your driveway and it comes to a stop in the grass. As a causally structured event, the movement of the ball starts the instant you kick it. The ball traverses several yards of objective space and a few seconds of objective time until it stops in another instant. You are convinced that the impact of the tip of your shoe on the ball initiated this movement, that the inertia of the ball caused it to continue on its course, that gravity prolonged the movement because the driveway slopes a bit, the ball's irregular course is due to the irregularities of the surface which also cause the ball to jump a little in a third dimension, and that the resistance of the grass caused the movement to stop. You witness sequences of cause-and-effect in objective time-space, without a doubt. Hume did not think it was that simple. Hume thought that our conviction that a cause is followed by its effect *of necessity* is not a logical necessity, nor that we can prove by experience that this sequence must always be the case in the future.

The description of my kicking the tennis ball as a causal event is incomplete, since one conceives of it as consisting of compartments occurring in objective time-space, separated by punctiform instants and geometric points. What is left out of this description are my personal, subjective experiences. When I see the tennis ball, a sudden intention to kick it grips me, which is reminiscent of my childhood experiences. I bend my knee and I stretch it with the firm conviction that the ball will go flying, and once it is moving I watch its irregular movements, and I wonder how far it will go with a bit of curiosity. Within the experience as a whole, the causally structured world and the purposively organized world interweave. But a natural scientist would tease the compartmentalized, causal aspect out of the holistic experience, since he concentrates on phenomena which happen in objective time-space, and hence he loses sight of my private mental processes.

Living in our world as total human beings, we do not separate causality from purposiveness. The way we use the word *because* bears testimony to the fact that these two concepts run into each other like two inkblots. The unreflective, merely acting person seems to be hardly aware of the difference between visiting a friend for a certain purpose and the flat tire which caused one to be late. It could very well be that both a natural scientist, as well as an introspectively inclined student of human behavior, might overlook this semantic finitude. A philosopher in his ivory tower is more aware of our Janusian outlook upon the world and upon ourselves and he will pick up the double meaning of the word *because*. He sees that the person who impetuously kicks the tennis ball and observes its motion, is immersed in his holistic experience of activities of body and mind without asking any analytic questions. The person of action finds gratification in his total experience, the scientist finds his satisfaction in quantifying a compartmentalized world. The philosopher tries to show each person that their outlook is one-sided and that we combine in daily, actual living the rational, compartmentalizing, extrospective approach with the intuitive, holistic, introspective approach.

Another question seeks an answer: Why did I kick the tennis ball in the first place? Here again, we come across a conjunction with two meanings, depending on whether you want to know more about the cause or more about the purpose of my act.

A neurophysiologist will try to answer your question by saying that some process in your frontal lobes caused activation of groups of neurones in the motor strip of your left cerebral hemisphere, which center subserves the region which controls movements of the right leg. These neurones cause, first, a group of neurones in the spinal cord to fire and they bring about contraction of the flexor muscles of your right leg, and next, another group of neurones in the spinal cord cause contraction of the extensor muscles, which neuromuscular events are the cause of your kicking the ball with your right foot. After these extremely complex neuromuscular events, chains of causes and effects start in the surrounding world. This is an answer to the question Why? within the causal context.

The other meaning of the question refers to the motivation and the purpose of your action. From the viewpoint of purpose-striving, one can think of two answers to the question why I kicked the ball. In part, the pleasure which I get from kicking it reminds me of similar pleasurable feelings I had in my childhood when I made things move: I regress to childhood for a short while. Huizinga suggests that a man, if he truly wants to play, must play as does a child.

But there is yet another motivation for my activity which on the surface looks so meaningless: it gives me a feeling of power over the causally structured events in the outside world and, under the proper circumstances, this feeling of power can grow to a feeling of omnipotence over nature. If we move the discussion into the complex situation of a tennis game, we become aware of the struggle for power that goes on ferociously not only within amateurs but even more so within professional players. These people do not hit the ball just for the fun of it; they put enormous effort, concentration, training, and determination into their game. And this is exactly the reason why they are deeply emotionally involved in it.

Johan Huizinga has studied the wide-ranging meaning of play in human life exhaustively. Aristotle extended the definition of man beyond *homo sapiens* to *homo ridens*; Huizinga added *homo ludens*, "man who plays."

> From the point of view of a world wholly determined by the operation of blind forces, play would be altogether superfluous. Play only becomes possible, thinkable and understandable when an influx of *mind* breaks down the absolute determinism of the cosmos. By this quality of freedom alone, play marks itself off from the course of natural process. Inside the playground an absolute and peculiar order reigns; it creates order, it *is* order. In an imperfect world and into the confusion of life it brings temporary and limited perfection. By this quality of freedom alone, play makes itself free from the course of the natural process. It is rather a stepping out of "real life" into a temporary sphere of an activity with a disposition all of its own. The joy inextricably bound up with playing can turn not only into tension, but into elation. Frivolity and ecstasy are the twin poles between which play moves.[8]

One sees a gradual transition between my kicking the tennis ball in a frivolous mood and the dead earnest in which the finals are being played in Wimbledon. These players not only exert themselves to the utmost to impart accurate control over their body movements, those of their racket, and over the motions of the balls flying away and toward them; they also strain to overcome their opponent. They are involved in a double power struggle by overcoming the determinism of natural forces and also by overpowering the purposeful activities of another human being. If successful, they partake of the ecstasy of feelings of temporary omnipotence.

Football provides a dramatic example of one's power over an object and over one's own movements. When the quarterback throws the football to the receiver, the thrower and the one who catches the ball integrate movements of the ball in objective time-space with movements of their bodies in objective and in subjective time-space over an enormous distance. The control which these two people have over the beginning of the movement of the ball, its trajectory, and the ending of the movement of the ball is a fascinating mystery: they integrate here-and-now for everyone to see the functioning and the mutual integration of two mental triads in two precious presents. This is an impressive and illuminating instance of man's capacity to synthesize subjective time-space with objective time-space and purpose-striving with causality.

To sum up, we can divide the world in two realms: one occurs in objective time-space, it is causally structured, and we can compartmentalize it. The other realm proceeds in subjective time-space, it is purposively structured, and its elements interpenetrate. It follows from this division that we must approach these two realms of our world with the two contraposed methods of observation of extrospection and introspection. Extrospection completes and forms a cohering series with objective time-space, causality, and with compartmentalization. Introspection completes and forms a cohering series with subjective time-space, with goal-striving activities, and with interpenetration.

MECHANICAL MOVEMENT AND PURPOSE-STRIVING MOVEMENT

Two modes of movement, mechanical and purpose-striving, exemplify how two temporal-spatial forms, two categories, and two methods of observation can occur disjointedly but can also be conjoined. The different properties of mechanical movement and of goal-striving movements rest upon their different spatio-temporal structures. Contraposing these two kinds of motion leads to the description of the even more fundamental dichotomy of causality versus purposiveness in concrete terms. Mechanical movements are compartmentalized events which occur in the form of objective time-space, which are structured by the category of causality and which we observe by extrospection. Purpose-striving movements are interpenetrating activities which are embedded in the form of subjective time-space, which are structured by the category of purposiveness and which we observe by introspection and its ancillaries. The most important contrast between these two kinds of movement is that mechanical movement is devoid of emotional coloring, and that purpose-striving movements are suffused with feeling tone.

Comparing these movements with each other we face several problems. First, mechanical movement such as water streaming down a hillside, or the rotation of the earth on its axis, occurs in pure form, but we never observe purpose-striving movements in the absence of mechanical movement, since the two are always conjoined. We use mechanical movement for specific purposes, and we are apt to assume that therefore these movements must be purposeful. For instance, the up and down movement of a piston in the cylinder block of a car is a mechanical movement, but since it is part of a total mechanism, the automobile, we think that the movements of the piston must be purposeful. And again, many of our body movements are of a mechanical nature, such as the rhythmic movements of the muscles of respiration, which are caused by the rhythmic firing of neurones in the respiratory center of the medulla oblongata; but we can slow down and increase the depth of our breathing movements on purpose, if we want to use deep breathing to relax, or, in the case of actors, to use slow or fast breathing to express emotion.

One can illustrate the differences and the interconnections between the two kinds of movement with several examples. For instance, by comparing how an artist draws a picture and what a student does who copies it. Before the artist starts to draw, he has a picture in mind of a real or of an imagined object, and he puts it very quickly on paper by performing a set of coordinated arm, hand, and finger movements, whereby he transforms his private mental image into the public, two-dimensional reality of the drawing. These movements are one cohering, holistic, mental and physical activity interpenetrating in subjective time-space and in objective time-space. The student follows step by step, stretch of line by stretch of line, of curves, shadings, and forms which were concretized previously by the artist. These movements are similar to mechanical, compartmentalized movements in objective time-space. The artist's drawing exemplifies purpose-striving movement, although the artist also executes mechanical movements. But mechanical aspects are in the foreground in the compartmentalized movements of the student, although his activity is not totally devoid of purposive movements.

Van Meegeren's imitation of the style of Vermeer caused a scandal in the art world. Van Meegeren wanted to prove that he could paint pictures in this style, to mislead the art critics whom he detested. Many top experts classified these imitations as authentic Vermeers. A Belgian curator finally admitted that he and his colleagues had arrived at wrong conclusions. He compared van Meegeren's brush strokes with those of Vermeer in minute detail, and he noticed that the master's brush strokes displayed a holistic, interpenetrating structure, while van Meegeren used compartmentalized, interrupted, short strokes which started and stopped in short spatial stretches, and therefore, van Meegeren's brush strokes displayed an entirely different temporal structure compared to those of Vermeer.

We are confronted with the problem of temporal organization in many contexts, whether we are contemplating a minor artistic activity like quickly drawing a sketch, or the slavish copying of it, or whether we think about the compelling mystery of the temporal-spatial organization of the work of great artists. It is said that Picasso had put up an empty canvas in his studio for several weeks, knowing that he was going to do something totally new and unique and then suddenly he started to work on *Les Demoiselles d'Avignon*. Our concern of what is going to happen to us and within us in the near and the distant future is always right underneath the surface, whether we work in a miniature temporal frame, or in a huge temporal panorama. Intentionality drives us into the future-in-the-fringe of a precious present and we calculate rationally how the objective future may shape up. The tensions between these constituent aspects of our living-in-the-world are a source of our suffering. An analysis of the temporal-spatial structure of mechanical movement and of purpose-striving movement brings these tensions into clearer focus. Since the mechanistic aspect of the world is continuous and deterministic, while purpose-striving activities are discontinuous and indeterministic, the analysis of the temporal-spatial structure of movement brings to light some of the essential attributes of human existence, especially emotionality and freedom of choice. These are constituent attributes of purpose-striving movement but they are missing in mechanical movement.[9]

Let us compare three modalities of locomotion.

1. A healthy looking young man seems to be walking along quite normally, until one examines this process a little closer: instead of swinging his arms in a pendulum fashion, the forward swing of the arm stops during a fraction of a second and the backward swing also stops for such a brief, clearly observable period. This gives the entire movement of walking a saccadic, interrupted appearance and it lacks, as Bergson would say, gracefulness. Normally, the forward swing as it stops is immediately reverted into the backward swing, while in this case there was a visible interruption of the flow of the normal pendulumlike movements.
2. A somewhat obese, stockily built man in his thirties is jogging; his face is red, the facial expression is tense and determined, the arms swing from the side of the body more or less in front of the torso, and one gets the impression of extraordinary effort being put into the self-imposed exercise. He strains to make his body go faster than he can make it go, and gravity obviously interferes with the man's efforts.
3. A lanky, tall, young black man is jogging and his arms, legs, and torso work in unison; he seems to float along effortlessly in an undulating, graceful manner of integrated body movements. He controls his body, whereas the thirty-year-old man is controlled by the heft of his body. The young man seems to overcome the pull of gravity, while the older man's movements are impeded by this force.

These three modalities of locomotion integrate time-space in a slightly different manner. The interrupted swinging of the arms in the first case lifts these movements out of normal interpenetration of body movements in subjective time-space; they remain stuck, so to say, in objective time-space. The stocky, straining jogger is encumbered by the recalcitrant body. The young jogger shows both interpenetration in subjective time-space and his overcoming of gravity to a certain extent.

THE BILLIARD GAME

A billiard game is an example of how these two kinds of movement can be both contraposed and integrated. There are three possible approaches to an analysis of this game. The players and the spectators accomplish the integration of the two types of motion by a subjective, nonanalytical route, since they do not differentiate between the movements of the player versus the movements of the balls because the game happens in subjective time-space for them. A physicist looks at the game objectively and he can choose to look only at the movements of the cue and the balls. He disregards the movements of the player since he wants to describe the movements on the felt quantitatively; he conceives of these movements as if they occur in objective time-space. A neurophysiologist attempts to explain how the movements of the player are initiated in his central nervous system, since he conjectures that these purpose-striving movements are produced within the nervous system and projected outward in terms of the movements of the cue.

THE PHYSICIST

The *physicist* describes movement as beginning in a geometric point-punctiform instant, traversing a straight or a curved trajectory, and ending in a geometric point-punctiform instant. The trajectory is made up of numerous short stretches, or deltas, which, in turn, are delimited by time-space points. Henri Bergson describes this treatment of movement either as unrolling in time or as unrolled in space. The unrolled movement and their trace in space can be divided, that is, we can compartmentalize them. Unrolling movements which we perform ourselves cannot be so represented and they cannot be subdivided, they interpenetrate. Bergson hints that the physicist describes the movements of the cue and of the billiard balls as having been unrolled, while the players and the spectators experience these movements as unrolling. We should go one step farther: we can measure objective time-space but we cannot measure subjective time-space. One can use Bergson's conceptions about movement and duration to compare the temporal-spatial structure of the movements of the billiard balls with the temporal-spatial structure of the movements of the player. As mechanical movements, the trajectories of the balls are made up of stretches which one delimits with geometric points and punctiform instants. That is, one represents the unrolling movements as having been unrolled.

> When we witness a very rapid movement, such as that of a shooting star, we distinguish the fiery line very distinctly, divisible at will from the indivisible mobility which it extends.... [10]

The moments of impact of the cue tip on the cue ball and of the mutual impacts of the balls with each other and with the rims of the billiard table are moments which actually occupy a fraction of a second of objective time, but we compress these moments into punctiform instants which do not occupy any stretch of time whatsoever. Hence, we can represent them with geometric points which lack any spatial dimension. A mechanistic description of the trajectories of the billiard balls conceives of these movements as if they are going from geometric point to geometric point, where they come to a standstill—*in abstracto*—in punctiform instants. One describes mechanical movement as if it proceeds in a saccadic manner, since one breaks it up into stretches of lines, which stretches begin and end in spatiotemporal point instants. This static, spatial representation of movement serves the goal of mathematical calculation of the trajectories.

The physicist aims to describe the trajectories of the balls in mathematical equations, in order to be able to predict the future course of the balls as accurately as possible. He may very well introduce the following artifacts: he marks the felt with a carefully measured grid of straight lines, which cross each other at ninety-degree angles, and he installs a moving picture camera above the table; he loads the camera with a film with time markers of seconds or fractions of seconds, which are synchronized with the speed of the camera's driving mechanism. Now he can reproduce on film a spatial representation of the movements of the balls against the background of two space coordinates and one time coordinate. These artifacts give him the means to visualize and to calculate accurately the beginnings, the extent, and the endings of the motions of the cue tip and of the billiard balls.

However, he still has to contend with one immensely variable set of factors, namely the postures and the movements of the player. If one could limit oneself to an analysis of the movement of the player's skeleton, his joints, and his muscles along mechanistic lines, this would be child's play compared to describing the neural mechanisms which activate and coordinate the contractions and relaxations of the player's muscles. The physicist will leave this investigation to the neurophysiologist who may hope to be able to explain, some day, goal-striving movements in mechanistic terms. But such understanding lies far in the future.

How to get around this impasse? I offer the following solution. The physicist sidesteps the motions of the player by constructing an electronically activated and controlled robot on which he installs a short cue. This instrument gives the scientist the means to apply a precalculated thrust in the necessary direction to the cue and therefore to the cue ball. The robot cue is the ultimate step toward an abstract, mechanistic interpretation of the game, since this gadget makes accurate, mathematical calculation and prediction of the future course of the balls possible. Now the physicist has "caught" these mechanical movements, which he initiates with a calculable device, in the "net" of a space-time coordinate system, and he can measure the force and the direction of impact, speed, and shape of the trajectories of the balls. What is most important, he can measure the amounts of clock time which are covered by the various stretches of movement. He measures the time factor of the movements of the billiard balls by counting the number of stretches of movement on the trajectories and by counting the number of intervals of objective time, provided by the time markers on the film. These figures quantify the time factor and he can use them in his mathematical formulas. These equations are cousins of the causal equation. This grounding principle of natural sciences states that each stretch of movement of the cue and of the balls is the cause of a subsequent stretch of movement, and, since one may overlook the loss of energy due to friction, compressions of the balls on impact, sounds caused by impact, one can say that each cause produces an equal effect. Mathematical equations and the causal equation work, because mechanical movement can be cut up in stretches by geometric points-punctiform instants. The application of these highly abstract concepts is etched upon one's mind if one remembers that NASA was able to land a space vehicle on the moon in a precisely predicted location with stunning accuracy.

Now let us see what the physicist is actually doing at the billiard table. After a few mathematically calculated practice shots with the robot, he learns to adjust the electronic cue according to the dictates of his mathematical formulas. Now he can pit his instrumented, mathematically transilluminated game against a skillful player. Since the robot and applied mathematics assure him of nearly one hundred percent accuracy of prediction of the future course of the balls, the scientist has far greater power of prediction over the movements on the felt than even the most skillful player can hope to match.

But what is left of the real game of billiards? The player and the spectators will marvel at the physicist's precision of prediction, but on the reverse side of this gain there is a decided loss: because prediction of the future has become so refined, anticipation of the future and the emotional tension which this uncertainty about the future creates, have been eliminated. If the superlative performance of the scientist is repeated very often, everyone will lose interest completely in the robot game. This extreme mechanistic approach to the game leads to the loss of the most important motivation for people to play games: players and spectators get an emotional charge out of playing and watching games. Emotionality is one of the fundamental characteristics of human existence.

It is mostly the robot which is responsible for the elimination of emotionality from the mechanized billiard game. Therefore it is relevant and enlightening to think what robots do and what they don't do.

Speaking with an electrical engineer, I asked him if he thought that a robot can replicate our purpose-striving movements. And we got into an impasse. We talked about an electronically controlled robot which I had seen performing on a TV program, operating certain functions in industry, and I described these movements as going from point-instant to point-instant in an interrupted fashion. The human hand and arm go in one spatially and temporally unitary movement from initial position to intended goal without interruptions in time-space.

The engineer took exception to this comparison of the two types of movement in the first place, because the human hand and arm also move from point-instant to point-instant, but we are not aware of these fractions of movement and their interruptedness since the periods of time involved are so small that we cannot perceive them; secondly, one could build a robot whose movements are divided into such minuscule stretches in time and in space that they would look exactly like our own purpose-striving movement. Although it is theoretically possible to achieve such a construct, the refined robot would be much too costly to be operative in industry.

The reason why we got into an impasse is that the engineer recognizes and uses only objective time-space, while I make a plea for the need to describe purpose-striving movements with both forms of time-space. The engineer sees no problem in compartmentalizing both the movement of the robot and those of the human arm. I see the first type of movement as compartmentalized and the second modality as interpenetrating. Therein lies precisely the emotional neutrality of mechanical movement in contrast with the emotional charge which is characteristic of purpose-striving movement. Here lie precisely the reasons for the inadequacy of mechanistic concepts in the field of biology as a whole and especially when one deals with the mind-body relationship. The mechanical billiard game sidesteps this relationship and its emotional impetus.

The Neurophysiologist

Perhaps the *neurophysiologist*, who seems to stand midway between the mechanistic, mathematical approach and the player's concrete, subjective approach, can explain why the playing of games is charged with inner tensions and a whole gamut of feelings. Might he rescue one of our most precious possessions, which is emotionality?

This fascinating science, neurophysiology, is built upon the spatial framework of gross and especially of microscopic anatomy and histology. Within this unbelievably complex labyrinth, electrical impulses speed from neuron, through axon, to end plate where they set chemical and physical processes in action within the interneuronal cleft. From there, secondary electrical processes enter the dendrite of the next neuron and reach its body. These electrical and chemical processes are cause-effect sequences which go forward in the before-and-after of objective time-space. A feedback system is superimposed upon this ongoing flow from neuron to neuron and dampens these processes down or accelerates them in the service of homeostasis.

This simplified and sketchy description of neurophysiological events shows that one deals with cause-effect sequences in objective time-space. But now an entirely different factor enters the workshop of the neurophysiologist. We have an innate tendency to identify with living organisms, their organs, and their cells, and this identification leads to our projecting our human attributes onto them and especially onto the central nervous system. Identification and projection add teleological admixtures to causally structured neural mechanisms. One shifts from objective natural scientific research to the intuitive introduction of mentalistic attributes into the nervous system.

Feedback mechanisms tend to reinforce the teleological coloring of neurophysiological concepts. These mechanisms are similar to man-made cybernetic servomechanisms which are intellectual constructs expressing man's purpose-striving activities. Reflex mechanisms with their superimposed feedback circuits are interpreted as purposeful, which seems the more plausible the more complex the concatenated reflex chains become.

Martin Nemirof has recently discovered the diving reflex in humans, which explains why some swimmers have survived for some thirty minutes after they had been submerged.

> The diving reflex was first identified in seagoing animals such as the whale and the porpoise. The porpoise can in an emergency remain submerged without breathing for twenty minutes, the whale for up to two hours. The reflex slows the heart rate and constricts the flow of blood to the skin, muscles and other tissues that are more resistant to oxygen-loss damage. At the same time, the remaining blood oxygen is directed to the heart and the brain. The cold water reduces the need for oxygen of tissues further and lengthens survival time without external oxygen.[11]

His description of this reflex is a typical example of the teleological interpretation of an exceedingly complex nexus of reflexes. The diving reflex suggests convincingly that its purpose must be to assure the animal's or the human being's survival under certain adverse conditions. However, one should not overlook the alternative interpretation of the sequence of events as a succession of a number of interrelated causal processes going on in objective time-space and without the need to add teleological admixtures. The other side of the coin is that some reflexes are downright harmful. A heart attack is caused by the occlusion of a branch of a coronary artery which causes necrosis of the portion of heart muscles which is severed from its blood supply. A reflex spasm of arterioles adjacent to the necrotic area may enlarge the area of already moribund muscle fibres thus making the situation worse then it was already.

John Dwyer has written a very lucid and informative paper on the immune system, to which a few random quotations hardly do justice. I shall not touch on the technical, detailed information which Dwyer presents so vividly, but I shall try to bring out that he follows a mixed causal-teleological approach.

> In the sixth to eighth week of fetal life, cells which given the appropriate educational experience [micro-environment] can mature to become 'B' lymphocytes [which] come from the bone marrow. The education process is directed by a complex genetic mechanism and we understand some of the skills taught these cells. The fundamental skill taught each of these cells is to recognize with exquisite sensitivity one and only one antigen. It recognizes a specific chemical and steric configuration. Some lymphocytes when triggered by antigens divide to produce clones of memory cells. The memory cells produced by the initial encounter with antigen ensures both a more rapid and more sophisticated secondary immune response. Another

problem was distinguishing between self and non-self. To make dangerous amounts of auto antibody, B cells that can recognize self-antigens require cooperation from the corresponding T cells. It seems that the dichotomy was a powerful evolutionary step as small amounts of auto-antibody may actually help in the maintenance of T cell non-reactivity to self. Lymphocytes do not aimlessly wander around, their homing characteristics are well defined if little understood. The major T cell function involves the control of the immune response. Such a powerful biological response must be actively controlled to prevent excesses that would do more harm than good. These suppressor T cells can control B cell responses as well as T cell response.[12]

On the one hand T lymphocytes and B lymphocytes are centers of physical, chemical, electrical forces, but on the other hand they are portrayed as displaying mentalistic and purpose-striving attributes. One projects human psychological properties onto these cells such as being taught, learning, developing skills, recognition, memory and in their homing characteristics they seem to have an anlage of foresight. They cooperate with each other and keep each other in check, which suggests a primitive form of interpersonal communication. Over and above these mentalistic factors they differentiate between self and nonself. In brief, T lymphocytes and B lymphocytes are made into homunculi.

Purposiveness comes clearly into view when evolution and nature are mentioned. To inhibit only the T lymphocytes "was a purposeful evolutionary step."[13] An anthropomorphic vision of "Nature" stands behind all these marvelous mechanisms and interrelations so that it is as if Nature exercises foresight and as if Nature has designed the structure and the function of organisms in a rational and purposeful manner. Nature, then, is the agent which projects human attributes upon the total organism, its organs, its cells and their organelles, and upon their mechanical, physical, chemical, and electrical functions.

Teleological interpretations are not worthless. They are valuable during the development of scientific theories because they often lead to causal explanations. It is as if causal explanation is lying in wait to take over from teleological interpretation. And this by the way, is an anthropomorphic way of speaking about causality and teleology!

I presume that when processes going on in the B lymphocytes and the T lymphocytes are better known in the future when physical, chemical, and electrical processes will explain their behaviors, then these mixed causal-teleological descriptions will shed their teleological components. Then the immunologist will see these cells as mechanistic centers of physical forces and they will be stripped of their mentalistic attributes. This is bound to be a slow process since the living cell is a very fragile structure indeed and life processes are enormously complex.

It is difficult to roll back teleological interpretations in biology for psychological reasons. Not only do we stand in awe of the functions which organisms display, but in addition the drive to identify with all that lives is a compelling emotional factor: we tend to see purpose-striving in phenomena which we ought to explain as cause-effect sequences because we ourselves are living organisms. In certain realms of biology we are quite justified to identify with some animals, for instance with the apes, although one should keep this identification within bounds. In the case of the structure and the functions of cells we must eliminate all vestiges of identification. It should not surprise us that this rollback is a slow process, considering that it has taken centuries to cleanse astronomy from teleological pseudoexplanations.

Neurophysiologists hope to solve the riddle of purpose-striving movements and, if possible, of consciousness, within a mechanistic framework. But, if they follow this approach, they overlook the facts that they conceive of the physiology of the nervous system as going on in objective time-space, that its functions are causally structured, and that they observe their subject matter, which is man, by extrospection. This form of time-space, this structuring category, and this method of observation are indicated when one deals with compartmentalizable phenomena. This approach is too restricted and one-sided if one wants to get insight into interpenetrating purpose-striving activities, and, even more so, if one wants to study consciousness. In these areas the form of subjective time-space, the category of intentionality, and the method of introspection with its ancillaries are obligatory. Matters are further complicated by the fact that purpose-striving endeavors are always conjoined with mechanistic events, and hence one is obliged to combine the two forms of time-space, the two categories, and the two methods of observation when one studies man as a whole human being.

One sees the same uneasy shifting back and forth between causal determinism and teleological indeterminism in neurophysiological interpretations of man, as we have seen to be the case in the theoretical problems which the immunologist faces. They get lost in an impasse which is not unlike the one in which the physicist gets ensnarled when he attempts to describe the real billiard game along mathematical principles with the aid of the electronically controlled robot. Natural science studies phenomena which can be compartmented, therefore, it cannot gain insight into interpenetrating, holistic, goal-striving activities and into mentation. The grounding attribute of mentation is intentionality, which projects our mental activities and our purpose-striving movements into the future fringe of a precious present. The billiard player lifts himself beyond the realm of causality and objective time-space, into the world of intentionality in subjective time-space, thanks to the activities of the mental triad. The physicist and the neurophysiologist cannot follow him into this purposively organized world, because as natural scientists they are bound down to the causally structured world in objective time-space.

In order to separate teleology from causality one should take heed of recent developments about the concept of causation. Professor Maurice Natanson has alerted me to the narrow scope of the foregoing discussion of causality. It is based on Hume's views upon causality and on Newton's physics. In recent times one speaks of sufficient conditions and necessary conditions instead of a linear cause-effect process. A most problematical renovation of the idea of causation is that cause and effect do not occur in the before-and-after in objective time-space, but that cause and effect are contemporaneous. This is a radically deviant temporal structure of this relation which demands careful scrutiny. However, it seems that the element of necessity between cause and effect, over which Hume agonized, is retained in these modern versions of the concept of causation.

> *Causes And Necessary Conditions.* Most contemporary philosophers agree with Hume... that a causal condition of any event is any condition which is such that, had it not occurred, the event in question would not have occurred.... *Causes And Sufficient Conditions.* According to the foregoing and now fairly widely held conception, a causal condition of an event is a *sine qua non* condition under which that event occurred or condition which was such that, had the condition in question not obtained, that the event (its effect) would not have occurred, and that the *cause* of the event is the totality of these conditions....*The Plurality Of Cause.* J. S. Mill mentioned that many events are such that they can be produced in a variety of ways. A match can be ignited by friction, but also by being heated, and perhaps in other ways too.... Most philosophers have supposed that causes could be distinguished from their effects in terms of time, the cause always occurring before the effect.... *Contemporaneous Causes And Effects...* Indeed it has often been maintained... that all causes and effects are contemporaneous, that there is never a real temporal succession of events that are causally connected.... Consider, for instance, a locomotive that is pulling a caboose, and to make it simple, that this is all it is pulling and that the two are tightly connected.... There seems however, to be no temporal gap between the motion of one object and the motion of the other.[14]

Returning to the motion of the billiard balls and the analysis of these movements by the physicist, one may say that the physicist conceives of the balls as ideally rigid bodies and of the interaction between the individual balls as occurring in punctiform instants. When we look at these motions and we hear their mutual impacts in reality, this transfer of movement and of kinetic energy takes place in a fraction of a second, that is, in an exceedingly brief interval of objective time-space. The physicist might install a moving picture camera above the billiard table with a timing mechanism built into the driving mechanism of the camera, which might deliver far more numerous space-time frames than a conventional moving picture camera. If he then should project the film of the motions of the balls in slow motion, he might be able to make visible the indentation of the active ball while it hits the passive ball and how it resumes its round shape almost immediately afterward. Simultaneously the passive ball is indented ever so little and recovers its original shape almost immediately. This indentation and dilatation does not happen in a punctiform instant but in a minute interval of objective time-space, measured in a fraction of a second. Although the physicist may abstractly conceive of the moment of impact of the two balls as happening in a punctiform instant, he could make this cause-and-effect transfer concretely visible with the film. Hence, cause and effect are not contemporaneous because the billiard balls are not absolutely, ideally rigid objects.

One can treat, along the same lines of thought, the causal relationship between locomotive and caboose which are tightly connected. The presupposition behind this example is not only that these two objects are rigidly connected but also that they are ideally rigid objects. Such bodies do not exist in our world of sensory impressions.

We can bring this example back into reality by watching how a freight train starts to move. It is powered by two locomotives which are connected with 125 freight cars interconnected by drawbars with about seven inches of "give" each. This adds up to eighty feet of slack along the length of the train. We have to remember that when the engineer did bring the train to a stop he applied the brakes of the locomotives very carefully so as to make certain that there would be a slack in all of the couplings. Actually, a still-standing freight train with all the couplings taut could not be moved by two locomotives. Now the train starts to move: the locomotives give a short jerk to the first few cars and they *almost* come to a stop. In the meantime one hears a rippling sound of the stretching couplings going down the trail of cars. Each moves over a short distance, giving a short jerk to its successor and *almost* comes to a stop while its successor goes through a similar compartmentalized motion. When the ripple of the stretching-out couplings reaches the last car, it gives its predecessor a brief push, and now one hears the ripple effect passing along the entire length of the train going forward, until it reaches the locomotives which receive a short forward push, and now the train begins to move smoothly without interruptions.

A freight train getting into motion is a convincing instance of the before-and-after relationship between cause and effect in objective time-space. Incidentally, the engineer's foresight, first by halting the train in the appropriate manner and then starting it up very carefully so that the wheels of the locomotives won't spin, adds man's goal-striving actions in subjective time-space to the causally structured machinery in objective time-space. One stands in awe of man's omnipotence who can manipulate, control, and calculate what goes on in the enormous mass of the freight train.

Now we should look in more detail at what happens in the coupling bars. First they tighten up and then they release. This stretching and contracting of the bars is possible because each drawbar can stretch slightly and then it contracts over the same distance. Now we step from the macrophysical world in which we see and hear the train move, into the microphysical world of molecular, atomic and subatomic movements, and the generation of heat. When we take these steps, we get farther and farther removed from the macrophysical world. The atomic physicist thinks of bodies imparting motion to each other in punctiform instants. This world lies beyond our capacities of direct sensory observation. It may be argued that cause and effect are contemporaneous in a punctiform instant in the microphysical world, but we have seen that only ideally rigid bodies can interact in a punctiform instant from which instant all motion has been eliminated. It is up to the atomic physicist to worry whether molecules, atoms, and subatomic particles are ideally rigid bodies.

One absolutely rigid body cannot transfer its motion and its kinetic energy to another absolutely rigid body. In the first place such objects are unknown to us in our lived-in world of sensory impressions and, secondly, if we imagined how such objects might interact they would probably break each other apart. Therefore the example of the tightly connected locomotive and caboose as an example of instantaneous causation has no parallel in the real, observed, seen-and-heard world of the billiard balls and of the freight train.

Mario Bunge discusses the following statement about causation.

> Sometimes a more sophisticated sentence is regarded as the correct formulation of the causal principle, namely, *The knowledge of the initial state of a [closed] system is sufficient for the prediction of its state at any other later time.* (11). However, (11) is first of all incomplete, as the law of motion (and eventually the force function) as well as the constraints and the specification of the surroundings (boundary conditions) are needed besides the initial state in order to perform some prediction. But even if these qualifications are added, the uniqueness that characterizes the causal bond is lost in (11), since empirical information is never quantitatively exact; instead of an unambiguous set of values specifying the initial state of the system concerned, observation and measurement usually yield statistical distributions of the relevant variables. This uncertainty in the initial information—which always shows a spread around average values—spoils the one-to-one correspondence among neatly defined states even if, as in classical physics, the theoretical values are supposed to be sharply defined. This is one of the reasons for regarding (11) as an inadequate formulation of the causality principle. Besides, it (11) is not an ontological statement, not a proposition concerning the world, but a sentence concerning our knowledge and prognosis of events. Certainly, the knowledge of a certain type of law of nature alongside a set of items of information . . . , make an almost unique prediction of future states often possible. . .; provided we know how to solve the mathematical problems eventually involved in the prognosis, our prediction may be adequate for limited periods of time. But the causal problem, far from having solely an epistemological side, is chiefly an ontological question. Hence, considerations about knowledge and foreknowledge are out of place in connection with statements of the causal principle;. . .[15]

I read this quotation as a warning against regarding the causal principle as resting solely on the possibility of prediction of a future state of events. From the viewpoint of physics and mathematics, prediction is only one of several aspects of the causal principle. Bunge's astute analysis does not annihilate the attempt to derive causality from the mental triad. He warns only against seeing causation exclusively as an ontological question, although it has indeed an epistemological side also. As a lay person, I may have overemphasized this side of the causal principle by laying stress on its future-attuned characteristics.

PURPOSE-STRIVING MOVEMENT

The physicist peers into the future through the telescope of the mathematical equation. How does the player look into the future? What kind of future does the scientist and what kind of future does the player look into?

Let us study the outer and the inner activities of the player in more detail. He surveys the relative positions of the balls, he gets his body and his cue into a proper directional position, he makes a few preparatory movements with his cue, and he sees before his mind's eye how the balls will interact with each other and with the rims of the table; that is, he calculates the future behavior of the balls. His calculations are both similar to and different from those of the physicist. He visualizes how the balls start in a point, go through a stretch of movement which is interrupted or decelerated in another point when the cue ball hits a second ball, or it changes direction in a point when it hits the rim of the table. He sees these movements in his fantasy both as an unrolling interpenetrating whole and as a compartmentalized unrolled set of stretches of movements. Though both the player's and the scientist's calculations are concerned with mechanical movement, the player's calculations are more intuitive, more judgmental, and therefore he is far less sure of the future course of the balls than the scientist is.

After the player has evaluated the future course of events which are immediately ahead, he now makes his shot. He follows the course of the balls with considerable, or even with very acute, mental tension and then, if he is successful, he experiences a feeling of relief or of triumph, but if he misses his shot he feels disappointed or possibly harassed.

Seen as a goal-striving movement, the shot begins before the player makes an external movement, and it does not end for the player when the movements of the balls are finished. He prepares before he hits, he judges while the movements are going on, and afterward he evaluates his performance critically. Preparation, shot, and evaluation are a time-binding process in which the player organizes his past-in-the-fringe with his act in a durational present and with his projecting the act into the future-in-the-fringe in subjective time-space. During this brief time span, while his mental triad is active, he experiences a whole gamut of feelings. The kaleidoscopic feeling tone of the mental triad functioning in a precious present covers a state of tension toward the future-in-the-fringe, apprehension during the period of evaluation when the balls are rolling, and as a result of the shot the player may feel satisfied, quiet, happy, or elated, or, on the other hand, he may feel a slight setback, or annoyance, or he may even feel depressed. Goal-striving movements carry very decided and variable feeling tones since they are the expression of the activities of the mental triad in a precious present.

During the few seconds, or sometimes even in the fraction of a second, during which the player calculates and shoots, his mind extends far backward in time. He condenses in the past fringe of this precious present what he has learned during his training period months or years ago and during recent practice sessions. These memories are now a real, active experience in the lived in here-and-now. Condensation transforms these specific memories into a useful conglomerate which the act now projects into the future fringe. His mind and body extend briefly forward into the future fringe and into subjective space by means of the final, quick, thrusting movements of arm and hand, which send the balls on their precalculated courses. In this brief moment he condenses an immense number of past experiences as present, past, and future come together in the precious present when he plans and executes the shot.

The player who learned this game in his personal past has his own history as a player, but the billiard game itself has also a long history which reaches back in objective time one hundred, sixteen hundred, even two thousand years. In this sense the player profits from some of the experiences of others who lived long ago. Purposeful activities in the present can relate to and can condense huge spans of calendar time in a discontinuous manner in a nonhomogeneous past fringe in a precious present. "Anarchis [travelling through Greece] 400 B.C., saw a game similar to billiards.[16] A king of Ireland, in the second century A.D., left behind him fifty-five billiard balls, of brass with the poles and the cues of the same metal. In the eighteen hundreds leather tips and rubber cushions along the side of the table permitted strong, accurate rebound, and from that time billiards developed into a game of scientific precision."[17]

For the player, present, past, and future are one interpenetrating whole, when the mental triad creates a new emergent out of the mechanical movement of the billiard balls which are going on in objective time-space. The mental triad, while creating purpose-striving movement in a precious present, superimposes goal-striving upon causally structured mechanical movement.

The integration of purpose-striving movement with mechanical movement is a mystery and all the more so since this integration does not disrupt causal structuring of movement. The player reaches into the world of purposiveness without leaving the causally structured world. His movements, which superimpose subjective time-space upon objective time-space, intentionality upon causality, and interpenetration upon compartmentalization, make the mystery manifest. As Wittgenstein says, "There are, indeed, things that cannot be put into words. They *make themselves manifest.* They are what is mystical. What we cannot speak about we must pass over in silence."[18] Psychologically speaking, the silence into which Wittgenstein falls is not a true statement about ourselves. We may be silent verbally and refrain from writing our thoughts down, for instance in the face of the mystery of the integration of intentionality with causality, but one can be sure that when Wittgenstein was faced with a mystery, he continued an internal dialogue with himself. Search for the solution of a problem, whatever its nature, goes on unabated, because we always reach only partial solutions. Further solutions come to us if we continue this search in silent dialogue with ourselves.

The player is placed squarely in between the world of subjective time-space in which the mental triad unfolds, and the world of objective time-space in which movement is causally structured. A mental and physical mystery is happening when the player by preparing, executing, and judging his shot interweaves and unifies these two levels of our spatiotemporal organization. This mystery is the creation of a mental and physical whole, of a new emergent.

When we analyze the game from an introspective, empathizing, identifying, and projecting viewpoint, we discover layers upon layers of feelings within the players and the spectators. The feelings of omnipotence which flow from the player's expert control of the trajectories of the balls are enhanced by overpowering his opponent. Scorekeeping, which introduces the magic of numbers into the game is another artifact that stimulates upsurging feelings. The spectators, identifying with the players, transfer his feelings of elation or disappointment to themselves by identification, and they express their inner tensions by derogatory exclamations or by applause. Trophies, ever present, encapsulate the brief afterglow of triumph as if they were three-dimensional, concrete memories. Gambling underlines and strengthens the spectators' identification with the players, and the thrown-in money artificially stimulates their anxiety level toward the unpredictable future.

This analysis of games is similar to the approach which Kurt Goldstein has followed in his extensive work with World War I veterans who had brain damage. "If I see things correctly, all attempts to understand life so far, have travelled the road from 'down below' toward 'up above.'... The following presentation starts out with man and tries to understand the behavior of other living beings from that vantage point."[19]

I fully agree that we must follow the road from "up above" toward "down below" in our study of human behavior and mentation. Expressed in my terminology, Goldstein starts his research with man as he exists in subjective time-space and in objective time-space; man who is structured by the category of causality and whose functions are compartmentalized. But man also lives in subjective time-space as a goal-striving being who makes contrasting elements interpenetrate. We have to approach man with the combined extrospective and introspective methods of observation. As we descend on the evolutionary scale toward "down below" we should progressively curtail purposiveness, subjective time-space, and introspection with its ancillaries. But if we remain "up above" and seek to understand man as a rational, emotional, communicating whole, we need to follow the combined routes of observation and the contrasting sets of concepts. We have to judge critically if these methods and concepts are appropriate for each particular level of reality we are studying, including the inanimate world. Or else we might regress to the animism of childhood. This may happen when we project mentalistic attributes onto a machine, such as Sinclair Lewis describes so beautifully in *Babbitt*. His neighbor has difficulty starting his car motor and Babbitt feels the same apprehension as his neighbor may be feeling. It is as if he urges the recalcitrant motor, please, to turn over.

> As he relaxed, he was pierced by the familiar irritating rattle of someone cranking his Ford: snap-ah-ah. Himself a pious motorist, Babbitt cranked with the unseen driver, with him waited through taut hours for the roar of the starting engine, with him agonized as the roar ceased and again began the infernal patient snap-ah-ah round, flat sound, a shivering cold-morning sound, a sound infuriating and inescapable. Not till the rising voice of the motor told him that the Ford was moving was he relaxed from the panting tension.[20]

If the billiard player has as lively an imagination as Babbitt, he may feel as if his foreknowledge and power of prediction are incorporated in the movement of the balls and as if they knew where they were going. This fantasy is a rather conscious regression to childhood animism, and it shows us how hard it is to develop a mechanistic conception of the inanimate world without inserting teleological admixtures which do not belong.

Our tendency to slide back into primitive, animistic identification with objects in our surroundings, makes us overlook the difference between the two kinds of movement since animism does not contrapose them. We must be on our guard against this regression because the fact that mechanical movements are always a component of our purpose-striving movements entices us to see mentalistic, purposive attributes in purely mechanical events.

We can hold the tendency toward animistic interpretation of mechanistic events in check if we are aware of the contrasting spatiotemporal structures of these two modalities of movement, of their opposite categorial structuring, and of their being compartmented versus being interpenetrating. Moreover, we should be aware that these contrasts require that we observe mechanical movement by extrospection and goal-striving movement by introspection and its ancillaries. The two sets of concepts and of methods of observation form two contrasting as well as cohering series, which give us a firm basis from which we can contrapose and integrate not only the two modes of motion, but to elucidate even more important dichotomies and fractures in our human existence.

NOTES

1. Aristotle, *Physics, bk. 2, The Basic Works of Aristotle*, ed. Richard McKeon (New York: Random House, 1966), chap. 8, p. 251; chap. 3; pp. 240-41.

2. G. B. Kerferd, *Encyclopedia of Philosophy*, 1967, s.v. "Aristotle."

3. David Hume, *A Treatise on Human Nature*, ed. L. A. Selby-Brigge (Oxford: Clarendon Press, 1965), pp. 10-11.

4. Alfred Jules Ayers, *Language, Truth and Logic* (New York: Dover Publications, 1935), pp. 54-55.

5. Georg Henrik von Wright, *Causality and Determinism* (New York: Columbia University Press, 1974), pp. 1-2, 39.

6. Jean Piaget, *The Child's Conception of Physical Causality* (London: Kegan, Trench, Trubner & Co.; New York: Harcourt, Brace & Co., 1930).

7. Peter Kunstadter, "Living with the Gentle Lua," *National Geographic* 130, no. 1 (1966): 122-52.

8. Johan Huizinga, *Homo Ludens: A Study of the Play Element in Culture* (Boston: Beacon Press, 1966), pp. 3, 7, 10, 8, 21 (Huizinga's italics).

9. Adrian C. Moulyn, "The Limitations of Mechanistic Methods in the Biological Sciences," *Scientific Monthly* 71 (July 1950): 44.

10. Henri Bergson, *Durée et Simultanéité*, 7th ed. (Paris: Presses Universitaires de France, 1968), pp. 47-48 (Author's translation).

11. Martin J. Nemiroff, "Reprieve from Drowning," *Scientific American* 237, no. 2 (August 1977), pp. 57-58.

12. John M. Dwyer, "Understanding Modern Immunology," *Connecticut Medicine* 39 (March 1975): 170-73; (April 1975): 216-18.

13. Ibid., p. 174.

14. *Encyclopedia of Philosophy*, vol. 2 (New York. MacMillan Publishing Co., 1967), s.v. "Causality."

15. Mario Bunge, *Causality: The Place of the Causal Principle in Modern Science* (Cleveland; New York: World Publishing Co. Meridian Books, 1963), pp. 71-72 (Bunge's italics).

16. *Encyclopedia Britannica*, 13th ed., s.v. "Billiards."

17. *Universal Standard Encyclopedia*, 1958, s.v. "Billiards."

18. Ludwig Wittgenstein, "Tractatus Logico-Philosophicus," trans. D. F. Pears and B. F. McGuinness (London: Rutledge & Kegan Paul, 1961), p. 74.

19. Kurt Goldstein, *Der Aufbau Des Organismus: Einleitung in die Biologie unter besonderer Berücksichtigung am kranken Menschen* (The Hague: Martinus Nijhoff, 1934), p. 1 (Author's translation).

20. Sinclair Lewis, *Babbitt* (New York: Harcourt Brace & World, 1922), p. 3.

6
Determinism versus Decision Making

THE DETERMINISTIC VIEWPOINT

The dichotomy, determinism versus freedom of decision making, troubles those who study man and the behavior of animals. Natural scientists are not concerned with this fissure since they have built up a causally grounded, thoroughly mechanistic conception of the inanimate sector of the universe. Heisenberg's indetermancy principle does not intend to challenge this deterministic viewpoint. The dichotomy, determinism versus freedom of decision-making, of choice, of the will, comes to the fore in the following definition:

> Determinism, a philosophical thesis that conditions control the course of events, a doctrine opposed to libertarianism, or belief in freedom of the will. Modern determinism is based on the type of psychology that sees the individual as controlled entirely by his history. The doctrine is opposed to the principle of emergence, which states that truly novel and unpredictable events may occur out of the composite forces of the situation.[1]

When we study the world of living organisms, the deterministic viewpoint is not as firmly grounded as it is in the sciences which study the behavior of inert matter. The behavior of lower organisms seems to be nearly deterministic in nature while the higher animals begin to show more and more evidence of the capacity to choose, of purpose-striving. In particular, the subhuman primates display intelligent behavior and are capable of learning; but these attributes are insufficient for the building of an indeterministic life-style. This life-style demands that the organism is capable of synthesizing objective time-space with subjective time-space. Only man is capable of striving toward this modality of living, since only man is capable of integrating the two forms of time-space.

The deterministic world is encapsulated in the form of objective time-space and it is causally structured. Indeterministic behavior and mentation move within the form of subjective time-space and the category of purpose-striving. But the form of subjective time-space and the category of intentionality do not exist separate from the form of objective time-space and the category of causality. Therefore, we shift back and forth between existing, mostly, in objective time-space and in a deterministic life-style, versus living, mostly, in subjective time-space and in an indeterministic life-style. Hence, we view the world in which we live and ourselves, now, as rigidly determined, and then again we are blessed with less or more freedom of choice. We cannot strip ourselves of the deterministic world of our body as a structural-functional unit, nor from our ambient world; all we can hope to do is to superimpose our indeterministic, goal-striving drives upon the deterministic aspects of ourselves. Determinism, which excludes any possibility of freedom of choice, leads to giving in and giving up, while indeterminism challenges a pessimistic view of ourselves and urges us on to liberate ourselves from our deterministic attributes to a certain extent. This process is never complete, and therefore we have to settle for a tension-ridden synthesis of this dichotomy. We should attempt to make the deterministic viewpoint and our limited measure of freedom intermingle and cooperate. However, one contraposition stands firm: indeterministic constituents of human, mundane existence do not exist apart and separate from the deterministically structured world. In practical, concrete daily living we do not exist either in objective time-space or in subjective time-space, but we do live in an uneasy balance in which now the one and then the other form of time-space predominates. If objective time-space tips the balance, then we live closer to a deterministic life-style and, if subjective time-space is in the ascendency, we partake of the indeterministic world of freedom of decision making.

Long before natural science established a mechanistic cosmology determinism was and still is a deeply rooted belief of mankind. Most mythologies abound with examples of this belief which is expressed in predictions about the distant objective future, knowledge about which is handed down to humans by superhuman deities. Carroll Stuhlmuller throws light on this subject in a quote from the author of Deutero-Isaiah who doubts that the pagan gods are really gods as he challenges them to foretell the future. "Let them come near and foretell us what it is that shall happen! What are the things of long ago? That we may know that you are gods!"[2]

Sophocles, speaking through Iocaste, states that man is bound to his Fate because he cannot foretell, nor can he influence his future.

> Why should anyone in this world be afraid,
> Since Fate rules us and nothing can be foreseen?
> A man should live only for the present day.[3]

Nevertheless, it was King Laios and Iocaste who tried to get around the predictions of the oracle and the power of Fate: they bound the baby Oedipus's feet and ordered a shepherd to leave him, with swollen *oede*matous ankles to perish in the mountains. Oedipus, warned of his future crimes, fled from Corinth to undo the words of the oracle and the power of Fate, which drove him exactly into the predicted tragic situation. He and his father and mother behaved as if they believed in the power of free will opposing the power of Fate, which free will turned out to be powerless against the deterministic power. A paralyzing view of human existence, which led Iocaste into taking her own life, because she saw that there *was* no future for her. Oedipus wrecks his own body, not because he feels guilty, but because he is mortally ashamed of what Fate has wrought upon him, and he cannot stand the sight of himself and of his children. His suffering is worse than death and yet the second part of the tragedy ends on an optimistic note. He did find some contentment in exile and King Theseus has him buried with a mysterious ritual as if he regarded Oedipus as a hero:

> For he was taken without lamentation,
> Illness or suffering, indeed his end
> Was wonderful if any mortal's ever was.[4]

The religious philosophers of India divided cyclical time into huge periods of astronomical time which were ordered by divine fiat. These tremendous cycles of time are decked out with emotionally loaded meaning since they cycle downward toward darkness and destruction and upward toward resurrection in a predetermined cosmic ordering.

> YUGA (skr., an age, a yoke) is one of the four ages of the world. The first is the *Krita Yuga*, or golden age, the duration of which is 1,728,000 years. The second is *Treta Yuga*, a fourth darker and less righteous than the preceding, and a fourth briefer, enduring 1,296,000 years; the third is the *Dvopara Yuga*, yet a fourth darker and briefer, 864,000 years; the last is the *Kali Yuga*, the present age, darkest and briefest of all, beginning 3,102 B.C., enduring 432,000 years. The total period of 4,320,000 years is a *Maha Yuga*, or *manvantara*. The close of a Manvantara is signaled by a pralaya, or apparent destruction of the world, which inaugurates the "night of Brahma," and endures until the end of thousand cycles of time. Brahma awakes and renews the cycle of ages, or "day of Brahma."[5]

We hear about a similar theme of the distant future being predetermined in the tale that Prometheus, chained to a rock in the Tartarus, tells to Io. His mother had predicted to him when he was a young man that Zeus would not be omnipotent forever. He would eventually come to his downfall and then Prometheus would be restored to his former power.

In Greek mythology the three Moirae, Clotho (spinner) who spins the thread of life, Lachesis (dispenser of Lots), and Atropos (Inflexible) who cuts it off, are personifications of determinism and so are the Nordic Norns, Past, Present and Future, who reappear in Shakespeare as the three weird sisters who foretell his future to Macbeth.

De Santillana has gathered an impressive body of materials on the concept of time in relation to the power of gods which rule man's destiny.

> One would wonder about this obsessive concern with the stars and their motions, were it not the case that those early thinkers thought that they had located the gods which rule the universe and with it also the destiny of the soul down here and after death.... Our forefathers concluded with speculations about the fate of man's soul in a cosmos in which present geography and the science of heaven are still woven together. Worse, maybe, they built them up on a conception of time which is utterly different from the modern, metric, linear, monotonous conception of time.[6]

Nevertheless, belief in freedom of the will, of decision making, of choice, has stubbornly held its own over against the power of time cycles or the relentless prediction by seers, oracles, or gods of the distant future. People have invented many types of behavior which they think will get around the all-powerful superhuman influence of Fate, of determinism. Amulets ward off spirits, offerings and prayers supposedly influence the decisions of the gods or may even lead them on the wrong track. For instance, the animistic world image of the Kpelle of Liberia gives the deceased ancestors great power over presently living members of the tribe, as John Gay tells us. A young man who has just heard that his wife has delivered two strong, healthy baby boys "thought that one of the newborn boys should be called "Good-for-Nothing" and the other "Dirt," so that the ancestors would not try to take them back again. When they were initiated in the Poro Society, then they could be given new names, proper Kpelle names."[7]

People find some ad hoc solution to curb the power of the feared gods or, in this instance, the power of the envious ancestors. Underlying these maneuverings is the conviction that we have a free will thanks to which we may be able to change the undesirable course of predetermined events to an outcome which is more to our liking.

Perhaps we can learn a lesson from the thinking of these members of foreign cultures: our existence is suspended between the two extremes of inflexible determinism and absolute freedom of choice and decision making. A purely indeterministic inner world in which we believe that we have absolute freedom of the will borders on grandiosity, and it goes against the grain of all that is given in experience. For instance, we are not free to decide to make ourselves free of gravity. Even the most sophisticated space vehicles do not release us from being earthbound human beings, living in our gravity-pervaded world. The concept of absolute freedom of the will is based on the conviction that we are able to leave the causally structured world in objective time-space behind us and that we can, on occasion, live exclusively in the "free" world of purpose-striving in subjective time-space. This is in my opinion a totally fruitless approach, inspired by fantasy, to the problem of freedom of the will.

As is so often the case, semantic unclarities confuse our thinking. The word *free* suggests that if we wish, we can do most anything at all. But what do we want to be free from? From our bodies? This is a nonsensical wish, because mentation never goes on without appropriate brain function. Take the case of "brain death" when the electroencephalogram is "flat" and when the dementalized body continues to "live" as a physiologically functioning, causally determined specimen in objective time-space. This nonhuman existence is "free" from its brain, but no one would want to strive consciously and willfully for this state of affairs, because this kind of quasi freedom is valueless. Not only can we not *wish* to be free from our body, we do not *want* to be free from our body, simply because we do not wish to die.

For certain, purpose-striving activities by means of which we accomplish our most cherished deeds and constructs are not free from the body, which is an integral cooperant in these activities. We do, however, acquire a certain degree of freedom over the mechanistic reflex mechanisms, by superimposing purpose-striving attributes upon these causally determined processes. This transformation of reflex mechanisms lifts us above the level of determinism. Far from getting rid of these sensory-motor reflexes and instead of "freeing" ourselves from this deterministic aspect of our existence, we integrate them within our purposive activities. We try to synthesize the dichotomy determinism versus indeterminism by integrating these two ways of living in the world and within this synthesis we capture a certain degree of freedom of the will.

Gravity, our body and its genes, our private history, our social environment, our level of education, are some of the deterministic elements in our mundane existence which are evidence of our having been 'thrown-into-the-world'. These elements exist side by side, and they impede as well as facilitate the decision-making process. Due to their sometimes cooperating, at other times mutually contradictory effects, we never capture freedom of the will in the absolute. But if we are fortunate enough, we grasp an uneasy compromise in which we taste a sense of victory over opposing deterministic and indeterministic constituents which we synthesize when we make a relatively free decision.

DECISION MAKING

Sometimes we make meaningful decisions, but most of the time we make meaningless decisions. While going about routine daily tasks, we pay attention only now and then, here and there, to what we are doing and usually we are thinking about something more interesting while we get the necessary chores out of the way. Routine activities tend to be deterministic, the thinking and planning that is going on contemporaneously veer toward indeterminism.

Heidegger makes the point that these activities reduce us to the life-style of the average man (*das Man*). "Existence is above all and mostly with the 'world' which we take care of. This being involved in has mostly the character of being lost in the publicness of the average person. The self-assurance and decisiveness of the average person spreads a growing needlessness in reference to insightful understanding."[8]

94   The Meaning of Suffering

Workers on a production line are graphic examples of this life-style. It seems as if they do not make any decisions at all. Visualize three workers, $W1, W2, W3$, sitting in front of a conveyor belt which stops every five minutes. When product P arrives in front of $W1$, it is his task to add or insert component $c1$ onto P; when it arrives in front of $W2$, he has to add $c2$ and $W3$ adds $c3$ five minutes later. These men are locked into the "windows" of the spatial compartments of their working areas and into the time slots of five minutes in which each repeats the same manipulation eight or more hours a day in objective time-space. As workers, they live in the stimulus-response pattern in which mentation hardly plays a role. Each reacts to the stimulus of the object appearing in front of him during the allotted interval of five minutes, and they have no interest in what happened to product P before the beginning of this interval, nor in what is going to happen to it at the end of the interval. Insofar as they are workers, they live mostly in objective time-space, in compartmentalized cause-effect chains. In this situation there is no opportunity to integrate past with future within the present. This is exactly the reason why piecework on the assembly line can have a dehumanizing effect. They are locked into the deterministic life-style in which they find minimal opportunity for the expression of selfhood, of decision making, of personality growth, and of feeling tone.

Heidegger enlarges upon this dreary picture of man as a robot by showing how average man rationalizes this existence as commendable, constructive, and comfortable.

> In this quiescent understanding of everything and comparing oneself with everyone, drives being into alienation in which its inmost capacity to be, hides itself.... But this fall remains hidden from it by public explanatory maneuvers, in such a way that it is explained as "ascension" (upward striving) and "concrete life."... This steady tearing loose from appropriateness, and yet always making believe of it, together with pulling into average man, characterizes the movement of falling away as *vortex*.[9]

We can tone down Heidegger's dizzying view of average man as living in some kind of a vortex by taking another look at the three workers. Because they are minimally involved with their actual work, they are free to indulge in daydreaming, in planning what to do after work, how to spend their paycheck at the end of the week, and most likely they will talk with each other about objects of common interest such as the news of the day and the victorious deeds of their favorite athletes. These thoughts and chats take them out of the narrow, meaningless confines of their work going on in objective time-space as they escape into the more meaningful world in subjective time-space even if it is only in fantasy. Besides being causally driven robots they are also purpose-striving, emotionally charged human beings. Within this spatiotemporal context Heidegger's vortex looks more like the teeter-totter movement of a pendulum. In a metaphorical sense the pendulum swings from one extreme of living totally in objective time-space, causality, and a compartmented life-style, to the other extreme of living totally in subjective time-space in which one would enjoy ideal freedom in making choices. Living in the reverberating flux between the two forms of time-space we experience emotion which is generated mostly within the activities of the mental triad in a precious present, which counteracts our falling away into average man.

The mental triad is a necessary constituent in the creation of a life-style in which we can make decisions and in which we have a measure of freedom of choice. Does this agency set us free from determinism? It certainly does not. The mental triad condenses memories which are "there" in our so-called memory bank which memories are deterministic elements in our mental makeup. But condensing the deterministic material with an eye to the future-in-the-fringe forms a new emergent. Thus the mental triad mediates between determinism and indeterminism since purpose-striving lifts our existing in objective time-space onto the region of indeterminism in subjective time-space. The actions of the mental triad seek to enlarge the deterministic life-style so that it can become part of and integrated with a life-style in which we enjoy freedom.

This trend of thought falls in line with Engelhardt's vision of the mind-body relationship as a categorial relation in which the lower category, body, "nests" within the higher category, mind. "The lower category is not only a moment of the higher category, it is also uniquely explained by the higher category.... The higher category is thus 'higher' in that it supplies a further dimension of significance in which the 'lower' category can be more fully significant. The higher category develops and completes the lower."[10]

One can graft the relationships between the lower and the higher categories upon my terminology by embracing a wider range of relatedness. The lower category of causality structures events in objective time-space which can be compartmentalized and which we observe by extrospection. The higher category of goal-striving structures events proceeding in subjective time-space. These experiences interpenetrate, and we become aware of them by introspection and its satellites. The lower category of causality is fulfilled and enriched by the higher category of purposiveness. Transferred to the dichotomy of the deterministic life-style versus the style in which we hunt for freedom, this proposition becomes: average man, fallen away into the deterministic life-style is lifted upon the higher plane of goal-striving; he is lifted up out of a poorer into a richer life-style.

It may seem as if modern physics curbs the deterministic conception of the universe in which trend Heisenberg's uncertainty principle is given an important role to limit all-pervasive causality. But warnings come from the community of physicists themselves, such as Schroedinger's remarks, quoted by Poortman (see below). Heisenberg's principle applies to the world of microphysics, and it does not apply to the macrophysical world of purpose-striving activities. Heisenberg's indetermancy principle states that:

Its restrictions are sufficient to prevent scientists to make absolute predictions about future states of the system being studied. The uncertainty principle has been elevated by some thinkers to the status of a philosophical principle, called this principle of indetermancy, which has been taken by some to limit causality in general.[11]

Schroedinger imagines as an example of microphysical uncertainty the following experiment. There exists an instrument, a so-called electron counter, which is able to give a signal when a radium atom explodes. If one allows a little hammer of the signal to shatter a little vial which contains cyanic acid gas, a cat, which is present in the room, will be poisoned and she will die. Now, one can put such an amount of radium in the counter vial, that the chance that an atom explodes within one hour, is of equal magnitude that this does not happen. However, one cannot know if it will happen or not. If the cat is actually poisoned, then, Schroedinger says, by means of the reinforcement installation with microphysical uncertainty, she has therefore died indeterminately.[12]

This principle is inappropriate for attempts to underpin freedom of choice, of decision making, of the will. This principle operates within the form of objective time-space and the category of causality, and it does not transcend into the world in subjective time-space which is purposively structured.

Does the drive to go from objective time-space toward subjective time-space undergird freedom of the will? Yes, with the limitation that freedom of choice and of decision making is modified and restricted by deterministic constituents in our psychophysical makeup. We can free ourselves from determinism to a certain extent by creating true propositions, beautiful things, and doing ethical deeds. But while we are involved in the world of freedom during these activities, we are also hemmed in by the deterministic, surrounding world and our body. Is decision making a real process which is of our own doing, or is it only make believe? Gordon Allport is convinced that decision making is for real, but he upholds that we have only limited options for exercising this freedom.

It is customary for the psychologist, as for other scientists to proceed within the framework of strict determinism and to build barriers between himself and common sense lest common sense infect psychology with the belief in freedom. The scientist's frame of reference is like the frame of an omniscient being: to him all things have time, place, and determined orbits. But this frame is definitely not the frame of the acting person. The situation is more like that of the watcher from the hilltop who sees a single oarsman on the river below. From his vantage point the watcher notes that around the bend of the river, unknown as yet to the oarsman, there are dangerous rapids. What is present to the watcher's eye still lies in the future for the oarsman. The superior being predicts that soon the boatman will be portaging his skiff,—a fact now wholly unknown to the boatman who is unfamiliar with the river's course. He will confront the obstacle when it comes, decide on his course of action and surmount the difficulty. In short, the actor is unable to view his deeds in a large space-time matrix as does the all-wise God, or the less wise demigods of science. From his point of view he is working within a frame of choice, not of destiny.[13]

Allport's beautiful imagery shows the split between man the watcher and man the doer. We can split the watcher into an extrospective observer who takes notice of the mechanistic forces which threaten the man below and a communicating human being who is introspectively aware of his concern for the dangers that face the oarsman while empathizing and identifying with him. He sees how the man skillfully overcomes the turbulent forces of the river and, next, watches how he portages his boat to start quietly rowing again beyond the rapids. The watcher may feel triumphant to have seen that a fellow human has not been overpowered by the potentially destructive mechanical forces. As an extrospective observer he is a determinist, as an introspective observer he is an indeterminist who rejoices in man's show of freedom of action over against deterministic nature, although this power is limited.

Does this prove that the oarsman shows us that he had freedom of will? Hardly. He lived, especially during the critical period, neither in extreme determinism nor thoroughly imbued with indeterminism, but he did make a successful synthesis "in between" these two extremes. And precisely in this "in between" lies our freedom. However, our freedom is never absolute and unencumbered.

To illustrate with an almost banal example of making a choice, consider that one may decide while going routinely from home to office to take route a instead of equidistant route b. If driving alone, one will hardly be aware that one actually made this decision between the two alternatives. If one travels with a companion, there might be a short discussion about which route one will take, and one will probably end up with a compromise such as: "We will go a and we will come back b." Because in this situation emotional tension is at a minimum and barely detectable, one hardly would think that this was a decision-making event since the outcome will be the same in either case.

As in so many cases, here again we can learn about normal behavior by comparing it with abnormal behavior. When a compulsive neurotic person has to make a decision of a simple nature, emotional tension may flood his consciousness and make it into a painful, upsetting experience. I knew a seriously ill, neurotic farmer who could go from his farm to the village along two alternative routes, and, when he came to the fork in the road where he had to decide to go right or left, he had to stop his car, and it took him a long time to make that decision. He felt utterly miserable, literally moved to tears and drained of energy. The pathology of decision making demonstrates that this process goes on even in simple situations.

Next, an example of a disastrous, destructive decision. The *Andrea Doria* was heading for New York surrounded by fog. The captain was on the bridge, tense and anxious because he could see on the radar screen that other ships were near him in the relatively narrow passage. He was glued to the radar screen, and his nervous tension may have blocked him from evaluating information that came through unaided visual and auditory channels that were also available to him. The information on the radar screen convinced him that a collision was near at hand, and he decided to order his helmsman: "Tutto sinistra," with the well-known disastrous results. This decision not only caused the death of several passengers, but it ruined the captain's career and he was a broken man. Some critics, long after the fact, felt that he should have compromised in making his decision by relying on information coming in from sources other than the radar screen.

Some decisions which we have made when we were young may structure the pattern of the rest of our life. William J. Jacobs describes such a decision which Hannibal made at age nine. His very last decision was to take his own life so as not to be killed by the hated Romans. Hannibal's life was one long series of making decisions to wreck the Roman hold on the Mediterranean. When he was close to success, his hopes were shattered by the political realities in Carthage over which he had no control.

> Legend has it that before accompanying his father to Spain, Hannibal swore a sacred oath in the Temple of Dido. With his hand on a sacrificial lamb, killed by Hannibal himself, young Hannibal pledged undying enmity to Rome. Never, he declared, would he submit to Roman rule. Never would he cease his efforts to crush the Roman State. The accomplishment of the ever-confident Carthagenian general was formidable; by the sheer strength of his will he had deposited a compact, superbly disciplined force of killers at the very doorstep of Rome. In the annals of military history the lightning march of Hannibal across the Alps is considered one of the great strokes of strategy. One of the Bythnian envoys to Rome let the carefully guarded secret slip—Hannibal was in Bythnia. The Senate immediately authorized a small expedition to seize the great Carthagenian. . . . Seeing the guards, Hannibal realized that this, at last, was the moment which he had known would come. "Now," he said, "it is time to end the worry of the Romans, impatient as they are for the death of an old and hated man." Drinking the poisoned wine, he died.[14]

The following story which describes the making of a decision of great complexity in midlife has this same quality of a vivid and meaningful experience in early life, which generates subsequent and related decisions made during the coming decades. A medical colleague told me about a series of decisions which he had made during his lifetime. His history shows that our explanations of this complex process are quite tentative and tenuous.

He was from a broken home and he was brought up by his grandmother. His problem began in high school—age sixteen. His English teacher appealed to him and instilled an abiding love of literature in the young student. This teacher took a unique place in his personal history. It was an intuitive impression which he finds hard to explain, but it gave him the yen to become an English teacher. His German grandmother, on the other hand, had a reverence for scholarship, and she encouraged him to depart from the family tradition. No one on either side of the family had ever gone to college, they barely got through high school.

The grandmother idealized doctors, and she treated the family doctor as if he were a demigod. In his second year in a liberal arts college the young man was torn between his desire to teach English, which he wanted to do very much, and his grandmother's gentle pressure on him to become a doctor. He solved the problem both ways by majoring in English and in biology. He admired especially his biology professor, but he postponed the final decision until his senior year when he got offers to do some teaching in biology, or to write for a newspaper. Instead, he applied to Yale Medical School, was accepted, and eventually became a specialist in surgery.

He was successful. Becoming chief of surgery in his local hospital where he loved his work in the operating room, he also made money. But there was always this nagging doubt whether he really should have become a teacher of English. He could not shake it, and in his early thirties he started thinking about giving up his practice. In his forties he saw some of his colleagues die from coronary attacks, and he began taking his surgical problems more and more to bed with him. At age fifty he felt that he *had* to see if he had chosen the right profession. He did go to a teachers college, loved the life on campus, visited home on weekends, and began to prepare his master's thesis in English literature on Mark Twain.

But then, he remembers clearly the moment when and where it happened: he was sitting on a pier jutting out into a lake, writing away at his thesis when suddenly he put down his pencil and thought, "I am not equipped to be a teacher of English." This doubt had been brewing in his mind for a long time, but he could not give thought to it until that moment. Being trained in medicine and surgery, he always tried to find the simplest possible solution to a problem. Teachers, on the other hand, make simple things more complex; he could not see himself standing before a class, reading a short poem, and then talking about it from various angles for an entire hour.

The next day he went to New York and got a job as a ship's surgeon on the *Independence* where he served for one-and-a-half years. Then he took up emergency room duty in his local hospital, which he is still doing to this day, and he loves it.

In conclusion he said, "Events often force you to make a decision for which you have to wait." He had to wait twenty years before the urge to be a teacher surfaced again. Another twenty years went by before he was properly prepared to take concrete steps. After four years in college the realization suddenly struck him that teaching did not fit into his personal history. He had to go through long gestation periods of self-evaluation and introspection before he could make these true, self-propelled, and self-motivated decisions, after much inward- and outward-turned preparation.

Because true decisions are time-binding processes, it seems as if we make them on the spur of the moment. When we make a true decision, we synthesize past and future in the present through the activities of the mental triad in a precious present, and it carries a vivid feeling tone precisely because it is a time-binding event.

These five examples of choosing and decision making have the characteristics of an exercise of free will. However, this freedom is hedged in and limited by external circumstances, personality attributes, and life history, factors over which the decision maker has only limited control. Nevertheless, each person experiences a certain degree of freedom of the will. Some people live closely within the deterministic life-style, while others partake much more of an indeterministic life-style. But since real life is a mixture of both styles, freedom of the will or of choice, of decision making, is variable, fluctuating, and never absolute but limited and relative. I find Robert Nozick's thoughts congenial to my own.

> Philosophers often treat the topic of free will as a problem of punishment and responsibility;...Without free will, we seem diminished, merely the playthings of external forces....There is an incompatibility or at least a tension between free will and determinism, raising the question: given that our actions are causally determined, how is free will possible?...if an uncaused action is a random happening, then this no more comports with human value than does determinism. Random acts and caused acts alike seem to leave us not as the valuable originators of action but as an arena, a place where things happen, whether through earlier causes or spontaneously.[15]

There may be a way out of this dilemma if we think of determinism as happening in objective time-space and freedom as belonging in the sphere of subjective time-space. By deriving objective time-space from the precious present one might be able to reconcile the incompatibility of causality and freedom.

Decision making is an expression of selfhood. The self constitutes continuity in the discontinuity of the stream of consciousness. "Stream" is an unfortunate metaphor since it suggests a continuously flowing river, while observed introspectively the stream of consciousness displays many aspects of discontinuity. The self establishes a measure of continuity within this discontinuity. One moment we may be fully alert, and during the next moment we may be distracted and our thinking may be decentered. Being awake versus being asleep are two markedly contrasting states of mentation, of alertness and activity versus near inactivity. While we are awake our thinking goes on rather slowly compared to the lightninglike speed of dream activities. Our emotional life shows the discontinuity of being cheerful and full of vim and vigor and then becoming full of ennui and boredom, of lackluster feeling tones, even of feeling downright depressed, and then the mood may swing upward again.

The neurophysiological interpretation of mentation correlates these discontinuities in the stream of consciousness with fluctuating brain functions which are going on continuously while we are alive, even in deep, dreamless sleep. For this viewpoint neither mentation nor brain function are discontinuous. Apparent discontinuities in mentation are degraded to a faulty reading of our subjective states, a conclusion which one reaches after introspection which the natural scientist rejects as a nonscientific method of observation. Unlike extrospection, which yields objectively verifiable and measurable data, introspection cannot be verified, and its results cannot be compartmentalized, and therefore cannot be measured. Extrospectively observed data of brain function, though variable, are continuous, and therefore mentation must also be a continuum because the neurophysiologist conceives of physical as well as mental phenomena as occurring in objective time-space. There is no need and indeed no room for the self in this view of human existence.

If one recognizes this viewpoint as one-sided, impoverished, and inadequate, then one has to come to grips with the fact of the discontinuity of mentation. This fact starts us on the search for an element which can counterbalance this discontinuity. This element is the self, which we elevate to the status of the guardian of continuity within the aspects of our discontinuous mental processes and our truncated living-in-the-world. But most human beings are not satisfied with such a concept of self because it does not promise continuity of our existence after death. Therefore, we have created the concept of the immortal soul.

THE SELF AND THE SOUL

I am convinced that I am a Self. This conviction stems in part from the structuring of memory images. I remember my early childhood experiences as discontinuous, truncated, and compartmentalized. So are my memories of my adult life discontinuous, truncated, and compartmentalized when they surface in the here-and-now. These early and adult images have in common that I relive them as my very own. I assign to both kinds of memories the attribute of belonging to "me," to "myself," that my Self is present in them. Self not as just another image of equal significance among the others, but Self as a specific function of mentation which stands out from the stream of consciousness. We experience this constituent of mentation as continuous throughout life, as a rock-bottom fact of being-in-the-world. My Self establishes a thread of continuity in the discontinuity, the interruptedness of mentation. It connects what is disconnected in our living in objective and subjective time-space.

It would seem that memory is hardly sufficient to establish the continuity of self. Sometimes we hardly recognize the participation of our present Self in our childhood activities, the Self which is the core of whatever I may be doing as an adult. When you happen to see a photograph made in your childhood and even those made in your teens or twenties, you barely recognize this face as your own.

It is as if you look at a stranger. This body image of yore does not fit into the now-remembered Self of those years, and therefore it is hard to recognize "me," 'my person,' 'myself'. This brings man as a whole mind and body, a psychophysical being into the discussion. Self-image and body-image ought not to be separated when one contemplates concrete experience, but we do differentiate between them in abstract thinking.

The concept of Self is hard to pinpoint because it is a member of a series of concepts which progress from relatively concrete to totally abstract. This series begins with the concrete psychophysical person, and it ends with the abstract concept of the Soul. The Self occupies a place somewhere in between these limiting points. The concepts of 'myself', 'the Person', 'mind', 'mentation', 'psyche', 'spirit', and 'Soul' are separated by barely definable and subtle transitions. Inasmuch as 'myself' is a total psychophysical being, and the soul is defined as an immaterial substance, the mind-body relationship is pushed more and more into the background as one progresses along this abstraction series, at the end of which the psychophysical individual evaporates. This progression is conceived as if it were free from any discontinuous leaps and bounds, and hence this series could be regarded as an ideal model of continuity.

Self is not a monolithic concept. One can differentiate within this idea the components of body image, self-image, awareness of self, and feelings of self. We can clarify these shadings of meaning within the concept of self somewhat by considering how a baby, an infant, and a young child gradually develop the foreshadowing of a self concept.

It begins with the *body-image*. The baby becomes aware that this foot, this hand, which he can touch and see and bite, is his own, that it belongs to him. His overall exploration of his own body is the experiential substrate upon which he develops an image of his own body. This process is going on while he also develops the idea that objects out there are separate from him.

Self-image requires that the baby and the young child have some kind of feeling for continuity in time. The continuity of his relations with objects 'out there' is the background for his feeling of temporal continuity. He learns that objects which he no longer hears, sees, tastes, have not simply vanished but that they are still there, somewhere. Self-image gives him the feeling that he belongs in this crib, to this mother, to these objects standing or hanging around, touchable, visible, and that they are available to him when he wants them. At this stage and for quite some time to come, self image is on a concrete level since it is tied into the body image and relations to surrounding objects.

Awareness of self which we grasp in our early teens, is a state of contemplation of self, embedded in the stream of consciousness. Awareness of self is a more subtle, variable, fragile, and introspectively oriented phase of development compared to the previous phases because body-image gradually retreats into the background.

Feeling of self sets us apart over against the others, whom we recognize and apprehend as strange selves 'out there', but one realizes that 'their selves' are similar to 'my self'. Feeling of self is the most abstract element in the concept of Self.

One can see within the development of the concept of Self a progression from more concrete—body-image—to more abstract—feeling of Self—which process of abstraction mirrors the progression from the psychophysical individual, through the concept of Self, to the concept of Soul in the larger series mentioned above. Attempts to define the concept of Self are hampered by the fact that we do not separate this term sharply from the other members of this abstraction series, especially over against the concept of the Soul. I propose that we attempt to separate the concept of Self from the concept of Soul on account of the differences between their temporal structures.

The soul exists in eviternity, that is, beyond objective time-space and beyond subjective time-space. Therefore, the concept of Soul promises to insure the continuity of our existence beyond death and gives relief of the dread of dying and of death. Self functions as a provider of a measure of continuity within the discontinuity of our mundane existence in objective time-space and in subjective time-space, this side of the grave. Soul offers continuity in a wider temporal expanse because it overarches the fracture living-versus-dying, and it executes this function precisely because it exists in eviternity. The different temporal structures and the different functions of Soul and Self forbid us from regarding Soul as if it were the most abstract endpoint of the series of abstraction from psychophysical subject to Soul. Its temporal structure places Soul outside the aforementioned series.

For these reasons we ought not to think that there is a gradual transition from more concrete toward more abstract in this series when we progress from the feeling of self to the concept of Soul. Since we have seen that a progression from more concrete to more abstract can be noted in the several elements of the concept of Self, we might be lured into inserting the abstract aspects of the concept of Soul into the concept of Self, which blurs the delineation of the two concepts vis-à-vis each other. The different temporal structures and the different functions of Self and of Soul ought to abolish the tendency to use these concepts as if they were quasi-interchangeable. The various conceptions of the Soul have been summarized as follows:

> The concept of soul is being interpreted so differently in different religions that it seems almost hopeless to comprise all under one word in a rational manner. a. *Power of the soul*, which is also named *soul substance*, *soul-stuff*. The soul is impersonally divided over the entirety of man, attaching itself to different parts of the body, according to their power. This structure is widely represented among primitive people. b. *Soul in the plural*. The Egyptians differentiated *ka*, *ba*, and yet a host of other souls. The Israelite *negesh* and *roeach*. Plurality is the cause that the soul does not have a fixed form at this point; it is a transitional stage between soul-power and soul conceived as a person. c. *The external soul* (after an expression of Frazer).

> This is a soul-power, which allows itself to be separated from its bearer, whether forever, as by death, or temporarily, as during sleep, dreaming, etc. d. *Soul*, conceived dualistically, *in contrast to the body*, the soul must leave materiality behind as much as possible. Hence, asceticism. Here, the representation of the immaterial soul can arise (immortality). e. *The soul as a central power* which gives direction, which presupposes the unity of body and soul. The soul is then life, according to its quintessence, that which is essential to life. Man speaks about his soul in so many different ways, because he finds something within himself, which is more than he is himself.[16]

Not unlike what Cicero said about the Soul. For the sake of clarity we should point out the differences between Soul and Self. Self does not overcome the dichotomies and the fractures in our existence, since it is flawed with contradictory aspects which stand in tension-filled relationship to each other. Because selfhood is a synthesizing agent it provides a limited measure of continuity in our life's experiences. Its synthesizing power is artificially increased if we allow attributes of Soul to penetrate into the concept of Self. Soul is conceived as indivisible and perfect, and therefore no aspects of Soul stand over against each other in tension-charged turmoil. Soul rises above destruction, death, and suffering. When the psychophysical individual dies, Soul is released from the body, from mentation and its imperfections, so that Soul exists in its appropriate state of heavenly bliss after the death of the human being who was temporarily the host of the Soul. Self is prohibited from this rise *outre tombe* on account of its spatiotemporal structure.

Within the dichotomy continuity versus discontinuity, the concept of self is distorted on two counts if one projects some of Soul's attributes onto Self: (1) This maneuver presupposes that Self can break out of the confines of objective and subjective time-space; (2) Self cannot provide immutable continuity nor can it abolish discontinuity. Keeping the different temporal structures of the functions of Self and of Soul in mind, let us scan some of the conceptualizations about Self and Soul presented by others.

Kierkegaard describes the self as a center in which our temporal organization builds up a concrete synthesis of the self by future-striving, decision making, and despair. He follows the method of contraposing the hedonist over against the intellectualist. He shows that each fails to attain selfhood because they assume one-sided attitudes toward time.

> His [the hedonists's] life becomes a discontinuous succession of passing from one moment to the next. His personality lacks unity and continuity. He has dispersed and lost himself in the present to the neglect of his past and his future. For the intellectualist all reality is dissolved in general categories but he forgets the individual who apprehends himself within the particular and concrete history.... Choice liberates the self from the im-

mediacy of pleasure and the immediacy of reflection of pure thought, and makes possible the discovery of pure selfhood. In the act of choice the past is taken up, the future is acknowledged and faced, and the self is centralized.... To become oneself, is to become concrete. But to become concrete is neither to become finite nor to become infinite, for that which is to become concrete is of course a synthesis.[17]

Due to the act of choice, we synthesize past with present with future, in which act the self becomes centralized and which generates the dread of the next day. One seeks to escape from this anxiety in a "wild narcosis." One sees how Kierkegaard vacillates between the concept of self as it exists in subjective time-space and the concept of soul existing in eviternity; he chooses a solution in one's becoming concrete, a compromise between the different and irreconcilable temporal structures of Self and of Soul.

Popper puts forth a quite convincing schema of the development of the self.

First, the category of persons; then the distinction between persons and things; then the discovery of one's own body; learning that it is one's own; and only then the awakening to the fact of being a self.... The self observes and takes action at the same time. It is acting and suffering, recalling the past and planning and programming the future.[18]

The third step in this schema stresses the body-image as an underlying stratum of the self-image. By interrelating body-image and self-image Popper brings the psychophysical problem to the fore. Popper does not see man as made up out of the two compartments of the physical and of the psychical. If this were so, then one could fit Self into the psychical compartment, with Self separate from the physical aspect of the total human being.

Popper's conceptualization of Self combines self-awareness with death-awareness, in which formula self-awareness stands for continuity, regardless of it also being discontinuous, while death-awareness stands for the absolute discontinuity of human terrestrial existence. As a Self we experience continuity within discontinuity since self-awareness interpenetrates with death-awareness. Living as whole human beings we adjoin these opposite poles of the Self and we synthesize continuity with discontinuity. We are able to accomplish this interpenetration of opposites because we have the capacity to synthesize objective time-space with subjective time-space. We *conceive* objective time-space rationally and abstractly as being continuous, while we *experience* subjective time-space concretely as being discontinuous.

Piaget looks at the development of self-awareness in children within a spatio-temporal perspective. The child has already some awareness of his body-image before he has some notion about temporal relationships in the conventional, adult manner. He judges the differences in age between himself and adults in terms of how they look compared to himself. "Older" means "bigger," "very old" means slow, stooping, grey-haired. A six-year-old thinks that when he will be as big as his parents then he will be their equal in number of years. Children think in concrete, material terms. The body-image is in the foreground during the development of the concept of time which, also, is conceived in concrete images. A self-image does not develop until eight-, ten-, and twelve-year-olds begin to feel that there are some connections between today, yesterday, last week, tomorrow, and next week; in other words, when they begin to integrate objective time with subjective time.[19]

At this point we can bring the precious present into the analysis of Self. The ephemeral characteristic of the precious present, here now and gone a short time after, exemplifies the discontinuity of subjective time. We may feel deeply moved by an experience during a precious present lasting perhaps half an hour, and then fall back to living mostly in objective time, an experience almost devoid of feeling tone. Then we are floating along in the routine existence of average man, and the awareness of Self seems rather neutral and in low key. During these phases of emotional and intellectual stagnation we see ourselves as barely different from the others and the feeling of selfhood is in the background. However, when the mental triad is vitally active in a precious present, then our feeling of being a Self is scintillating. Like most other aspects of mentation, self-awareness fluctuates.

Charles Morgan says that it is as if we die in each instant, not because time abandons us, but because we do not fully live in such instants. Therefore the feeling and memory of ourselves as we were is fleeting and ghostly so that we can hardly communicate with this former self.

> Never before had he been so strongly aware that, in each instant of their lives, men die to that instant. It is not time that passes away from them but they recede from the constancy, the immutability of time, that when afterwards they look back upon themselves they see, not even—as it is customary to say,—themselves as they formerly were, but strange ghosts made in their image with whom they have no communication.[20]

Wisely spoken, but here again, we see a lack of differentiation between objective and subjective time. Morgan shifts from the durational present of subjective time to the immutability of objective time without seeming to be aware of this transition from one form of time to the other. Hence his description of self lacks sharp contours.

Theodore Lidz approaches the problem of self-awareness from a psychological and a societal angle.

> In finding his place in a society of peers, the child develops and forms a self concept that serves to regulate his ambitions and ways of relating to others. The ego, so to speak, contemplates and evaluates the self, but in so doing

the ego is considering the reactions of the others to him—to his self.... He learns to see himself as others see him and according to rather relentless standards.[21]

The self-concept develops in the rough-and-tumble of the child's society of his peers where relentless and often cruel standards are imposed on him. His ego has to evaluate his own self in order to relate to the others and to hold his own within this aggressive environment.

Jung speaks of the tension between the present Self with its personal and its communal history and the Self which is attuned to its personal future and the ego which contemplates the objective future. The split between ego and Self points up the emotional tension which is constitutive of the mental triad as it projects its act into the future-in-the-fringe of a precious present, versus the ego's reality principle which is concerned with the future in objective time-space and the distant future. The dichotomy Self versus ego, which Jung describes as tension, is one of the fundamental sources of suffering. More fundamental, however, is the copresence of self-awareness and death-awareness within the same human being. These two modes of awareness are forceful reminders of the fracture living-versus-dying.[22]

Heinrich Rickert's exceedingly abstract concept of the epistemological subject lies in contrast to the foregoing psychologically oriented and more practical definitions of the concept of self. This is the endpoint of an abstraction series which starts with the body-image and which goes beyond the feeling of selfhood. This series can also be regarded as a stepping-stone toward the discussion of the concept of Soul. The epistemological subject is more abstract than the concept of Self, but it is less abstract than the concept of Soul. It stands on the borderline between the several variations of the Self concept and the concept of Soul. Heinrich Rickert's presentation succinctly states:

> I want to talk about myself, not as an object, but as a subject.... There is something, that certainly belongs to me, but which cannot be referred to as that which I must call I-object.... I am a knowing subject and *at the same time*, known subject.... now we dare not even any longer speak of a subject which perceives the perceptions, which feels the feelings, and which wills the will, since all specific determination must be kept far removed from the epistemological subject.... If I myself were not this specific individual and at the same time a subject in the epistemological sense, or if I could not dissect myself conceptually in subject form and subject content, then I could never know what the word subject exactly means.... This concept does not let itself define any further.[23]

Rickert forms the exceedingly abstract concept of the epistemological subject by eliminating from the concrete psychophysical subject inch-by-inch elements from one's body until one has only the psychological subject left. Then the abstraction process goes forward until one faces a subject which neither thinks nor feels nor wills. This is the epistemological subject from which all determinants have been removed. It is exactly this ephemeral epistemological subject which is the endpoint against which we are able to have awareness of self, the source from which we can separate subject *form* from subject *content*, and from which our insight emerges into the meaning of the word *subject*. Very significantly, Rickert says, "We are therefore far removed from any kind of soul substance."[24]

Although Rickert often uses the words *self*, *I*, *individual* in this very extensive discussion, he does not attempt to give a definition of the concept of Self. Nevertheless, I include this study in advancing abstraction since the epistemological subject is the end stage of the aforementioned series just short of the even more abstract concept of the soul.

The concept of Soul takes on various forms in the religions of the world, but for us Westerners the Greek conceptions seem most important. Here too we see many variations on the same subject which has in common with the others the idea of an immortal substance. Homer represented soul as the breath of life, something airy or ethereal. Achilles, who was allowed to visit the underworld, did see the spirit of Patrocles, and he could recognize him but he could not embrace him.

Speaking through Socrates, Plato presents quite a different image of the soul, in the *Phaedrus*.

> Of the nature of the soul...let me speak...in a figure....a pair of winged horses and a charioteer...one of them is noble and of noble breed, and the other is ignoble and of ignoble breed; and the driving of them of necessity gives a great deal of trouble to him [the charioteer]...;when perfect and fully winged she [the soul] soars upward, and orders the whole world; whereas the imperfect soul, losing her wings and drooping in her flight at last settles on solid ground—there finding a home she receives an earthly frame which appears to be self-moved, but is really moved by her power; and this composition of soul and body is called a living and mortal creature....The wing is the corporeal element which is most akin to the divine and which by nature tends to soar aloft and carry that which gravitates downward into the upper region, which is the habitation of the gods.[25]

Plato describes man as a mystical being composed of the human and of two winged steeds. If the soul is animated by reason, she ascends toward the upper regions where the gods reside and where she will see truth and beauty. If she succumbs to passion, the ignoble steed drags her down, back to earth where she becomes a soul and a mortal body. The striving to overcome gravity by means of wings is common to both gods and man, but only the gods who have steeds with perfect wings can ascend to the very summit of the spheres and they look beyond. Man's soul with its imperfectly winged steed often tumbles down to earth and is again enclosed in a body which causes him to be mortal. Up toward heaven we become immortal if we can overcome the downward pull of gravity. If we can't we die, and we end up in the underworld of the shadowy dead.

As far as the soul is concerned, no matter how light and transparent it may be depicted, it is a corporeal entity in Greek mythology.

In Christianity the concept of the soul is all important. But since the Bible does not give a definition of the soul, Christian concepts about it vary greatly. The origin of the soul has been a controversial subject in the history of Christian theology. Two opposite views may be discerned: "creationism" which holds that God creates each individual soul in a special act of creation (at the time of conception according to some) or at the moment of birth (according to others) and "traducianism" which holds that the parents in begetting the child beget the soul too.

· Saint Augustine sees Soul as a spiritual substance intimately united with the body. The soul is a completely immaterial substance, *res spiritualis, res incorporea.* The Thomistic explanation is that the body is dependent on the soul and exists by virtue of the soul's existence. The body appears as an instrument of the soul, in its language, in gestures, in a smile, or in a tear. According to Maurice de Wulf, the scholastics taught that the soul did not partake of immortal life as merely a cold intelligence in isolation but included conscious and personal life and enjoyment therein.[26]

For many Western philosophers the term *soul* is synonymous with *mind*, for instance, for René Descartes. Others, although asserting its indefinability, have seen it as a useful element in a system of ethics, Immanuel Kant, for example. This indefinability has led others to reject the idea of soul and to postulate ethical systems based upon a different conception of man's nature, such as, William James. Feibleman holds that the concepts of psyche or soul are consistent as well as contradictory and that they represent the *nous*, the upper level of consciousness.[27]

Three themes stand out in almost all of these interpretations of Self and Soul, namely that psyche and soul, mind and soul, Self and Soul hardly differ in essence, that Soul is immortal and that intelligence is rated as the most valuable attribute of man, while emotionality seems hardly worth mentioning.

We can gather together the several strands that run through these discussions from the viewpoint of time. This approach promises to differentiate Soul from Self and to prevent these two concepts from fusing so that neither can be defined. Both Self and Soul are traditionally regarded as being immaterial, but they are contraposed to each other on account of their time structure. Soul exists in eviternity; Self exists in objective time-space integrated with subjective time-space. Another common attribute is that both Soul and Self are concepts stemming from our attempts to liberate ourselves from determinism. Soul strives for total liberation from determinism because it overcomes the dread of death. Self accepts that we can secure only a limited degree of freedom from determinism.

Limiting our horizon to the Self, we can sketch an image of human terrestrial existence. With a view toward the past, body-image, self-image, awareness of self, feeling of selfhood, are built upon our memories of who and how we were and how we acted in years past; these memories carry an aura of "me-ness," and they contribute to a sense of continuity and selfhood. With a view toward the future, Self is built up as we express ourselves in present deeds and feelings, which we project into the future in the fringe-of-a-precious-present and which, we trust, we will be able to realize in the distant, objective future. Not only our past, but also our present and our future are experienced and evaluated as being intimately our own; therefore, we are confident that we will experience a similar feeling of selfhood in the objective future as we are presently aware of in the here-and-now. In brief, selfhood is undergirded by our capacity to synthesize objective time-space with subjective time-space. In turn, selfhood makes our acts of decision making possible, which give us limited freedom.

Two theories concerning determinism, decision making, and selfhood remain to be considered. First, man is not an animal because our spatiotemporal structure is essentially different compared to the spatiotemporal structures of animals. Second, Freud's metapsychology assumes a distinctly deterministic stance because he bases his theory about mentation on the concept of objective time.

MAN IS NOT AN ANIMAL

From the viewpoint of common sense, the negating proposition "man is not an animal" seems to be self-evident. In the world of science however, one seeks to prove that there is a gradual transition from the animal kingdom, through the realm of the subhuman primates, on to man. The theories of evolution and biochemistry present powerful arguments which seem to prove this conclusion.

The theory of evolution opts for cohering descriptions of the structural differences and similarities between lower and middle stages of development and the highest stage of evolution, which is man. The development of eohippus through several forms into the contemporary hippus (horse) is a convincing example of structural coherence, without breaks in the chain of intermediary stages of development. The structure of the brain of apes is more complex than that of monkeys, and the structure of the human brain is more complex than that of the apes. Because of this step-by-step increment in complexity, one is convinced that there cannot exist any essential differences between the behavior of monkeys, of apes, and of man.

However, when Hugo de Vries discovered mutations in certain plants, he inserted a note of discontinuity in the evolutionary process. But although mutation does show sudden "jumps," it does not disrupt the causal structure of these biological events.

> In 1901 the observation of mutants among evening primrose plants led the Dutch botanist Hugo de Vries to present his theory that new characteristics may appear suddenly and these characteristics are inheritable. The work of de Vries has shown the importance of mutation in the mechanism of evolution. Most mutants are lethal, since any change in the delicate bal-

ance of an organism having a high level of adaptation to its environment tends to be disruptive.... This process is now believed to be a chief agent in the process of evolution and in the extinction of species that fail to mutate in a changing environment.[28]

De Santillana sounds an ill-boding warning against extending the theory of evolution beyond the field of biology in the restricted sense into the areas of psychology, psychiatry, history, and cultural endeavors.

> The simple idea of evolution, which it is no longer thought necessary to examine, spreads like a tent over all those ages that lead from primitivism into civilisation.... Those soporific words, "gradually" and "step by step", repeated incessantly, are aimed at covering ignorance which is both vast and surprising.... The lazy word "evolution" has blinded us to the real complexities of the past.... As far as human "fate" is concerned, organic evolution ceased long before the time when history, or even prehistory, began. We are on another time scale. This is no longer nature acting on man, but man on nature.... Rudolph Virchow warned time and again of an evil "monkey wind" blowing around; he reminded his colleagues of the index of excavated "prehistoric" skulls and pointed to the unchanged quantity of brain owned by the species Homo Sapiens. Mistaking cultural history for a process of gradual evolution, we have deprived ourselves of every reasonable insight into the nature of culture.[29]

De Santillana's warning is of basic importance because he shows that development in the biological sense and development in the psychological, cultural, and historical realms are going on in contrasting, incomparable, spatiotemporal milieus. Evolution stretches over periods of millions of years; it is causally structured and compartmentalized; and we study this subject matter by extrospection. Cultural, historical, and personal development happens in subjective time-space and in much shorter periods of objective time-space, and it is purposively structured and interpenetrating, wherefore we study these undertakings by the combined extrospective as well as introspective methods of observation.

Claude Bernard has put organic function on a firm physico-chemical basis with the principle of the *milieu intérieur*.

> [He] explicitly recognized that the general laws of physics and chemistry are valid in living systems. In the process of organization, living systems acquire new properties that are not easily predictable from simple application of the laws of chemistry and physics. The new properties, however, can only be studied by the methods of physics and chemistry. The stability of the *milieu intérieur* is the primary condition for freedom and independence of existence; the mechanism which allows of this is that which ensures in the *milieu intérieur* the maintenance of all the conditions neces-

sary for the life of its elements. The concept of the *milieu intérieur* provides the first scientific integrative approach to biology.[30]

Physico-chemical methods have been successfully applied to investigations of the functions of the animal brain and of the brain of apes. Biochemistry of the brain of subhuman primates teaches us a great deal about the biochemistry of the human brain. In the opinion of some neurophysiologists, some psychiatrists, and some psychologists these sciences will offer mechanistic explanations of human behavior and of human mentation with the aid of our increasing knowledge of the functions of the human brain. The more accurately we learn to predict these physical, chemical, electrical processes which go on in the brain, the better will we be able to predict and hence to understand human behavior. Armed with this knowledge we will need less and less such mentalistic concepts as goal-striving activities, intentionality, the mental triad, freedom of choice, the self, to understand human behavior. When we finally possess an overall causal explanation of human brain function, then we will possess a mechanistic view of man. This natural-scientific outlook upon man denies the existence of our essential constituents of purpose-striving and freedom of decision making. I see this trend as truly ominous.

We can constrain this one-sided and impoverished view of man by showing that man is not an animal.

One must set limits to a mechanistic view of man because as an intelligent, emoting, communicating, playing, and laughing individual he rises above the confines of the field of causal relations. Having conquered this limited measure of freedom we should not devaluate or even try to eliminate the mechanistic theories of cellular and organ functions. On the contrary, the biochemical investigation of compartmentalized cell and organ functions is of great practical, medical, and social value. In brief, from a deterministic viewpoint there is no essential difference between man and animals, man so far being the highest evolved animal. Looking at man and animals from the viewpoint of goal-striving activities, man is not an animal since his spatiotemporal structure is essentially different compared to that of animals. I propose that we climb up the ladder of evolution in large blocks of organismic divisions to show this essential difference by comparing the movements of plants, of lower organisms, of animals, with movements that only man is capable of performing. We shall see an increasing complexity and widening of scope of spatiotemporal structures, but finally there is an unleapable gap between the subhuman primates and man.

Plants display two types of movement: the slow movements of growth and the much faster nastic movements. Growth such as a tree trunk growing upward and its roots growing downward is caused by hormone-like substances, the auxins. Nastic movements are caused by sudden changes in the turgor of cells due to changes in the water and electrolyte balances. One is justified in classifying and explaining these movements as causally determined and mechanistic.

Tropistic movements of growth are believed to be triggered by the presence of hormones (Auxins) that promote cell growth.... Generalized plant responses to a stimulus are called nastic movements or nastics. These include the opening of bud scales and flower petals.... Turgor movements are effected by changes of water content in cells and are often quite rapid. Examples are the "sleep movements" of clover, sudden dropping leaves of the sensitive plant (mimosa), and the reactions of insectivorous plants to the response of their prey.[31]

I do not see any need to ponder whether or not these movements might also have purpose-striving tendencies. Clearly they are causally structured and they happen in objective time-space.

The movements of *protozoa* are explained with the concept of tropism to indicate that the entire organism reacts to stimuli such as light, heat, conditions of a surface.

See especially Jennings' fascinating account of the pursuit of Amoeba by a larger specimen—the persistent flight of the smaller and the persistent pursuit by the larger.... Amoeba sometimes detaches from the solid surface on which it normally crawls: it sends out long pseudopodia in all directions, until one of them comes in contact with and adheres to a solid body; the other pseudopods are then quickly withdrawn and the whole substance flows towards the point of attachment.[32]

Jennings's detailed study of the behavior of amoeba causes one to wonder whether there is already an inkling of purpose-striving in this monocellular living thing. The movements of plants are well within the realm of causality without question and they can be explained on the basis of the structure and function of their cells. But protozoa which are less restricted in their motility compared to sessile plants, display movements which some investigators explain by the causally structured concept of tropism, but on which they project goal-striving tendencies. In other words, one cannot be sure that tropistic movements are exclusively causally structured.

When one studies the behavior of *worms*, clear-cut deterministic views are inadequate. Yerkes performed a classical experiment when he placed an earthworm in a T tube with an ingenious impediment against escape: he placed a piece of sandpaper on the floor of one of the escape routes and immediately behind the paper an electrically charged copper wire.

An earthworm is placed in a T shape labyrinth...[in which] a strip...of sandpaper extends across one arm of the T....just beyond the outer edge of the sandpaper...lie copper electrodes....The worm apparently profited from experience with surprising quickness....the gradually increasing avoidance of the sandpaper, which was meant as a warning against the electrical stimulus; the acquired tendency to avoid contact with the electrodes...[shows that] the worm is capable of profiting from experience in a simple maze.[33]

Whether one is justified to say that the worm showed some kind of capacity to choose is a moot question, but it certainly looks as if the worm liberated itself to some degree from rigidly determined behavior.

The behavior of *insects*, though on the whole causally determined shows the ability to change in the direction of indeterminism. A species of solitary wasp displays a rigid, ritualistic behavior: it carries its paralyzed prey to the entrance of the nest, puts the prey down, disappears in the nest, comes out after a few seconds and drags the prey into the nest. Two experimenters moved the prey a few centimeters away from the entrance of the nest, while the wasp was underground, and when the insect reappeared it searched for its prey, dragged it to the entrance, and repeated the ritual of going into the nest, reappearing, replacing the prey in its original position, numerous times. Finally, after the experimenters had repeated their interference with the ritual a great many times, the wasp dragged the prey down into the nest, without first replacing it at the entrance. A truly astonishing change from deterministic to indeterministic behavior.

However, is one justified in adding indeterministic, teleological elements to the causal explanation of the wasp's undisturbed ritual? Does its behavior show a glimmer of intelligence, of judgment about the whole situation, or even an inkling of freedom of choice? A *caveat* is in order here: if we attribute purpose-striving abilities to the wasp, we give evidence of identifying with the insect, of empathizing with it, and of projecting human attributes of foresight, judgment, decision making upon it. This goes beyond what is critically and scientifically permissible. The wasp's rigid ritualistic behavior is evidence of causally structured neural mechanisms, interlocked in objective space-time. These mechanisms have become nonoperational due to the interference by the investigaters because the source of chemical neural transmitters in the neuronal gaps of the insect's nervous system have become depleted. This neuro-chemical explanation sidesteps our empathizing with the wasp and projecting human attributes onto it. This is where applied chemistry performs a cleansing operation by keeping the concept of instinct uncontaminated by teleological admixtures.

Maurice Maeterlinck describes an experiment which one of his entomologist friends executed with the *Processionary Caterpillar*.[34] These creatures move in a long line, each holding on with its jaw to the terminal hairs of its predecessor. The experimenter placed a long line of processionaires on the rim of a round vessel and watched for three days and three nights to see what would happen. The insects went around and around in the circle until one after the other dropped dead after some seventy-two hours. This experiment and the observation of insects in their natural habitat without man interfering with their behaviors, support the thesis that most of their behavior is fixed in prescribed phases, each previous phase acting as a cause which elicits the next phase as its effect.

The honey dance of the honeybee, discovered by von Frisch, seems patently purposeful and makes one doubt whether this highly sophisticated behavior can possibly be causally determined. However, the variations in the details of the honey dance from one species of honeybee to another, gives support to the idea that it is indeed the structure and function of the nervous system of the bee, and that physical, electrical, chemical processes which are going on in this already quite complex network are the cause of the honey dance.

> Honeybees frequently dance with a view to the sky, orienting themselves to the sun or natural patterns of polarized light. Three new conventions have been discovered in the dance language which are used in these circumstances to eliminate potential ambiguity in the dance message.... They ensure that both sender and receiver are using the same reference system. The rules are presumably a consequence of neural wiring. They are not, however, necessary or even desirable for any of the vast number of social and non-social animals that perform the same feats of navigation but lack symbolic language.[35]

The territorial behavior of *fish* is a well-established fact which makes it possible for a big fish and a small fish to survive in the same aquarium: the smaller fish stays within a smaller territory than the territory of the big fish, but the small one defends its territory ferociously by chasing the big one out of it. I saw an example of territoriality in a big aquarium in which the owner housed three fish. The biggest and most spectacular fish occupied the length and the breadth of the tank; a smaller one stayed mostly in the left one-third, while the smallest fish was seldom seen since it stayed near the rear wall and behind rocks.

Konrad Lorenz specializes in the study of the behavior of *geese*. The gander of one goose disappeared one day, and the mate began to show most unusual behavior: she was slow in her movements, withdrew from the group, and then restlessly began to fly away from her territory obviously looking for the lost mate. She covered larger and larger distances and emitted a wailful cry.

> The first response to the disappearance of the partner consists in the anxious attempts to find him again. The goose moves about restlessly by day and night, flying great distances and visiting all places where the partner might be found, uttering all the while the penetrating, tri-syllabic distance call.... Just as [in] the human face, it is the neighbourhood of the eyes that in geese bears the permanent mark of deep grief. The lowering of the tonus of the sympathicus causes the eye to sink deeply inside the socket, at the same time, decreases the tension of the outer facial muscles supporting the eye region from below.... My dear old graylag Ada, several times a

widow, was particularly easy to recognize because [of] the grief-marked expression of her eyes.[36]

Lorenz says that those who live with these birds and study their behavior systematically become very definitely aware of the feeling tone of the inner state of their subjects, a fact that would be lost on a casual observer. I am sure that he suspects no problem at all in empathizing and identifying with his geese nor does he hesitate to project some humanlike attributes onto them.

The *beaver* is a truly amazing animal. Especially in this case of obviously purposive behavior of an animal one has to be on his guard against going too far in identifying with the beaver by projecting human capacities of foresight and judgment. Leonard Lee Rue writes very warmly and also very critically about beavers in a balanced evaluation of their accomplishments and their failures.

> When trees [that stand next to the water] are cut, they are heavier on the stream side and fall in that direction, a fact for which the beaver gets all the credit.... When a tree becomes hung up on another, the beaver does not know enough to cut the supporting tree, but may make repeated cuts on the hung-up tree itself.... [They] often construct dams that are 100 feet long, whereas if they had sought the ideal spot they would have held the same amount of water with a dam 25 feet long.... In a man-made pond, where the water is already held by a good solid concrete dam, they may still not be satisfied until they improve on the job by attempting to cover the concrete with mud and sticks.... During times of flood the pressure on the dam may become so great that it will be washed out. If there is time enough, the beavers will usually cut a spillway in the dam itself to relieve the pressure. After the floodwaters have subsided, the beavers have only to repair the hole they have cut.[37]

Most of the trees which they cut to build their dams do not fall to the ground and are not used by them. They do not "choose" the most efficient place in a stream to build a new dam. Lee Rue gives a clear picture that the life of the beaver is of a predominantly deterministic quality and instinct-driven. But he also gives them credit for sporadic, indeterministic, purpose-striving activities.

Our *domestic pets* react to many household noises, such as the opening of the door of a refrigerator, or of a can. They also send signals to us when they want something. They give evidence of pleasure and displeasure, they communicate with us and with each other on a concrete, here-and-now level. Dogs learn to open doors without explicitly being told to do so, but I have never seen or heard about a dog who would "spontaneously" close a door with one exception. Two young brothers who had observed that their dog opened the front door by pushing the latch down, taught him to close the door after he was in the house. While they exclaimed loudly, "Shut the door," several times, they also pushed the dog's shoulder against the open door and thus made the dog's body shut it. After five or six repetitions of this maneuver, the dog closed the door on the loud, repeated, verbal command without physical exertion by the boys. By that time, reinforcement by a reward of food was no longer necessary.

Animals do look ahead into the immediate future; as Sherrington says, "The act puts forth, as it were, a little bud of futurity. Then also there is lent to it something of the past."[38] But, and this is the important point, the dog opens the door almost without training so that he can be in the warm house in the immediate future. But once there, he does not look back into the space behind him and the cold environment he has just now escaped from in the recent past, and so he leaves the door open.

Our Siamese cat treated us to an elegant instance of her being attuned to the immediate future. After the first snowfall in early winter, we let our dog out and the cat was ready to follow her, but paused, looked out at the snow, and while on the perfectly dry carpet, she shook first one, then the other front paw, as we often saw her do on a wet surface, and turned around as if she "decided" to stay where it was dry. Living in daily contact with animals, one is astounded every now and then by how similar they seem to us during their brief periods of purposive behaviors.

Dolphins have been studied intensively and seem to be closer to man than the subhuman primates.

> If one of their number is injured or sick they make every effort to rescue it, holding it above water for air.... It displays considerable tool-making and tool-using and manipulative ability.... One female may watch over several calves, while the others hunt, or during battle.... Each dolphin has a signature whistle; a calf soon learns to recognize its mother's whistle. They are observed to converse, and it has been repeatedly shown that one animal can convey instructions to another.... They demonstrate foresight, learning from observation, perform elaborate tasks, and learn multiple procedures simultaneously.... The U.S. Navy has trained dolphins to act as messengers to underwater stations, to rescue wounded frogmen and protect them from sharks, and to seek and destroy submarines.[39]

Their subtly different sound production is an interdolphin communications system which is somehow reminiscent of human language. Their social organization and foresight, learning from experience and performing complex tasks, testify to a level of mentation which is quite similar to that of the subhuman primates. Interestingly, people who study and work with dolphins become emotionally attached to them, quite like those who study and train the higher apes. These studies depend to a large extent on our empathizing and identifying with dolphins and apes and on projecting human attributes onto them.

The behavior of *monkeys* and, even more so, that of the *apes* is in some respect remarkably similar to our own behavior. On the basis of the differences between the spatiotemporal structure of simian behavior and the spatiotemporal structure of human behavior, an essential contrast between the penultimate highest level and the highest level of evolution can be posited.

Portielje conducted experiments with *monkeys* in the zoo in Amsterdam. He put some nuts in their cage and also a hammer. If the hammer happened to be in their field of vision, they would use it to crack the nuts, but as soon as they had eaten all the nuts, they did not rest until they had loosened the head of the hammer from its handle. Here we see the first difference between the simian's use of a tool and man who not only uses it but also saves it for use in a distant future.

Chimpanzees build comfortable and remarkably efficient nests up high in the trees in which they sleep during the night. But they use the same bed very seldom for another night. Baron Hugo van Lawick has made moving pictures of the toolmaking of these apes. They are very fond of eating termites, so they break off a small branch, strip it of its leaves, and use it to fish for the termites. They lick the insects from the little stick, and, when they have eaten their fill, they either throw the stick away, or, at most, they carry it with them for a short distance, but they do not use the same stick for future termite fishing expeditions.[40]

Now let us compare this toolmaking with the toolmaking of Stone Age man, who fashioned arrowheads during a time when no fighting was in sight, nor any planning, here-and-now, for hunting trips. His toolmaking anticipated a future fight or hunt and, moreover, he did not throw unused arrowheads away but saved them carefully.

In recent years several investigators have been successful in teaching chimpanzees and gorillas sign language (ASL), such as speech therapists teach the deaf. If subhuman primates can use this language, then man is no longer the only "talking animal" and the chasm between man and the animal kingdom seems to have been overbridged. This research adds a powerful argument in favor of interpreting the animal kingdom, man included, as a gradually, step-by-step evolving, unitary group of living creatures.

The Gardners decided to teach Washoe, a one-year-old female chimpanzee, ASL. Previously, Yerkes had found that his apes, Chimp and Panzee, "would imitate many of my movements but rarely made a sound peculiarly their own in response to mine."[41] Other experimenters, the Hayeses, had succeeded in teaching their adopted chimpanzee three words in more than one year, by molding its mouth and lips. Following on this, the Gardners taught Washoe ASL by shaping her hands and then giving the hands and arms the appropriate movements. She knew thirty signs by the end of twenty-two months of training. After she had learned to make the sign for "open door," she used this same sign to request that they open the icebox, or a briefcase, or a can, which meant that Washoe was capable of generalizing sign language. Instead of the sign "cold box" for refrigerator, which was the sign she had been taught, she constructed the sign "food drink" for refrigerator.

> In terms of the general level of communication that a chimpanzee might be able to attain, the most promising results have been spontaneous naming,

spontaneous transfer to new referents, and spontaneous combination and recombination of signs. Already Washoe has provided an example of chimp-to-chimp communication. One day when she and other chimps were playing together outside, they spotted a snake. They all fled, except Washoe and a youngster who didn't notice the snake. Apparently by reflex, Washoe signaled in ASL, "Come, hurry, hug." Only when the youngster did not respond, did Washoe drag him off to safety.[42]

Some researchers are expectantly looking forward to the possibility that a chimpanzee might teach ASL to another chimpanzee. The Gardners have described how Washoe used sign language trying to communicate with a younger playmate, and Patterson has recorded how Koko tried to make Michael tickle her by signing to him. On the other hand, after Nim had been returned to his original colony, Terrace reports that he used ASL to a minimal degree.

David Premack has refined communication technique with chimpanzees with the aid of plastic signs, attached to a magnetic slate, representing objects and words which the chimp was taught.

Each word of the language used with the chimp (Sarah, 6 years old) is a piece of plastic, backed with metal, that adheres to a magnetic slate. Sarah was induced to make a prescribed response with the language element, after which she was given some fruit. The chimp was almost immediately proficient in this act.[43] The exemplars I am dealing with here concern selected aspects of (1) words; (2) sentences; (3) questions; (4) metalinguistics (using language to teach language); (5) class concepts such as color, shape and size; (6) the copula; (7) the quantifiers all, none, one, and several; (8) the logical connection if-then. SYMBOLIZATION: When is a piece of plastic a word? The subject assigned the same properties to the plastic that she had earlier assigned to the apple. The properties she assigned to the word "apple" show that her analysis of the word was based not on the physical form of the blue piece of plastic, but on the object that the plastic represents.[44]

Sarah was able to perform such a fairly complex task as putting an apple in one colored dish and a banana in a dish of a different color when Premack attached these commands in plastic language onto the slate. But Premack, the meticulous researcher, is reluctant to answer the question whether a chimpanzee learns a language. The basic difference between human language and our communicating with a chimpanzee either by ASL signing or by means of the pieces of plastic is that these experiments are conducted in total silence. Human language is a two-way, spoken-and-heard route of communication; communication between human and chimp is seen and motorized. It seems that we have here an essential contrast between internalized, intra-audited human language and externalized, visualized, and motorized communication. Moreover, Washoe's and Sarah's language productions seem to operate entirely with concrete objects, notwith-

120   The Meaning of Suffering

standing the astonishing capacity to generalize from one object or "word" to another object or "word." Human language proceeds from concrete to abstract when we operate in some universes of discourse. The question whether a chimpanzee or a gorilla can travel this road remains unanswered.

Francine Patterson, working with a gorilla whom she names Koko, has added modern electronic equipment which allows Koko to enlarge her repertoire beyond ASL to auditory productions through a speech synthesizer.

> Now, with the auditory keyboard, which produces spoken words when she presses keys, Koko can talk back as well as listen. As she signs, she can type out an identical or complementary phrase, and the synthesizer will vocalize her message.... Koko... has taken it upon herself to coach Michael's execution of the signs for "Koko" and "tickle."... One bright morning that followed weeks of rain, I told Koko that if it was still sunny during the afternoon, I would take her out. When I arrived at three o'clock, she looked out at the still bright weather and collected her gear to go outside.... My try at cross-examination three days after the event [Koko had bitten Francine] went [in part] as follows: "What did you do to Penny [Francine's nickname]?" "Bite. Sorry bite scratch—because mad." "Don't know." Of striking import to me was that Koko knew she could not remember or express whatever it was that prompted her bite.[45]

The temporal sequences in these two instances are exceedingly interesting. At three P.M., some six hours after being promised to go out for a walk, Koko checks the weather and gets ready for the event. Three days after biting Francine she remembers what she has done, she verbalizes her anger, and one is led to believe that she is self-critical of her behavior, at any rate, so she signs. Three days in the first instance and six hours in the second seem long periods of time for an ape to remember, in rather complex situations. Apes seem able to interconnect activities and occurrences separated by fairly long intervals of objective time-space.

Herbert S. Terrace describes in detail his *Project Nim* in which he provided a socializing environment for a two-weeks-old chimpanzee for research which lasted about four years.[46] His technique differed from that of previous researchers in that he did not reward Nim for correct responses with food or drink but by an approving attitude of the teachers.

The detailed analysis of the videotapes made during the years of studying Nim after he had been returned to Dr. Lemmon's chimpanzee colony revealed, rather to the astonishment of the people who had worked with Nim, that he reacted most of the time to the preceding signing of the teachers. Also, when he was seen signing alone, he was looking at a picture book, and he signed to each individual picture appropriately. In both contexts, Nim's signing remained within the stimulus-response sequence in objective time-space. This is characteristic of the subhuman primate's near deterministic, compartmentalized behavior and segmented signing. However, he was seen to sign for an object that he wanted and which at that moment was not in sight.

Nim used to love to wash dishes, but, since these dishes just had been cleaned by one of the teachers, this behavior shows that he was not interested in getting the dishes clean, but that he wanted to perform the compartmented movements in interrelated sequences which he had observed being executed many times by his teachers. Unlike a human being, he did not go through this tedious task with a goal in mind, projected into the near future. On first sight, Nim's movements appear to be goal directed, but on further examination they really aren't: these compartmented movements themselves are his "goals" and they are not structured by the more distant purpose, "clean dishes." Nim's washing dishes exemplifies how he lives in a concrete, deterministic "life-style." Nim's movements are segmented and they are not time-binding operations.

In the main, the behavior of simians is activated by stimuli which originate in their surrounding world. But occasionally they can break out of this deterministic mode and they seem to behave indeterministically, such as when Nim invents rubbing his hands together as a sign for "handcream," clapping his hands for "play," and when Washoe invents "food drink" instead of "cold box." A most dramatic and emotion-laden instance of breaking out of the deterministic style is the occasion when Nim, instead of biting Laura, signs "bite" and quiets right down; when Washoe signs to a youngster who does not notice the presence of a snake "come hurry hug." But on the whole one gets the impression that subhuman primates live fairly locked into the immediate present.

It is the very efficient memory of chimpanzees which induces us to think that they have the same access to their past as we have to our past, for instance, when Koko remembered what she had done three days previously. These memories were revived in response to Francine's insistent questioning. Francine was the activator of her memories. Actively remembering presupposes the faculty of imagination, of internal dialogue with oneself, of listening to intra-audited past conversations. This faculty requires that one possess a very complex temporal structure, but this is the exclusive prerogative of man, who synthesizes objective time-space with subjective time-space. Subhuman primates are not capable of this synthesis, since they live in the immediate, here-and-now present. An example of this living in a truncated present instant is Nim's "changing from a cuddly creature to a menacing little monster" within a few seconds.[47]

Touching on the problem of the temporal structure of a chimpanzee's mentation, Terrace asks a most pertinent question: "When Washoe signed *time*, did she do so out of a sense of time, or had she simply learned a gesture to request food, as in the sequence *time eat*? Time was never contrasted with other related signs, such as *now, later, before*. Instead she seemed to be imitating her teacher, who had just asked her *time eat*?"[48] Laura knew what Nim had done two minutes ago and what he was going to do two minutes into the future, which is an example of human temporal organization, which was a source of the great power Laura had over Nim. It is as if Laura structured present, past, and future, with four minutes of clock time *for* Nim.

One can clarify one's insight into the language structure of simians by considering this structure in the light of their living in the immediate present. After chimpanzees learn individual signs, they soon begin to combine two or three signs spontaneously, which impresses us as the dawning of sentence construction. But when the strings of signs become longer, the initial, short string of signs is not elucidated nor is it enriched by the following members of the longer string of signs. The several segments do not proceed from simpler to more complex context, they only emphasize the initial short string of two or three signs. This elucidation and enrichment of one utterance by a following one, requires the capacity of integrating present with past with future within a larger context than the immediate present, of this subhuman primates are not capable. Nim's seemingly more complicated language productions, up to sixteen signs in a string, remain compartmentalized and they do not interpenetrate like utterances in human language.

Washoe's sign for swan was "water bird." Terrace points out that this is no more than the juxtaposition of two signs, just as the bird is juxtaposed to, or superimposed upon, water in reality. She does not create a new emergent as human beings do who think of a water bird as a bird who has a functional relation with water. Chimpanzees are locked within the immediate present; therefore their manner of living is concrete. Man moves from a more concrete level of living and thinking to an abstract level when he creates the concept of "water bird," in which concepts "bird" and "water" interpenetrate. Washoe handles the two signs as segmented, compartmentalized, concrete entities. The compartmented aspect of Nim's longer strings of signs points up this characteristic of animal mentation. They are unable to rise from the concrete, individual sign to the more abstract, generalizing sign within an interpenetrating sentence, replete with meaning. Terrace shows with most effective graphs how Nim's vocabulary grows, sometimes by leaps and bounds, but how the number of combinations of signs remains more or less stationary.

I take exception to Premack's statement that Sarah makes the logical connection "if-then" and to Terrace who thinks that Nim learns to use abstractions such as "big," "give," "look," as children do. In the first place, it is questionable if children have the intellectual capacity to think abstractly, and, second, the temporal structure of mentation of subhuman primates prevents them from progressing from concrete to abstract. The supposition that chimps can proceed from the concrete use of concrete signs to the construction of abstract sentences amounts to our projecting human attributes onto them which go beyond their living in the immediate present. Especially when one communicates with animals which are as close to man as chimpanzees, orangutans, and gorillas, one must constantly examine critically how far one is justified in identifying and empathizing with them. Terrace is well aware how important this is when he gives us a glimpse into the emotional meaning which Nim had for him personally, when he describes the saddening scenes of their necessary separation. "One of the reasons this parting was so painful was that there was no way to talk with him about it. Nim and I were able to sign about simple events in his world and mine. But how could I explain why I suddenly abandoned him?"[49]

Terrace is sad because he is confronted with the chasm that divides us from nonhuman primates, no matter how close a relationship we may feel with one of them. Terrace and Nim could sign about "simple events" that were concrete, compartmented things and happenings in the immediate here-and-now. But for Nim to understand the highly abstract state of affairs of the planned separation would demand that he had the capability of looking ahead into the distant future in objective time, far beyond the 'now'. Nim would be surd to such a discussion, and Terrace experiences this limitation of his simian friend painfully. Animals cannot move from concrete living in objective time-space toward considering the more abstract, personal past and distant future in far-away calendar time.

One of the most deeply probing questions that Terrace had hoped to solve with *Project Nim* is, Can a chimpanzee create a sentence? He leaves the answer open to further research, but he seems to waiver between a positive and a negative answer. In my opinion a subhuman primate is incapable of creating anything whatsoever, including sentences, because they live concretely in a segmented, compartmentalized world in the immediate present. Creativity presupposes the capacity to synthesize objective time-space with subjective time-space, cooperation between causality and intentionality, shifting between the outward-bound and the inward-centered methods of observation, a shift which implies imagination and interweaving of compartmentalization with interpenetration.

Another fundamental question urgently demands an answer: Do chimpanzees suffer? The scenes of Nim's separation from Terrace give a tentative answer. Nim was upset, angry, frightened, and depressed during the afternoon and evening of final separation, mostly on account of his new environment in which he met several chimpanzees for the first time in his life. But next morning he showed none of these signs of emotional upset. At the time of the actual parting he immediately and effortlessly shifted from being with Terrace to playing with his new chimpanzee pals without any signs of pain or tension.

Terrace, on the other hand, felt the separation very deeply because he knew that this was the end of a most stimulating, revealing, and enlightening relationship with an animal. Nim felt a negative effect but he did not suffer. Terrace not only felt a negative effect, but he also suffered because he saw before his mind's eye an enormous temporal panorama of four previous years of research and all its vicissitudes and its gratifying accomplishments. Now there was an empty spot in his life which he knew would last for a long time. I feel indebted to Professor Terrace for the frank descriptions of his introspectively observed personal reactions to a dramatic period in his life.

Animals experience pain in the here-and-now. Humans also experience pain in the here-and-now, but, in addition, humans suffer because they integrate objective time-space with subjective time-space.

The basic schism between man and animals lies in man's awareness of his future death. Animals are not aware of that impending fracture in their existence. Our awareness of the inevitability of our dying some time in the future is the source of the most painful suffering. Animals who do not look ahead into the distant, objective future do not suffer from this foreknowledge. This is the major reason why an unleapable chasm exists between man and animals and why man is not merely an animal.

Kierkegaard derives the difference between man and animals from two modes of dualism: (1) Self versus nonself; (2) Man's existence in the complex, intermingling of present, past, and future over against animal's existence in the present. "Man is precisely defined by this difference from animals in being able to relate himself to himself. Were he neither future, nor past, then man would be bound in slavery like the animal, whose head is bowed down to the ground and whose mind is captive in the service of the instant."[50]

Man is not an animal on account of his complex spatiotemporal structure. His complex structure entails that man suffers from neurotic and mental illnesses. Therefore it is obligatory for us to build psychiatric theories from the viewpoint of time. I see fundamental flaws in Freud's metapsychology because he failed to recognize the importance of the problem of man's temporal organization. From this oversight flows the impediment in Freudian theory: it is a one-sided deterministic outlook upon human existence in which there is no room for freedom of decision making.

FREUD'S METAPSYCHOLOGY FROM THE VIEWPOINT OF TIME

Freud the theoretician versus Freud the practicing psychoanalyst is an enlightening instance of one whose deterministic, theoretical outlook upon human existence vied for dominance over the common sense conviction that we are able to exercise a certain measure of freedom of choice and decision making. As a theoretician he followed in the footsteps of natural science, but as a practitioner he was guided by his intuition. In his metapsychological writings he conceived of mentation as existing in objective time, as causally structured and compartmentalized, so that extrospection was the indicated method of observation. But in his practice he saw his patients as rational, emotional, holistic individuals who communicate with their fellow humans, in particular with the psychoanalyst. Intuitively he felt that his patients lived in their world in subjective time-space, structured by purpose-striving tendencies. He observed them by introspection, empathy, identification, and projection. Starting out within the framework of neurophysiology, an offshoot of natural science, his technique grew into an intuitively conceived, artistic, mythological edifice. Freud superimposed free association and dream analysis upon extrospection, which insures an emotionally neutral, distant relationship between observer and observed subject. In combination with the couch-technique, he developed an emotionally, deeply charged relationship with his patients, which he kept strictly under control from his side.

This doctor-patient relationship blossomed forth into transference. This paved the way for his patients to produce hidden experiences, feelings, and memories which at first made no sense, but which Freud structured systematically, first for himself and then, when he felt they were ready for interpretations, for his patients. As a result, he said that his case histories read more like psychological novels than objective, scientific reports.

He postulated in his metapsychology that two kinds of forces impinge on a human being: sensory perceptions coming from the ambient, outside world and subconscious instinctual drives which energize the mental apparatus from within ourselves and which seem to remain unchanged from their archaic beginnings in early childhood throughout adult life. The weight which he gives to external physical influences and to internal subconscious processes, over both of which forces we have little control, impresses one as a deterministic conceptualization of mentation. Psychoanalysis has uncovered the formative influence which our private past has over our private, here-and-now present. The crux of stressing the power of the past over the present is that it crowds out projection into the future fringe of a precious present.

Notwithstanding these strong deterministic trends in Freud's metapsychology, he believed intuitively in his patients' capacity to change their outlook upon themselves and upon life and that, like all other human beings, they *can* outgrow the influence of their early, past experiences. He helped his patients to release themselves from past-determinacy and to develop a future-oriented outlook, which released them from a stifling, deterministic life-style and which made room for indeterministic activities and a more normal life-style. In his practice he went beyond determinism. Deeply influenced by Goethe, who believed in our ability to capture freedom, Freud, the indeterministic practitioner, overcame natural-scientific determinism as he developed a combined deterministic-indeterministic view of human life with its various tensions, dichotomies, and fractures which are the sources of our suffering.

In short, the inadequacy of a deterministic outlook upon man is that it overlooks the basic importance of suffering, as it synthesizes the contrasting elements in life, such as determinism versus indeterminism. The goal of psychiatric treatment is not to eliminate suffering, but to turn neurotic, destructive, meaning*less* suffering around toward normal, constructive, meaning*ful* suffering. We are striving to help a truncated individual who lives in a compartmentalized, deterministic life-style, to become a holistic human being who lives in an interpenetrating, deterministic as well as indeterministic life-style. Released from the one-sided, overwhelming influence of their private, early past, they develop a future-striving life-style, becoming capable of integrating objective time-space with subjective time-space, causality with intentionality, an outward-bound with an inward-looking attitude and compartmentalization with interpenetration.

I do not pretend to challenge Freud's genius and the perduring greatness of his work, yet it makes sense to take a critical stance vis-à-vis certain shortcomings in his metapsychology. Major flaws exist in this aspect of his writings because he adopted Newtonian time, the time of natural science, in his descriptions of the psychic apparatus. The master himself has given us cues for a critical analysis of the grounding principles of his theories, by his ceaseless building and rebuilding of his hypotheses.

A theory of mentation should attempt to answer the question: How do we human beings live in time? We cannot mold an answer within the confines of linear, Newtonian time, since the punctiform instant is the central concept within this form of time. In a punctiform instant we could not move a finger, we could not think one thought, we could not experience one feeling, and we could not utter one word, which means that we could not communicate with each other in such an instant, in the 'now' of Newtonian time. We need to conceive of a present which has temporal extension, so that we get liberated from the imprisonment in the punctiform present instant. The precious present, with its durational nucleus, its future fringe, and its past fringe, releases us from the confines of the punctiform instant. The precious present accommodates the activities of the mental triad, the dynamic aspects of mentation. The mental triad would not be able to exercise its functions of condensation, action, and future projection in linear time. The precious present and the mental triad are of help in making us aware of the limitations of a deterministic framework and the one-sided, extrospective approach of natural science.

In his metapsychology, Freud *thought* like a natural scientist since he used the form of Newtonian time, but he *worked* as a psychoanalyst in the ambience of objective as well as subjective time. While observing his patients, he went beyond extrospection and followed the introspective, empathizing, identifying, and projecting routes to communicate with them.

Since Freud built his metapsychology around the concept of spatialized, physical time, his thinking about the structure of the psychic apparatus was decidedly spatially oriented. This trend of thinking was in line with medical theory and practice during the nineteenth century, during which time medicine and surgery were put on the firm footing of applied natural science. Before graduating from the *gymnasium*, Freud vacillated for some time between several choices of career, until he was deeply moved by Carl Brühl's presentation of Goethe's essay on nature which made him settle on the study of medicine.[51] The vast scientific panorama that Darwin's theory of evolution opened up and the use of the methods of chemistry and physics in medicine stimulated his natural scientific bent. In addition, being steeped in anatomy, in Virchow's cellular pathology, and in the histology of the nervous system it is little wonder that Freud designed spatial schemata to explain the workings of the psychic apparatus. Newtonian objective time fitted right into these schemata. In various publications he used the telescope, the embryonic blastula, and most often the sensory-motor reflex to explain what was going on in the psychic apparatus. The theory of association of ideas, which played a large role in the then-developing experimental psychology, was modeled after the natural sciences. Freud used this theory of mentation which contains strong deterministic trends.

> Hence we represent the psychic apparatus as a complicated instrument that perhaps, the systems have a certain spatial orientation in relation to each other, somewhat like the various lense systems of a telescope are situated behind the other...that the psychic processes in these systems are being traversed by the excitation in a specific temporal succession. All our psychic activity is initiated by inner and outer stimuli and ends in innervation. Therefore we ascribe a sensory and a motor side to the apparatus.[52]

When he compares the psychic apparatus with a telescope, Freud spells out that external stimuli are passively received by the system preception-conscious and proceed to the subsequent instances of memory-1, then to memory-2, and so on toward the system subconscious and from there to the system preconscious until they are finally discharged as motor activity. These are clearly cause-effect sequences in the before-and-after in objective time-space. This conception of the time structure of the psychic apparatus is reinforced by Freud's interpretation of the hallucinatory dream state in which stimuli traverse the psychic apparatus in the opposite direction and again, in the before-and-after of objective time-space.[53]

Freud declares that the sensory-motor reflex is the ideal model from which the psychic apparatus must be conceived. Here we have a sensory stimulus which elicits electrical, chemical disturbances in the receptor organ which are transmitted through nerve fibers from neurone to neurone and finally to the motor endplate terminating around muscle fibers which respond by contraction or relaxation. The reflex, though basically a mechanical, causally structured process in objective time-space, is contaminated as it were by goal-striving attributes since there is an obstinate belief afoot that biological functions in general and the sensory-motor reflexes in particular are purposive. This conventional, mixed causal as well as teleological method of explanation appears in the forefront when Freud explains how the psychic apparatus becomes active in the service of a wish fulfillment which will reduce pain and promote pleasure. But then, it is as if he drives the teleological components of his model underground by stating that it is basically mechanical conditions which regulate this complex process. Although Freud often uses the concepts of emotionality and goal-striving in his writings, he does not relinquish a mechanistically oriented interpretation of the psychic apparatus, as a result of his overlooking the concept of subjective time-space. This leads toward a confusing theory of drives.

> Hence, a drive might be an urge to restitute an earlier condition, which this living thing had to relinquish under the influence of outside disturbing forces....Differences between peace which was found and peace which was requested through gratification, delivers up the driving moment which does not permit us to remain in the fulfilled situation....The path backward is, as a rule, blocked by the resistances, which maintain repressions, and so, nothing else remains but to proceed in the other direction of development which is still free, although without any hopeful outlook that one will be able to terminate the process nor to reach the goal.[54]

The psychic apparatus seems to be oriented more toward the past than toward the future, since drives and wishes are not future-striving functions but arise from the urge to return to a past condition of lesser tension and greater comfort compared to what a living thing may be experiencing at present.

> Nothing but a wish could be capable of bringing the apparatus in movement, and the course of the excitation could be automatically regulated by the perceptions of pleasure and pain. Therefore, I make the assumption that the course of the excitement under the domination of the second system [secondary process thinking] is connected with totally different mechanical conditions than under the domination of the first [primary process thinking].[55]

Little hope is held out for our ability to regain the conditions of peace and gratification which we enjoyed during the periods of our earlier life. It appears in this view of man's temporal structure as if he is caught in a tight vise between past and future, unable to move freely in either direction in time.

This constricted view of man's temporal structure is expressed in the dictum "The subconscious is timeless."

> We have experienced that the subconscious psychic processes in themselves, are "timeless." That means first of all that they are not temporally ordered, that time changes nothing in them, that one cannot accommodate the representation of time to them; rather, it seems that our abstract representation is obtained from the manner in which the system perception-conscious works, and it seems to correspond to a self-perception of these.[56]

Here Freud falls back on the classical representation of time as an agent which slowly destroys all that exists, especially human life. Ordering in time means to Freud the linear ordering of the before-and-after in Newtonian time. As a natural scientist and as a medical doctor, objective time was for Freud the only thinkable and usable form of time. He assumed tacitly that conscious processes go on in objective time. This structure accounts for their changing rapidly from one moment to the next. Subconscious processes either do not change at all, or else so slowly that this change cannot be perceived by us. Hence, the subconscious cannot be placed in objective time.

Freud's mechanistic model of the psychic apparatus, encapsulated in objective time, leads unavoidably to a deterministic metapsychology of the ego. Therefore, the privilege of freedom of choice, of decision-making, is denied to the ego.

> That what we call our ego, behaves in life essentially passively, that we are "being lived" by unknown, uncontrollable powers. Now, an individual is for us a psychic Id, unknown and subconscious, upon which the ego sits superficially astride, developed as a nucleus out of the Perception-system. ...The Ego is before all things somatic, it is not only a being of the surface, but it is itself the projection of the surface.[57]

Here we touch upon Freud's spatially oriented theories which are underpinned by spatialized objective time. This leads him toward trying to localize mental processes within the brain, notwithstanding his warning, "We want to evade carefully the temptation to determine the psychic locality somehow anatomically."[58] He claims that consciousness must lie in between the incoming outer stimuli, thus facing the external world and the inner processes of pleasure and pain which consciousness must face in the opposite direction. This is exactly the locality of the cerebral cortex which is the outermost layer of the brain.

Freud's space-oriented thinking comes to the fore in yet another model of the psychic apparatus when he compares embryological development of primitive organisms with psychological development. He postulates that the embryonic blastula had to form a protective layer or crust around its vulnerable interior against traumatic influences which threaten its delicate substance. This physical defense against physical stimuli is compared to the infantile psyche developing its defenses against painful stimuli threatening it, especially authoritative, punishing parents. Freud transposed over into the psychic realm, which exists in subjective time, that which happens within the physical organism during early stages of its development, which happen in objective time.

Having located the subconscious drives outside of time and not changing in time, Freud considers that these drives are a thoroughly conservative influence within the psychic apparatus. Moreover, he postulates that they are the driving forces behind the activities of our ego, and, since we do not know these drives and cannot control them, the ego is a passive play ball without freedom of action, so that we are "being lived" by these subconscious processes. The passivity of the ego prevents it from striving into the future, and hence Freud assigns an overriding influence to past experiences. The definitive power of the subconscious and its atemporal structure becloud the fact that we are purpose-striving beings who project our acts into the future.

Although Freud thought in his theoretical conceptualizations as a natural scientist, in his practice he introspected, empathized, and identified with his patients. He projected human attributes onto them, and he saw them living in subjective time-space as interpenetrating whole human beings. If it were so, that the subconscious is timeless and that we are preconditioned by memories locked within the subconscious, while the external world rules the psychic apparatus from without, and, if it were true that man exists in objective time-space, then introspection, empathy, identification, and projection would have no business intruding into the work of the analyst and the psychotherapist. Clearly, these methods of observation and of interpersonal communication are the tools in psychoanalysis and in psychotherapy. As a practitioner, Freud believed that his patients could make the decision to battle against their neurotic life-style and that they had the freedom to develop a normal life-style.

In order to bring theory in line with practice we need to develop a time-oriented metapsychology which holds that man exists in the combined forms of objective and subjective time-space, that he is structured by the two categories of causality and purposiveness, using both extrospection and introspection with its ancillaries as methods of observation. Such a metapsychology can be free of deterministic one-sidedness. From the viewpoint of a dualistic, temporal-spatial structure of mentation and our existence, we can look both inward and outward so that we gain insight into our mentation, our rationality, and our emotionality, while by moving outward toward the others, by empathy, identification, and projection, we get a wider theater of operations for interpersonal communication. Thus we can build toward a comprehensive view of man's mundane existence which we can bring to bear profitably upon the theory and practice of psychiatry.

This line of thought can stem the intrusion of determinism into the realm of human mentation, for which intrusion psychiatry is in part responsible. We can raise a bulwark against unwelcome and stifling trends with the aid of the mental triad and the precious present. These two concepts set us free to build an indeterministic life-style in which we pursue purpose-striving activities and in which we make decisions. A Janusian survey of ourselves in both directions in timespace harmonizes the deterministic with the indeterministic life-styles. We can partake of an interpenetrating, holistic mode of living in the world which is causally structured and compartmentalized by balancing our existence between extreme determinism and extreme indeterminism. This balancing act is bound to be painful, tension ridden, discouraging, or exhilarating since attempts to heal blemishes, dichotomies, and fractures in our existence draw us through a complex gamut of pleasant and unpleasant feeling tones. Daring to heal some of life's contrasts and fissures we have to accept the path of suffering.

MARIE BONAPARTE

There are trends in recent decades to build a time-oriented metapsychology which liberates us from Freud's space-oriented metapsychology. Marie Bonaparte moves the time problem into center stage. She bases her wide-ranging and deeply probing paper on "Time and the Unconscious" on four areas of experience: the world of childhood, of dreams, of ecstasies, and of the arts and literature. Two basic themes run through this paper: she differentiates between time and eternity. I take it that the term *timelessness* refers to events which do not occur in linear, irreversible, objective, Newtonian time. In contrast to structured limited time, eternity is unstructured and without limits.

Bonaparte states that the child lives in a timeless world which she interprets as the paradisian existence of childhood. However, this existence is neither timeless nor is it a blissful paradise.

> He simply regards the attempts made by adults to impose their "time" on him, still in its essence of virtually infinite duration, as an intrusion on the part of a strange and hostile world. In spite of clocks, the child long retains his sense of "time" of infinite duration. The child's mind is permeated with the limitless unconscious.[59]

The child lives in a present which does not fit into the world ruled by the objective time of adults. The young child cannot fit linear, compartmentalized, measurable, objective time into his world which is structured within interpenetrating subjective time. Childhood existence is neither timeless, nor is it a blissful paradise. The child lives in the here-and-now, immediate present which is highly charged with a mostly positive feeling tone during most of the day. These happy periods are interrupted by brief crying spells, pain, disappointments, and frustrations which tarnish the so-called paradisian bliss. His existence is more like a mosaic of brief periods of time alternating between euphoric and sad feelings. This state of mind is not unlike the case in adult life, except that the child's happy periods usually last longer than the depressed phases.

Bonaparte thinks that we adults hanker back after the lost paradise of our childhood in various states of altered consciousness such as dreams and ecstasies. "While we sleep we are immersed once more in the nebulous, almost indefinite time we knew in childhood."[60]

I interpret the cessation of time while we sleep as a destructuring of our complex temporal structure in the direction of subjective time-space, so that objective time relinquishes the dominant role it plays while we are awake. During sleep, ego boundaries lose their sharp boundaries and hence objective time loses its compartmentalized characteristic. It is as if we live in subjective time exclusively while we sleep and during dreams with their intense emotional charge.

Bonaparte believes that we experience eternity during various forms of ecstasies induced by drugs or alcohol, during sexual orgasm, during mystical experiences, all of which bring about many varieties of altered states of consciousness.

> Ecstasies (in all its forms) succeed in destroying the subject's sense of time.... To lose one's sense of time is to forget the existence of death.... Man vaults the barriers of death and prolongs his earthly existence in an imaginary world beyond the grave.[61]

I think that we can describe these experiences more aptly and without invoking the concept of eternity, if we interpret these rapturous states of mind as the mental triad coming alive in a precious present. The precious present is certainly not eternal. Moreover, there is a deeper hidden motivation behind the attempts to grasp eternity through ecstasies: we try to overcome the destructive qualities which we ascribe, wrongly, to objective time which we equate with the destructiveness of death.

Incidentally, the huge spatial panoramas which the opium smoker seems to enjoy, together with the loss of the boundaries of objective time, are an indication that we should not separate time and space from one another. Fairy tales also point toward this trend of thought about time-space.

Bonaparte's extreme position that literature goes beyond time and space can be expressed more felicitously by saying that the artist does not grasp eternity in poetry, painting, music, but that he synthesizes what exists in objective time-space with the world of his ideals which exist in subjective time-space. We grasp beautiful things within experiences which have lasting value and meaning for ourselves and for others through the mental triad functioning in a precious present. The precious present is neither eternal nor is it timeless.

The concept of eternity overshadows the forms of objective and of subjective time-space in Bonaparte's paper. The insight that we live in both forms of time-space leads to the understanding that the horror of time is really our dread of dying and of death.

Surprisingly, Bonaparte raises doubts about her own and Freud's thesis that the subconscious is timeless.

> One can scarcely imagine any living thing...being immune from the effects of time.... Thus even Freud is prepared to admit that repressed psychic content undergoes *some* modification.... *Time does not appear to form an integral part of our fundamental nature.*[62]

Bonaparte and Freud derive the idea of the timelessness of the subconscious from the different velocities of conscious and subconscious processes. The latter change at an exceedingly slow pace, while conscious processes vary sometimes with great velocity. Since we hardly perceive changes in subconscious processes we may think the subconscious is timeless. But this more flexible stance does not change Bonaparte's conviction that we have no power over changing our subconscious. Since the neurotic life-style is distorted by being determined by a person's past experiences which he relives as if they were the here-and-now present, one can understand Freud's and Bonaparte's confused thoughts about the problem of objective and subjective time-space.

Bonaparte does not differentiate between the time in which we live and the time in which our surrounding world exists, which leads her to the obscure pronouncement: "Time does not appear to form an integral part of our fundamental nature."[63] Overlooking the fact that time is indeed fundamental to our nature flows from recognizing only the form of objective time and of eternity. It is quite true that objective time is an inadequate form in which to impress mentation, but this is precisely the reason why we need to develop the concept of subjective time-space as the appropriate milieu in which we human beings live.

Heinz Hartmann

Compared to Freud and Bonaparte, Hartmann adopts a very different approach toward the problem of time. He centers upon the importance of our lived-in present, our personal past, and our living into our personal future. The ego evaluates not only present circumstances in which we find ourselves, but it also foresees the configuration of the future.[64] Hartmann enlarges on Freud's proposition that the process of maturation depends on the young child's evolving the capacity to postpone gratification, by describing it as a future-attuned process of development.

Between the third and the fifth months of life, he learns to anticipate the feeding situation without crying. Some kind of anticipation of the future events play a part in each of these operations....The reality principle implies something entirely new, namely, the familiar function of anticipating the future, orienting our actions according to it and correctly relating means and ends to each other.[65]

Hartmann enlarges the scope of the pain-pleasure principle and the reality principle by the attribute of anticipating the future which releases the baby about six-months-old from the confinement within the immediate here-and-now. In parallel with the baby's enlarging spatiotemporal perspective and with ego development, there is a change in the baby's reaction to danger and in the configuration of anxiety.

Originally, fear is a reflex-like reaction to danger...later, it acts as a signal of changes to come....Anxiety as a signal can operate only when the child has learned to anticipate the future....[This makes] the transfer from the pleasure principle to the reality principle possible, the one seeking immediate gratification, the other postponing gratification.[66]

Whereas anxiety is originally part and parcel of a reflexlike somatic reaction to danger, it becomes a signal of danger lurking in the farther-removed future beyond the immediate here-and-now of a reflex movement. Anxiety takes on a new physiognomy, and at the same time the reality principle takes over from the pleasure principle, hence, the baby learns to postpone immediate gratification. This putting off flows from projecting into the future-in-the-fringe which interlocks with a dawning awareness of the objective future. The baby becomes capable of applying his past experiences of dangerous situations in his stance toward the future within the present. The progression of anxiety as a reflex reaction to anxiety as a signal of future possible occurrences involves the capability to superimpose objective time-space upon subjective time-space. A subtle, yet formidable shift is taking place in the baby's relationship to the future, and there is already an inkling of a two-level temporal organization at this early stage of development. The baby of six months undoubtedly wants immediate relief from his hunger pangs, but now he can put off getting relief in the objective future for a few minutes, a long time span for a baby. The deeper meaning of this adaptation process is that it sets the young baby on the road toward becoming a human being. Its highest significance is that already at this age man is not merely an animal.

Hartmann differentiates between anticipating the future and scientific prognosticating of the future. Therefore our actions are a mixture of rational actions and irrational decision making. "The first signs of intentionality appear around the third month of life...in crucial areas of life only extremely rarely can we prognosticate with scientific certainty....In the meanwhile, action—at least in part—remains a matter of irrational decision.... The rational plan must include the irrational as a fact."[67] Therefore our actions are a mixture of rational actions and irrational decision making.

Intelligence is only one of many activities of the ego. The dichotomy of rational versus irrational constituents of our existence is one of the basic sources of anxiety, which in Hartmann's interpretation has risen from the status of a concomitant to reflex mechanism to a constructive and valuable element of the psychic apparatus.

Our living into the future should be separated into projecting our acts into the future fringe of a precious present and calculating and predicting the distant, objective future. Projection into the future fringe is the irrational, noncalculable aspect of our activities; planning for and calculation of the probable results of our activities is the rational aspect of mentation. The integration of these two different attitudes toward the future which always work in tandem and which synthesize objective time-space with subjective time-space, is the hallmark of our maturity. This sophisticated spatiotemporal organization takes a long time to develop; in this respect we mature at about age eight or ten, or even later.

Our dyadic spatiotemporal structure underpins the rational and the irrational dichotomy in our human existence. We think that our acts are rational if we can fairly accurately predict their outcome in the future in objective time-space. But the operations of the mental triad contain irrational elements, since we cannot accurately predict events that are going on in the not compartmentalizable and not measurable precious present. Projection of our acts into the future-in-the-fringe of a precious present is of a mysterious nature. The ego's reality principle stands for the rational side of our nature. The reality principle is concerned with events that happen in objective time-space, which are calculable to a certain extent and which are structured by the causal category. The rational and the irrational sides of our person are integrated when we synthesize objective time-space with the precious present and causality with the mental triad.

Hartmann's ideas on adaptation reveal that this process points toward the future. However, the term *anticipation of the future* carries the notion that the future is something concrete, real, ready-made, already there, somewhere, waiting for us to grasp hold of it. This concretization of the future is in line with a forceful tendency in theorizing about time, which conceives time as if it were thinglike, an entity, an agency, a power, instead of thinking of time and of space as the forms of our inner and outer experience of reality. The mental triad does not get hold of something already there in the future-in-the-fringe; to the contrary, the mental triad creates new emergents which are not waiting for us; these emergents are of our own making.

Hartmann's future-attuned thinking about time develops a metapsychology which counterbalances the originally prevailing theories about instinctual drives. It liberates the ego from domination by the conservative, past-oriented, instinctual drives and creates the future-tending climate in which the mental triad can fulfill its functions. Present-day ego theory is set free to describe man's terrestrial existence in terms of integrating our past with our future in our present. This integration underwrites trends toward phenomenological descriptions of how we human beings actually live in time-space. The mental triad and the precious present work toward such detailed description since these two concepts enlarge upon Hartmann's assertion that conflicts are part of the human condition. Conflicts create mental tension and emotional turmoils which manifest themselves within the operations of the mental triad.

Differentiating between the process of adaptation and the resulting state of adaptedness, Hartmann takes this process out of the conventional context of being static in space and transfers it into a temporal milieu, so that it becomes a future-attuned event.

> We may distinguish between a state of adaptedness and the process of adaptation which brings this state about. A state of adaptedness refers to the present. The process of adaptation always implies a future condition.... an evolution peculiar to man, namely, the influence of tradition and the survival of the works of man. Man lives, so to speak, in past generations as well as in his own.... The superego becomes the vehicle of tradition and of all the age-long values.... The ego too, has its share in the building of tradition.[68]

Hartmann includes man's cultural achievements in the dual process of adaptation and adaptedness, and he steps therewith into a huge spatiotemporal, historical perspective. The self-enclosed three-tiered Freudian model of the psychic apparatus is liberated from its constrictions, as ego and superego partake of large blocks of objective past periods and of anticipated wide leaps into the objective future. The Freudian superego is bound to tradition and to the individual's past, but now it recognizes and reevaluates not only our personal but also our communal past. Ego, superego, and mental triad transform the past into a newly emerging living past which can be adapted to the personal and the distant future. As a time-binding and time-synthesizing agent of the ego, the mental triad plays a powerful role in our adapting our past to our future.

Hans Loewald

Loewald places the time problem right in the foreground. Significantly, he holds that the interaction between the three temporal modes is basic to the theory and practice of psychoanalysis. He makes a forceful plea to see human life not as being determined by one's past, but to interpret it as being also attuned toward one's personal and communal future. "With the ascendency of ego psychology this time perspective has changed.... Life is not altogether motivated by forces of the past but is partially motivated by an attraction coming from somewhere ahead of us."[69] Loewald also says: "...man can own up to his past and gain thus mastery of his present and shape his future."[70] To rephrase: Man owns up to his past by condensing selective memories, and he gains mastery over his present by means of an act which uses this condensed material. This act shapes our future by projecting itself into our personal future. Loewald emphasizes our tending toward the future by stating that modern ego psychology no longer sees the ego as motivated by ties to the past but, as previously stated, "by an attraction coming from somewhere ahead of us." We can talk about this rather nebulous last statement more felicitously if we use the mental triad and the precious present as descriptive terms. The motivation for condensing specific memories is not inspired by our past but is initiated by the act which projects itself into the future fringe. To be sure, the future-in-the-fringe is somewhere nebulously ahead of us. This lack of demarcation places the not quantifiable future fringe in contraposition to the measurable future in objective time.

Loewald speaks of psychic time as distinct from clock time which is similar to the contrast between the forms of objective and subjective time.

> Past, present and future present themselves in psychic life not primarily as one preceding or following the other, but as modes of time which determine and shape each other, which differentiate out of and articulate a pure now. There is no irreversibility on a linear continuum, as in the common conception of time as succession, but a reciprocal relationship whereby one time mode cannot be experienced without the other and whereby they continually modify each other. As experiential phenomena they interpenetrate.[71]

He departs from the conventional conception of irreversible, linear time and posits psychic time as the three modes of past, present, and future interrelating and mutually interpenetrating. Interpenetration is the key concept that places compartmentalized objective time over against subjective time. Man as a future-striving being synthesizes present, past, and future through the process of interpenetration of the three temporal modes.

Loewald bases a clear-cut differentiation between the physical world and the world of mentation on their different spatiotemporal structures. He holds that physical processes are typified by spatial relations, but he does not go so far as to say that their structuring in objective time amounts to their occurring in spatialized time. He attributes a psychic present to the ego and a psychic future to the superego, so that the cooperation between these two instances creates a higher mental organization.

> Physical structures are arrangements characterized by spatial arrangements....Psychic structures are temporal in nature....Psychic time implies an active relation between the temporal modes of past, present, and future....The awareness of past, present, and future stands on the threshold of higher mental organization. In terms of psychic time, the relation between the ego and the superego can be seen as a mutual relation between psychic present and psychic future.[72]

Loewald localizes our future-tending activities in the superego. He makes a great leap forward from the classical Freudian conception of the superego which holds our past mistakes and misdeeds before the mind's eye in a punitive gesture.

Memories, instead of being regarded as in dead storage in our memory bank, partake of the lived-in present because from Loewald's point of view past and present interpenetrate within the present. Very incisively, he says that "memory...makes the past present."[73] I prefer to regard this statement in terms of the mental triad operating in a precious present. Memories, condensed by the act operating in a durational present, project memories as new emergents into the future fringe of a precious present.

Loewald takes issue with Freud's concept of the timeless subconscious by endowing the Id with the temporal attribute of futurity.

> The Id has a future insofar as we make it ours by acquiring it, by imprinting on it the stamp of ego organization....The ego envisages an inner future for itself, the superego being the representative of the ego's futurity....The superego functions from the viewpoint of a future ego....Conscience speaks to us from the viewpoint of an inner future.[74]

We imprint aspects of the ego upon the Id by seeing it as future-related and thus make the Id our own possession. Conscience therefore becomes a factor which moves us ahead into psychic time, rather than keeping us bogged down in our past. Loewald's metapsychology revamps our outlook on our private past, "In the early stages of psychoanalysis, the future was nothing but a time when a past state would be attained again."[75]

Loewald mentions the two states of fragmentation and ecstasy in which we experience time in a most unusual manner. "In the experience of fragmentation, meaning, i.e., connectedness, has disappeared, each instant is only its empty self, a nothing....In the experience of eternity (during ecstasy) all meaning is condensed in the undifferentiated global unity of the abiding instant, the *nunc stans*."[76]

During fragmentation we live close to an 'empty' interval of objective time-space, devoid of meaning and of emotionality. The *nunc stans* and the precious present have in common that past, present, and future interpenetrate and yet there is a vast difference between these two concepts which refer to the present. Religious mystics are driven toward bringing about the experience of the *nunc stans* by the motivation to immerse themselves in a metaphysical, cosmic ambience, or being, or God. This experience seems to lift one beyond the awareness of being a particular human being. The mind seems to be released from the body, the experience of self counts for very little. The hidden motivation for mystics seeking the experience of a *nunc stans* is that through the loss of experience of self they hope to heal the dread of their own death.

Living in a precious present, to the contrary, is the epitomy of the here-and-now acting, emoting, cogitating, planning, communicating, creating, human being who is vividly aware of his selfhood, especially in a precious present. In a precious present the mental triad particularizes ourselves as this and no other person. Ecstasy in a *nunc stans* attempts to outreach the constraints of objective and subjective time-space in its emphasis on eternity; this experience strains to outstrip our enclosure in the holism of body and mind.

Loewald contributes to building a revitalized metapsychology since he emphasizes man's future-attuned life-style. This emphasis on the future cleanses Freud's metapsychology of past determinacy, which creates a pessimistic outlook on the human condition in which we appear powerless playthings of forces over which we have almost no control. Loewald's work initiates a desirable rollback of determinism in the field of psychoanalysis and psychotherapy in favor of indeterminism, in keeping with the shift toward the future and away from placing excessive weight on the power of the past. A time-oriented metapsychology is a constructive force in the practice of psychoanalysis and psychotherapy because one can interpret the goals for which we strive, as releasing our patients from being determined by their past and setting them free to develop an indeterministic life-style. Unencumbered by the weight of their past, they can move into their personal future.

J. T. FRASER

Fraser's interdisciplinary approach elucidates how deeply and widely psychoanalysis is affected by the problem of time. Three areas of particular interest in the rich content of Fraser's contribution involve theoretical concerns, the development of temporal structure in infancy and early childhood, and the practice of analytic treatment. In addition, Fraser provides a model of our temporal organization.

Fraser organizes time on five levels of decreasing complexity and fits them into an overall evolutionary conception of reality. According to Uexküll, for each animal the world-as-perceived is determined by the potential functions of the totality of its receptors and effectors. Its receptors determine the world of all possible stimuli that the animal may experience; he calls this *Merkwelt*, that is, the animal's universe of signals. The sum of all possible responses as determined by the effectors of the animal form its *Wirkwelt*, or universe of possible actions. The dynamic combination of the *Merkwelt* and *Wirkwelt* makes up the animal's *Umwelt*, best rendered into English as the animal's "specific universe."[77] He describes the five levels as follows: (1) a *nootemporal* umwelt is the most complex and the highest level of temporal organization, which is structured by symbolic causation, where man suffers from the conflict of self versus nonself, from guilt and anxiety, and where man reaches for freedom; (2) the *biotemporal* umwelt is the next lower level, characterized by immediate future and immediate past, devoid of long-term memory and expectation, which is a continuum, organized by intentionality, or final causation; (3) the *eotemporal* umwelt is even more abstract, lacking the directional quality of time, and which is also a continuous aspect of reality, with deterministic qualities; (4) The *prototemporal* umwelt occurs in fragmented time, in which statistical law holds sway; and finally, (5) an *atemporal* umwelt, which is chaotic and in which time blends into space.[78]

These five levels are not at all comparable to, for example, the stratification of layers of rocks, or to the multiple layers of tissues which envelop the brain. First, they stand in a relationship of "nesting" to each other, and second, the lower strata are contained within the next higher stratum, so that the nootemporal umwelt is a complex integration of all five levels of temporality.

Although Fraser says that a dichotomy is too simplistic to describe intricacies of our temporal structure, I feel that a dyadic, dualistic conceptualization has many features in common with Fraser's five-layered model, so that the two viewpoints can be quite harmoniously integrated. Very similar to Fraser's "nesting" of his mutually correlated levels of temporal integration, are the many intermediate transitional forms of temporal structure between objective time and subjective time regarded as the two limiting points of a series.

Fraser emphasizes that psychiatrists assume tacitly that time is "man-independent" and that it flows by itself, which preconceptions block our deepening of insight into our own and our patients' temporal structure. He brings to our attention that futurity is a powerful constituent of our temporal organization and shows how important this tendency is in the child's growing into a temporal hierarchy. He takes up Ferenczi's idea that within the initial phase of development of the human fetus in utero are feelings of omnipotence. Fraser describes how this process changes and broadens in scope after birth and that the verbal manipulations of time concepts cap this process off. After feelings of omnipotence have been tamed, the baby begins to project his own purpose-striving tendencies onto the inanimate objects of his umwelt and enters an animistic stage of development. Now he lives in a biotemporal umwelt. The study of the child's language capacity reveals a progression from an earlier phase when time-words referred mostly to the present, toward his using mostly future-words around age two.

> The first category of time-words which the child can verbally manipulate refers to the present.... At around 24 months the child commands future-words, at around 30 months past-time words, but a greater variety of future-words than past-time words.... development of time is from chaos toward conflicts of increasingly better defined articulations, along a hierarchy of temporal umwelts.[79]

By age three the child is versatile with all three varieties of time-words, but future-words are more frequent in this mixture than words which refer to the past. By now, the child lives in a nootemporal umwelt. Summing up, Fraser states that our temporal structure develops from the chaos of the prototemporal umwelt to the ordered perspective of the nootemporal umwelt.

Speaking of psychoanalysis, Fraser says:

> We cannot go to the future without clearing a path to it through the removal of obstacles in the past.... in which umwelt does the patient place his

major sense of identity?. . . in case of psychopathology, progress in treatment includes reestablishment in the mind of a dynamic balance among hierarchical, unconscious levels of time. Judged from this perspective, the reliability of reality testing is a measure of the degree to which the patient is able to accommodate to the archaic reality of primitive causations and temporalities, lodged in his fantasy, memory and dreams and integrate them with the ambiguities of human freedom and the certainty of the transience of human life.[80]

In contrast to the generative qualities of the future in the child's development, classical psychoanalysis has laid excessive value on unearthing the patient's past. Fraser does not deny in the least that this approach is necessary, but he stresses the future-attuned tendencies within the psychoanalytic process. In fact, progress in therapy signals that the patient is releasing himself from past dominance. Fraser contraposes an ideational self, operating neither in the future nor in the past, versus an observing self which lives in an eotemporal umwelt. This division is reminiscent of Allport's self as a doer and the self as an observer. The patient has to learn to shift from one to another temporal umwelt and if he lives mostly in an eotemporal umwelt he lives in a fantasy world where perceptions are confused with memories and expectations. This travelling among the "nesting" temporal levels brings about qualitative changes within the patient's temporal structure, and he begins to taste freedom as well as coming to grips with his dread of dying and of death.

Fraser's conceptualizations about time, causality, and goal-striving activities aid in the search for the proper temporal milieu in which we can place the subconscious.

We have given the name prototemporal to all umwelts which comprise indistinguishable entities. . . . In a prototemporal umwelt time and space are not sufficiently differentiated.[81]

Since the subconscious is the least structured of the three levels of the psychic apparatus in Freud's descriptions, we are looking for a temporal milieu which is also barely structured. I submit that we find this temporal ambience in Fraser's prototemporal umwelt. There is no temporal ordering in this umwelt, and we cannot count events which take place in it. The shadowy world of the subconscious seems to fit quite well into this temporally and spatially, barely structured umwelt. From this vantage point one can sketch a temporal schema in which the three-leveled psychic apparatus can fulfill its various functions.

When the ego acts according to the reality principle, it is concerned with objective time-space in which it places orderable and countable events. Since I see the mental triad as a cooperating power within the ego, the ego operates in subjective time-space when the mental triad takes the lead in the ego's pursuits. The superego is even more committed to subjective time-space than the ego, which proposition enlarges upon Loewald's view of the superego as the ego's futurity. So far, we lay stress on our future-directed activities of the highest level of temporal organization in a nootemporal unwelt, or, in a precious present.

The superego, the ego, and the mental triad are not limited to operating in this rarified atmosphere since they can penetrate through the lower temporal strata into the prototemporal umwelt of the subconscious. Because these temporal levels are "nesting," or interpenetrating structures, the four agencies (including the mental triad) of the psychic apparatus need not remain separate entities but can create holistic new emergents. These activities are undergirded by subjective time-space interweaving with objective time-space.

Time-oriented interpretations of human existence take the rigidity out of Freud's spatially dominated descriptions of the psychic apparatus. A combined deterministic as well as indeterministic description of human mentation approximates closely to the realities of our living-in-the-world and also to the flexibilities of psychiatric theories, which flexibility is essential in psychoanalytic and psychotherapeutic practice. In brief: a temporal schema of the workings of mentation constrains the domination of determinism over freedom of decision making in the field of psychiatry.

The constraint of determinism is quite important within the limited field of psychiatry. It is even more urgent to keep determinism within proper bounds when we theorize about the mind-body problem. If one is convinced that there exists a deterministic relationship of body over mind, then goal-striving activities are a figment of our imagination and we should have no freedom of decision making.

NOTES

1. *New Columbia Encyclopedia*, 1975, s.v. "Determinism."

2. Carroll Stuhlmueller, "Biblical Voices of Suffering and Prayer" (Paper read at the First Congress on Human Suffering), Notre Dame University, Notre Dame, Ind., April 1979, p. 21, by permission of Stauros International, West New York, N.J.

3. Sophocles, *The Oedipus Cycle. Oedipus Rex. An English Version*, Dudley Fitz and Robert Fitzgerald (New York: Harcourt Brace & Co., 1939), p. 49.

4. Ibid, *Oedipus at Colonus*, p. 163.

5. *Webster's New International Dictionary*, 2d ed., s.v. "Yuga."

6. Giorgio de Santillana and Hertha von Dechend, *Hamlet's Mill: An Essay on Myth and the Frame of Time* (Boston: Gambit, 1969), p. 150.

7. John Gay, *Red Dust on the Green Leaves: A Kpelle Twin's Childhood* (Thompson, Conn.: Inter Cultural Associates, 1973), p. 13.

8. Martin Heidegger, *Sein und Zeit* (Halle a.d.S.: Max Niemeyer Verlag, 1929), p. 175.

9. Ibid., p. 178.

10. H. Tristram Engelhardt, Jr., *Mind-Body: A Categorial Relation* (The Hague: Martinus Nijhoff, 1973), p. 91.

11. *New Columbia Encyclopedia*, 1975, s.v. "Uncertainty Principle."

12. J. J. Poortman, *Indeterminism or Determinism* (Assen, The Netherlands: van Gorcum & Co., 1949), p. 12.

13. Gordon W. Allport, *Becoming: Basic Considerations for a Psychology of Personality* (New Haven: Yale University Press, 1955), pp. 83-84.
14. William Jacobs, *Hannibal* (New York, McGraw-Hill, 1973), pp. 19-20, 32, 83-84.
15. Robert Nozick, *Philosophical Explanations* (Cambridge, Mass.: Harvard University Press, Belknap Press, 1981), pp. 291-92.
16. *Winkler Prins Encyclopedia*, 6th ed. s.v. "Soul" (Author's translation).
17. Søren Kierkegaard, "Either/Or," in *Masterpieces of World Philosophy in Summary Form*, ed. Frank N. Magill and Ian McGreal (New York: Salem Press 1961), pp. 614-19.
18. Karl R. Popper and John C. Eccles, *The Self and Its Brain* (Berlin: Springer Verlag, 1977), pp. 109-11.
19. Jean Piaget, *The Child's Conception of Time*, trans. A. J. Pomerans (New York: Basic Books, 1969), p. 198.
20. Charles Morgan, *The Fountain* (New York: Alfred A. Knopf, 1932), p. 49.
21. Theodore Lidz, *The Person: The Development throughout the Life Cycle* (New York: Basic Books, 1968), pp. 504-5.
22. Carl Jung, *Self*, as quoted in Alfred Freedman and Harold L. Kaplan, *Comprehensive Textbook of Psychiatry* (Baltimore: Williams & Wilkins Co., 1967), p. 369.
23. Heinrich Rickert, *Der Gegenstand Der Erkenntnis: Einführung In Die Transcendentalphilosophie*, 5th ed. (Tübingen: Verlag von J.C.B. Mohr, 1921), pp. 44-51.
24. Ibid., p. 47.
25. Plato, *"Phaedrus,"* in *Works of Plato*, selected and edited by Irwin Edman (New York: Modern Library, 1928), pp. 286-87.
26. Maurice de Wulf, *An Introduction to Scholastic Philosophy*, trans. P. Coffey (New York: Dover Publications, 1956), p. 126.
27. James K. Feibleman, *Ontology* (Baltimore: Johns Hopkins University Press, 1951), p. 345.
28. *New Columbia Encyclopedia*, 1975, s.v. "Mutation."
29. Santillana and Dechend, *Hamlet's Mill*, pp. 69-71.
30. Eugene D. Robin, "Claude Bernard: Pioneer of Regulatory Biology," *Journal of the American Medical Association* 242, no. 12 (September 21, 1979): 1283-84.
31. *New Columbia Encyclopedia*, 1975, s.v. "Tropisms and Auxins."
32. Herbert S. Jennings, "Tropisms," in *Body and Mind*, ed. William McDougall (London: Methuen & Co., 1911), p. 259.
33. Robert M. Yerkes, "The Intelligence of Earthworms," *Journal of Animal Behavior* (1912): 334-52.
34. Maurice Maeterlinck, *Les Sentiers Dans Le Montagne: Monde Des Insectes* (Paris: Bibliothèque Charpentier, 1919), p. 92.
35. Michael L. Brines, "Bees Have Rules," *Science* 206, no. 4418 (November 2, 1979): 571-73.
36. Konrad Lorenz, *On Aggression*, trans. M. K. Wilson (New York: Harcourt, Brace & World, 1966), pp. 207-9.
37. Leonard Lee Rue, *The World of the Beaver* (Philadelphia; New York, J. B. Lippincott Co., 1964), pp. 56-81.
38. Charles Sherrington, *Man on His Nature*. The Gifford Lectures, 1937-1938 (New York: Macmillan Co., 1941), p. 205.
39. *New Columbia Encyclopedia*, 1975, s.v. "Dolphins."
40. Jane van Lawick-Goodall, *My Friends: The Wild Chimpanzees* (Washington D.C.: National Geographic Society, 1967), pp. 48-51.

41. Robert M. and Ada W. Yerkes, *The Great Apes* (New Haven: Yale University Press, 1929), p. 306.
42. Allen R. Gardner and Beatrice T. Gardner, "Teaching Sign Language to a Chimpanzee," *Science* 165 (1969): 664-72.
43. David Premack, "The Education of Sarah: A Chimp Learns Sign Language," *Psychology Today* 4, no. 4 (1970): 55-58.
44. David Premack, "Language in a Chimpanzee?" *Science* 172 (1971): 808-22.
45. Francine Patterson, "Conversations with a Gorilla," *National Geographic*, 154, no. 4 (October 1978): 438-65.
46. Herbert S. Terrace, *Nim* (New York: Alfred A. Knopf, 1979).
47. Ibid., p. 98.
48. Ibid., p. 18.
49. Ibid., p. 203.
50. Kresten Nordentoft, *Kierkegaard's Psychology*, trans. Bruce H. Krimmse (Pittsburgh: Duquesne University Press, 1972), pp. 6, 129, 130.
51. Ernest Jones, *The Life and Work of Sigmund Freud* (New York: Basic Books, 1961), p. 22.
52. Sigmund Freud, *Die Traumdeutung*, 3d rev. ed. (Leipzig; Vienna: Franz Deuticke, 1911), pp. 358-59.
53. Ibid., p. 362.
54. Sigmund Freud, *Jenseits Des Lustprinzips* (London: Imago Publishing Co., 1940), pp. 38, 45.
55. Freud, *Die Traumdeutung*, pp. 398-99.
56. Freud, *Jenseits Des Lustprinzips*, p. 28.
57. Sigmund Freud, *Das Ich und Das Es* (London: Imago Publishing Co., 1940), p. 251.
58. Freud, *Die Traumdeutung*, p. 358.
59. Marie Bonaparte, "Time and the Unconscious," *International Journal of Psychoanalysis* 21 (October 1940): 427-28.
60. Ibid., p. 432.
61. Ibid., pp. 437, 436, 451.
62. Ibid., pp. 438, 465 (emphasis added).
63. Ibid., p. 465.
64. Heinz Hartmann, "Comments on the Formation of Psychic Structures," in *The Psychoanalytic Study of the Child*, ed. Ruth S. Eisler, et. al., vol. 2 (New York: International Universities Press, 1948), p. 15.
65. Ibid., p. 21.
66. Ibid., p. 28.
67. Heinz Hartmann, *Ego Psychology and the Problem of Adaptation*, trans. David Rapaport (New York: International Universities Press, 1958), pp. 51, 68, 71.
68. Ibid., pp. 24, 30.
69. Hans Loewald, "The Experience of Time," in *The Psychoanalytic Study of the Child*, vol. 27 (New York: International Universities Press, 1948), p. 404.
70. Ibid., p. 403.
71. Ibid., p. 407.
72. Hans Loewald, "The Superego and the Ego-Ideal: Superego and Time," in *International Journal of Psychoanalysis* 43 (1962): 264, 268.
73. Ibid., p. 264.

74. Ibid., p. 266.
75. Loewald, "The Experience of Time," p. 404.
76. Ibid., p. 406.
77. J. T. Fraser, *Of Time, Passion and Knowledge: Reflections on the Strategy of Existence* (New York: George Braziller, 1975), p. 75.
78. J. T. Fraser, "Temporal Levels and Reality Testing," *International Journal of Psychoanalysis* 62, no. 3 (1981): 10, 13, 18.
79. Ibid., p. 13.
80. Ibid., p. 78.
81. Ibid., pp. 78, 436.

7
The Mind-Body Relationship

Conventionally, one places the body in three-dimensional space and the mind in linear time; one uses the category of causality; one observes by extrospection; and one compartmentalizes body and compartmentalizes mind. Thus one is caught on the horns of the classical dilemma which is created by separating time from space. How can mind, which is spatially nonextended and which exists in linear time, interact with body which exists in three-dimensional space and in linear time?

By contrast, an unconventional exposition starts out with the more complex combination of objective time-space and subjective time-space; the categories of causality and intentionality; extrospection as well as introspection. An unconventional exposition also stresses the contrasts between emotionally neutral versus emotionally charged events. It starts out "from above" with man as a holistic mind-and-body and it proceeds "downward" to arrive at a conception of man as a causally organized physical entity.

Since mind-body interaction is fraught with several of the fissures, dichotomies, and fractures which are inherent in man's mundane existence, this challenging and unresolved problem belongs in my thesis about the meaning of suffering. Sometimes, interaction between mind and body progresses harmoniously and pleasantly; at other times the opposition of mind set over against body is a source of vexation and irritation, while a more serious discordance between mentation and bodily functions may cause us to suffer deeply.

The basic question is: Why do we split mind from body in the first place? And here we have to look beyond the dichotomy mind versus body to the most severe fracture in our existence, which is that of living versus dying. Mentation, inseparable from body, is just as much a victim of the ravages of death as is the body. Therefore, one has conceived of the immaterial soul, existing, not in limited subjective time but in eternity (eviternity: Saint Thomas Aquinas). This metaphysical leap contributes to the pervasive vagueness of definitions of mentation. I hold that the concepts of the mental triad and the precious present can be of help to define the concept of mentation from the viewpoint of time and space. Hopefully, this definition may lead toward a tentative solution of the mind-body problem.

CONVENTIONAL EXPOSITIONS

The conventional exposition of the mind-body problem stems from the legacy of René Descartes. His starting point is *cogito ergo sum*. His discussion of the mind-body problem emerges from this proposition. Descartes compartmentalized the whole human being into a *res extensa*, or body, and a *res cogitans*, or mind. That is, he started out from the concept of man as a compartmented rational being who lives in isolation from his fellow human beings. One can extrapolate Descartes' view of man as an island unto himself from his motto: *bene vixit qui bene transtuit* ("he has lived well who has hidden well").

In the split between *res cogitans* and *res extensa* we stumble upon the split between space and time. Space was a most congenial concept for Descartes, who made a homogeneous medium out of space by means of the three-dimensional coordinate system. His concepts of time and duration are derivatives of Aristotle's definition of time as "the number of motion." Although Descartes did differentiate between duration and time, he did not seem to apply these concepts in his exposition of the mind-body relation.

In earlier works, Descartes hardly differentiated between time and duration, but later he based their differences upon the way we perceive things and the way we think about things.

> Changes and motion, then, were conceived primarily in geometric terms as changes of place and position, not in terms of process. As a result, the concept of time plays a minor role.... He worried how we can possibly get from one sovereign moment to the next and he saw these independent moments as necessitating the repeated creative efforts of God in each moment.[1]

There is no inkling of the fact that we synthesize past, with future, with present in lived-in time.

Descartes thought in terms of temporal atomicity. Movement was for him the changing of an object from one geometric point in space to another point. His thinking was dominated by a concern about spatial relationships which kept temporal relationships almost out of sight. The atomistic structure of moments of time raised the problem of how they in fact do cohere. To save the continuity of time he invoked the miraculous intervention of acts of God.

Descartes presumed that we understand duration by comparing the day-night cycle and the progression of years in time with duration. Time then is nothing more than how we think about duration.

> Thus time, for example, which we distinguish from duration taken in general, and which we say is the number of motion, is nothing other than a

certain *manner* in which we think about duration. But in order to comprehend the duration of all things under the same measure, we ordinarily make use of the duration of *certain* regular motions which form the days and the years, and having thus compared it, we call it time; although in effect what we mean in this way is nothing, *over and above* the true duration of things, except a *manner* of thinking.[2]

One might paraphrase Descartes by saying, only the Creator is capable of making compartmentalized intervals of objective time separated from one another by punctiform instants into interpenetrating precious presents. Linear time prevents Descartes from defining mentation because the punctiform instant is surd to mentation's capability to synthesize our past and our future within our present. He split mind from body because he placed the *res cogitans* in objective, linear time and the *res extensa* in three-dimensional, homogeneous objective space.

Regarding movement as an object leaving one point in space and arriving at another point in space, the possibility was given Descartes to construct a mechanistic explanation of body movements of animals and of our own body movements. However, he had recourse to the ancient concept of "animal spirits" or "vital spirits" which he conceived as tiny particles of matter, something like a thin fog or minute water droplets. With these concepts he entered animistic and teleological elements into his mechanistic conceptions. He found inspiration for this approach when he saw how water, streaming in hidden pipes, could produce movements in animal or human figures which imitate movements of living beings. "Descartes...had been impressed by the fountains in the royal gardens which were so constructed that, actuated by water, clay figures moved, made sounds, and played instruments. He conceived the animal body to be actuated on the same principle: instead of pipes and water, there are nerves and animal spirits."[3]

In keeping with the convictions of contemporary physiologists, Descartes thought of the nerves as hollow tubes, and he imagined that vital spirits roamed around in these nerves and, ending in muscles, made them swell, that is, they were the cause of muscle contractions.

This mechanistic model molded Descartes's theories about the mind-body relationship. This was quite congenial to him, because he admired Harvey's mechanistic interpretation of circulation and of the heart as a pump device which sets the blood in motion. However, Descartes invoked the concept of the immaterial soul as an agent which directs the course of the animal spirits within the nerve channels. For Descartes, a mechanistic process of shock and countershock went on in the pineal gland. This aroused certain mental states in the soul when it was hit by animal spirits, while in reverse order the soul also influenced the pineal gland.

Descartes extended to the physiology of the nervous system the mechanistic conception which Harvey had made triumphant for the circulation of the blood. The soul is in immediate communication with one single part of the brain, the pineal gland. The "animal spirits" impinge on the pineal gland, thus exciting the soul and give birth to sensation, sentiment and appetite in it; the soul responds to it, by an equal shock (!)—against the pineal gland—and directs the animal spirits in certain directions.[4]

Stuart Spicker gives further evidence how Descartes arrived at the antithesis of body versus mind as meaning the divisible body over against the indivisible mind.

That is, there are properties of mind and properties of body, and none of the properties of either can be a property of the other.... And in Meditation Six, it appears that the last word has been said.... Descartes writes to the Princess [Elizabeth of Bohemia] that "everybody always has the awareness of the union of soul and body in himself without doing philosophy."[5]

It is as if Descartes's own doubts about his mind-body theory come to light in this admonition. It is reminiscent of Hume's suspending his critical attitude as soon as he leaves the ivory tower of abstract thinking and immerses himself in everyday, concrete social life.

The Cartesian dualism of mind versus body has not been resolved because of the one-sided use of the concepts of three-dimensional space and measurable, linear time; the lack of differentiation between causality and intentionality; the failure to define mentation; and the interchangeable use of the concepts of mind and soul. In addition, the off-handed remark to Princess Elizabeth about switching at will from critical thinking about the union of mind and body to intuitive immersion in their holism, overlooks the contrast between compartmentalization and interpenetration.

Descartes elevates spatial relationships to the place of honor while temporal relationships in man's terrestrial existence are relegated to the continuous, creative act of God who keeps body and mind together. Opting for a mechanistic description and explanation of the mind-body relationship, Descartes ends up with an outright teleological interpretation.

Descartes's problem is how to synthesize mind (compartmentalized in objective time) with body (compartmented in three-dimensional, objective space) into a holistic, interpenetrating entity. Recognizing only the forms of objective time and objective space, Descartes could not release himself from the extrospective method of observation, especially since he valued mathematics and geometry as the ideal, true sciences. The "vast difference" between the compartmentalized *res extensa* and the *res cogitans* splits body from mind. Rationality cannot heal this dichotomy in our existence, but Descartes could hardly have been enamored with the intuitive, nonquantifying observations of introspection and its ancillaries that are obtuse to mathematical treatment and cannot be pressed into geometric constructs.

Whereas Descartes did not develop a theory of mentation, David Hume brought forth his theory of the association of ideas and impressions. Simple ideas are linked together into more complex ones, not too different from the way in which atoms are linked together to form molecules. The temporal element of contiguity in the process of association is mentioned in one breath by Hume with the spatial element. This seems to give first rank to space over time. One gathers from the emphasis he gives to the power of cause-effect sequences, that association of ideas is of a causal nature, and one may conclude that this process occurs in objective time.

In contradistinction to Descartes, Hume speaks extensively about time and space. Time is equated with succession in the before-and-after of objective time. Hume seems to have a premonition that we do not experience time as such, but that we are aware of the temporal structuring of individual impressions. A melody of five notes is more than the sum of these five impressions, but it is not as if a sixth impression, that of time, is being added when we experience the succession of the five notes. For Hume, duration flows from the succession of individual and indivisible moments of time. He comes close to saying that time is a formative agent in sensory experience.

> As 'tis from the disposition of visible and tangible objects we receive the ideas of space, so from succession of ideas and impressions we form the idea of time, nor is it possible for time alone ever to make its appearance, or be taken notice of by the mind.... Five notes played on a flute give us the idea and impression of time; tho' time be not a sixth impression, which presents itself to the hearing or any other senses, nor is it a sixth impression which the mind by reflexion finds in itself.... From the tone of voice the dog infers the master's anger, and foresees his own punishment. From a certain sensation affecting his smell, he judges his game not to be far distant from him.[6]

One misses in Hume's discussion of time any mention of the mind's extension into the future, except when he is talking about the behavior of the dog.

To this day we live under the shadow of Humes's theory of association of ideas and impressions, which reinforces our tendency to compartmentalize mind. His attempts at a localization theory of mentation in the brain have a distinctly modern flavor. Instead of localizing the soul, or the "animal spirits" in the narrow confines of the pineal gland, Hume thought that the process of association is active in the entire brain. A sort of game is going on between the mind, the "animal spirits" and the traces in the brain. The mind does not always get its way in directing the spirits to the cell to which the idea is associated. The physical structure of the brain presents obstacles to the mind in keeping the associations on the paths which it intends them to follow.

> 'T wou'd have been to have made an imaginary dissection of the brain, and have shewn, why upon our conception of an idea, the animal spirits run

into all the contiguous traces, and rouse up the other ideas, that are related to it.... Wherever it [the mind] dispatches the spirits into that region of the brain, when they run precisely into the proper traces, and rummage that cell, which belongs to the idea. But as their motion is seldom direct, and naturally turn a little to the one side or the other; for this reason, the animal spirits, falling into contiguous traces, present other ideas in lieu of that, which the mind desired at first to survey.[7]

Hume gives the association of ideas breathing space, as it were, since he allows them to "rummage" around in the wide expanse of the brain. But Descartes's conception of the body as if it were a machine, and Hume's theory of association, fail to differentiate between causality and intentionality. Their thinking is dominated by the form of objective time in which one compartmentalizes animal spirits, and the other compartmentalizes ideas and impressions. Although both philosophers break ground for the introspective method of observation, they do not pry this method lose from the extrospective method, which prevails in the natural sciences.

Hume's sketch of the localization of the associative process in the brain, down to the cellular level, leads to the question how the mind and the brain interact, instead of how the mind and the entire body interact. This physiological theory puts the question of how mentation and cerebral functioning interact on the doorstep of neurophysiology. The theory of association of ideas and impressions, which relies on objective time, does not proceed beyond the category of causality and therefore its does not widen our purview of the interaction between mentation and brain physiology. To enter into the realm of intentionality and of interpenetration, we ought to rely on the service of the mental triad and the precious present, of objective time-space and subjective time-space.

McDougall reviews the historical development of the mind-body problem extensively from ancient times through the turn of the nineteenth century.[8] McDougall seems to get sidetracked into a defence of animism, as the subtitle of the book announces, but also in the concept of the immaterial, immortal soul. He attributes several biological functions to the soul in the fields of evolution, growth, heredity, and psychology. In short, he extends Descartes's teleological conception of the soul into the fields of biology and psychology.

McDougall shows that it is indeed of crucial importance to foster teleological trends in biology and in psychology, but his treatment of the differentiation of causality from goal-striving activities is far from acceptable because he does not introduce the difference between the spatiotemporal structure of purpose-striving phenomena versus the spatiotemporal structure of causally constituted events.

Wundt proposes a solution to the mind-body problem by psychophysical parallelism. "The difference between teleological and causal conception is that to every purposive reaction there belongs a causal connection."[9] Wundt does not see causality versus teleology as two incompatible principles of explanation since they depend on the manner in which we approach events. A purposive relationship can be completed by a causal connection and a causal formulation of the same event can be enlarged by insight into its teleological aspects. Thus, natural scientific explanation based on causation does not exclude philosophical interpretation which posits mentation as a goal-striving dynamism.

Henri Bergson has opened up new perspectives toward an interpretation and solution of the mind-body problem. He goes beyond Descartes's *res extensa* versus *res cogitans* by taking issue with conceiving mind as if it were thinglike. Bergson also goes beyond Hume's theory of association of ideas in an extensive and illuminating analysis of memory, recollection, and mentation. Specifically, Bergson denies that memory is a *res extensa*, and he places it as a nonextended process within the domain of mentation.

He places the time problem in the center of his discussion of the mind-body problem, as he has done in reference to other problems in his major works. He emphasizes the importance of understanding the temporal structure of movement since he sees the most concrete instance of mind-and-body interaction as a relationship between movement and cerebral and spinal-cord mechanisms. His *Matter and Memory* is a tightly argued, masterly interweaving of trends of thoughts which interconnect epistemology, psychology, neurophysiology of speech, and the phenomenological examination of our internal mental processes.

Bergson poses the question how cerebral processes relate to the body movements that we execute. The brain is a connecting agency between sensations and movements, but the brain does not cause these movements. It links movements which we perceive mentally with those we eventually perform. "We see in the brain no more then an intermediary between sensations and movements, which would make the extreme points of sensations and movements, a point which is incessantly inserted in the tissue of events."[10]

The more complex cerebral processes become, the greater the opportunity for choice which one can observe. But the affections which stand in between perception and movement are not visible. These remarks derive from discussions of the relation between memory and matter in the light of several neurological diseases such as apraxia, motor and sensory aphasia, and psychic deafness. Bergson also analyzes the process by which we learn something and when visual and auditory images initiated imagined body movements which cannot yet be perceived externally. The mind directs past recollections toward the present. But he is firmly convinced that memory is independent of matter.

> Our cerebral state contains more or less of our mental state, depending on whether we tend to exteriorize our psychological life in action, or to interiorize it in pure knowledge. If we take perception in its concrete form, as a synthesis of recollection with perception, that is to say of mind with matter, then we will compress the problem of the union of body and mind within the narrowest possible limits. The distinction between body and mind must not be established as a function of space, but of time.[11]

Mind versus body cannot be described in terms of space but must be conceived in terms of time. The duration, in which we observe ourselves act, consists of elements which are juxtaposed to one another. The duration in which we act allows our mental states to melt into each other. In brief, Bergson sees brain structure and function as a locus for sensory-motor possibilities. Mentation activates this site by choosing between those activities that are useful and that interest us at present and those that prolong these neural mechanical functions into the lived-in present and into the immediate future. Mentation concretizes memories that are available as well as nonspatially organized entities. He asserts time and again that the brain is not a storage place for memories. It is as if Bergson foresaw Stent's critical remarks about the "grandmother cell."[12]

Stent describes an abstraction process which reduces step by step the avalanche of incoming perceptual stimuli, as the excitation traverses the numerous levels of neural organization ending in the highest cerebral centers. Although certain nerve cells in the brain respond to exquisitely specific light and dark patterns, he concludes that it is inconceivable that these cells are capable of recognizing meaningful structures.

Stent's question reminds us that spatial localization in the brain impedes rather than furthers our understanding of mind-body interaction. We ought to keep the time problem in the forefront of the discussion. Notwithstanding Bergson's liberation from Descartes's spatially oriented approach to the mind-body relation, there is a fundamental similarity between the thinking of these two philosophers: both fail to analyze the differences between objective time and subjective time. But Bergson's contrasting concepts of *temps espace* and *temps duré* (spatialized time and durational time) do anticipate the concepts of objective time and subjective time. This is so, because he talks frequently of the power of mentation to integrate past, present, and future into an interpenetrating whole, when he describes how mentation exists in durational time.

But due to the lack of firm distinction between the two forms of time, one cannot establish a firm distinction between mechanical movement and purpose-striving movement, a contrast which must be anchored on the two contrasting time concepts. In many examples Bergson brings this difference to light very clearly from the phenomenological viewpoint of the immediately given data of consciousness. So long as Bergson discusses movement, he is on firm ground, but the lack of differentiation between the two forms of time leads into the pitfall of unclear distinction between the whole array of concepts of mind, mentation, memory, affectivity, spirit and soul, which other writers allow to run into each other like inkblots on tissue paper.

Going one step beyond Bergson, I propose that we start the examination of the mind-body problem with the holism of mind and body, which we experience as an immediately given datum of consciousness. The next questions are how and why we do split this holism of mind and body and how can we synthesize this antithesis? In order to take a few steps farther down the road, I propose that we do not contrapose linear time versus three-dimensional space as the two basic forms of our experience and of our existence, but that instead, we must conceive of the contrast between objective time-space versus subjective time-space. These two forms of time-space allow us to go beyond Descartes's concretizing the mind versus body split, beyond Hume's association theory, and beyond Bergson's failure to describe how mechanical perturbations in the brain relate to purpose-striving movements which we execute with our body. Instead of separating time from space, contraposing objective time-space versus subjective time-space, as well as the interrelation between these two forms, there is an alternative approach to the study of the mind-body relationship. This approach is in keeping with the fact that we do not experience time nor space, but movement in time and space. Therefore, time, space, and movement ought to be kept unified as a cohering threesome. And on this basis we can differentiate and synthesize mechanical events and purpose-striving activities.

Since we are surrounded by and propelled forward by mechanical movement, such as the rotation of the earth, gravity, outside our body, and by respiration, circulation of the blood, reflex mechanisms, inside our body, we think of these movements as the most important and primary ones, while occasionally performed purpose-striving movements remain in the background. However, within the framework of our personal and our intellectual development, we start out with purpose-striving movements, as immediately given data of consciousness as the primary kind of movement. Out of these data we conceive by a slow, time-consuming process of abstraction the idea of mechanical movement and its appropriate forms of time and space. Against the experienced splits of mechanical from purposive movement and the contrasting concepts of objective versus subjective time-space, the dichotomy mind versus body grows.

Popper and Eccles address the mind-body problem from the vantage points of philosophy and neurophysiology. Popper divides reality into three realms: "The universe of physical entities I will call World *1*; the world of mental states I will call World 2; the world of the contents of thoughts, and, indeed of the human mind, I will call World *3*.[13] This schema progresses from the causally structured World *1*, which I should like to conceive as occurring in objective space, toward the purposively structured World *2* and World *3*, which seem to be nonextended and which are presumably structured within subjective time. Within this central thesis, Popper develops the concept of the self, but he does not attempt to define mentation.

Eccles attempts to overbridge the mind-body split with his concept of the self-conscious mind which he presents as the main constituent of World *2*. World *3*, creates the self-conscious mind with the aid of cultural achievements and particularly with the aid of language.

Within World *1* the central nervous system is the area of main interest for the neurophysiologist and Eccles describes and explains an enormous amount of scientific material in lucid but compressed format. In his exposition of the mind-body relationship the modules in the dominant hemisphere carry the weight of his argument. They are more or less cylindrical units which contain up to

10,000 neurones each interconnected by a complex network of nerve fibers, measuring 3 mm in length and 0.5 mm in width. The modules possess input channels and output channels, an arrangement reminiscent of the sensory-motor reflex structure. These two billion modules are in constant conflict with each other by excitation and inhibition.

Eccles localizes mind-body interaction in the interface between World *1* and World *2*, on which plane the self-conscious mind "scans and reads out" what is going on in the modules of the "liaison brain" of the dominant hemisphere. Placing the self-conscious mind in World 2, which both Popper and Eccles conceive to be nonextended, Eccles is faced with the classical problem of how the nonspatial self-conscious mind can be in a giving and receiving interaction with the surface of open modules which are, patently, spatially structured elements. The drawback of these authors' exposition of the mind-body problem is that both treat the time problem only on occasion and with an almost off-handed attitude. Moreover, they do not contrapose causality over against purpose-striving activities. It is my impression that Popper thinks that gradual transitions occur between his three worlds. I glean this from his pondering that inasmuch as we can understand interaction between World *3* and World *2* fairly well, we may have a model for interaction between World *2* and World *1*, which is similar to mind-body interaction. But again, he allows mentation to infuse neural processes by gradual increments. It seems as if he does not see that there exists an essential difference between mentation and mechanism.

I propose that the transition from World *2* to World *1* is not a step-by-step, gradual progression from more concrete toward more abstract: I propose that there is an unleapable chasm between interpenetrating World *2* and compartmentalized World *1*.

Eccles follows a similar route of gradual transitions between body and mind as Popper does in interpretations of World *1* and World *2*. He points to the unfathomable complexity of the central nervous system in general and of the cerebral cortex in particular. He implies that there may be only a relatively gradual increased degree of complexity when we progress from individual neurones, to module, to the cerebral hemispheres, and finally to the self-conscious mind. However, he does mention several times that the cerebral machinery cannot explain the unity of the self-conscious mind.

He is quite positive about the impotence of neurophysiology as it is confronted with the phenomena of preprogramming and antedating. He invokes the capacity of the self-conscious mind to "play tricks with time," but leaves the question as to its specific temporal structure dangling. Obviously, he does not differentiate clearly between objective time and subjective time.

> This antedating procedure does not seem to be explainable by any neurophysiological process. The antedating of the sensory experience is attributable to the ability of the self-conscious mind to make slight temporal adjustments, i.e., to play tricks with time....Does the self-conscious mind have some specific temporal properties?[14]

Eccles is on firmer ground when he discusses memory. Histological, biochemical, and surgical procedures have accumulated a great deal of factual knowledge about the memory function. One of the important subcortical structures which plays an important part in memory processes is the hippocampal gyrus. Patients who have undergone hippocampectomy suffer from severe memory defects. They are left with short-term memory of only a few seconds which they have to constantly rehearse verbally in order to retain. The operation has also interfered with long-term memory.

The effects of sectioning the corpus callosum, which consists of millions of fibres crossing the midline in opposite directions, and which connect the dominant hemisphere with the minor hemisphere, reveal many heretofore hidden functions of the brain.

> The dominant hemisphere is predominantly symbolic and propositional in its function, having specialization for language with syntactical, semantical and logical abilities. By contrast, the minor hemisphere [is] appositional, with the property of apposing or comparing perceptions and schemas in some Gestalt sense, which is far beyond our present understanding.[15]

Eccles claims that the results of these operations intimate that the dominant hemisphere is and that the minor hemisphere is not in liaison with the self-conscious mind. The dominant hemisphere imagines fine detail, it is analytic and sequential; it is propositional and symbolic, logical and mathematical. The minor hemisphere specializes in pictorial and musical patterns and synthesizes gestalts since it brings contrasting elements together.

It seems as if the commissurotomy patient is made up out of two different human beings, each with his own mental, rational, and emotional attributes. Each separated hemisphere is part of the neural machinery which is causally structured and operates as a computerlike instrument. However, even the separated hemispheres outreach by far the limited functions of any purposeful man-made machine. The left hemisphere functions in a rational, compartmentalizing vein, while the minor hemisphere operates on a holistic, interpenetrating level. We ought not to compare either hemisphere with a computerlike entity because both hemispheres are necessary conditions for conscious experience. Eccles retracts somewhat his original statement that the self-conscious mind is only in liaison with the dominant hemisphere and its liasion modules, but that the minor hemisphere possesses this quality also but to a lesser degree.

We cannot construct the whole human being by starting out from the causally structured neural machinery of the brain which occurs in objective time-space and by introducing purposiveness by small increments into the increasingly complex neural structures and functions as we ascend from the lower, simpler regions of the nervous system to the higher and more complex areas of the cerebral cortex. The cerebral cortex is supposed to display purpose-striving capacities which proceed in subjective time-space. Popper tries to accomplish this feat by sketching a gradual transition between his three worlds. Eccles tries to overcome the lack of goal-striving in the neural machinery with the aid of the self-conscious mind. But neither the philosopher nor the neurophysiologist state unequivocally how they conceive of the temporal structure of the three worlds nor of the self-conscious mind. One has to be satisfied with the negative statement that World 2 and World 3 are nonspatial and so is the self-conscious mind. What kind of temporal structure the self-conscious mind may have is left unanswered. I trust that the mental triad and the precious present can be of service at this point.

Popper and Eccles both follow the route from "down below" toward "up above." Popper intimates a gradual transition from World *1*, to World *2*, to World *3*; Eccles starts out with the neurone, proceeds to the module, thence to the larger centers such as the cerebellum, hippocampus, thalamus, prefrontal lobes and advances ultimately to the highest level of the cerebral hemispheres. In the transition from World *1* to World *2*, Popper covers over the leap from causally structured events to intentionally structured activities. Eccles minimizes the fact that not only neurones and modules, but also large centers, including the hemispheres, are compartmented structures within the total nervous system; therefore, one overlooks the essential differences between compartmentalized structure and function compared with interpenetrating mentation and purpose-striving. Causality and compartmentalization are lumped together with intentionality and interpenetration. As a result, neurones, modules, hemispheres, are not physical entities of structure and function but psychophysical hybrids with distinctly mentalistic attributes that come more and more to the fore, the higher one ascends on the route from "down below" toward "up above." One ends up with a horde of homunculi populating the central nervous system, simpler ones in the neurones, highly complex homunculi in the hemispheres. One portrays the separated hemispheres of the commissurotomy patient as two, not-quite-human persons, and one buries the mind-body relationship almost completely under this large array of varyingly constituted homunculi.

Sperry has tested commissurotomy patients exhaustively. In one modality of these tests he projects an image on a transparent screen during one-tenth of a second, or less. These flashed images are so arranged that they stimulate either the left half-field or the right half-field of vision of one eye. The images in the left half-field are elaborated by the right visual cortex and the right hemisphere, the images in the right half-field of vision are elaborated by the left visual cortex and the left hemispheres.

> One hemisphere does not know what the other is doing. It is like two separate individuals working over the collection of test items, with no cooperation between them. This presence of two minds in one body, as it were, is manifested in a large number and variety of test responses. Whereas interference and extra delays are seen in normal subjects with the introduction of the second task, these patients with the two hemispheres working in parallel simultaneously perform the double tasks as rapidly as the single

tasks. The most remarkable effect of sectioning the neocortical commissures is the apparent lack of effect so far as ordinary behavior is concerned.[16]

Sperry deduces from these and many other test results that the left and the right hemispheres subserve distinctly contrasting mental functions which seem to be compartmentalized. He sees two individuals working at different tasks with no cooperation between them. When a normal subject is asked to perform two different tasks at the same time, he deliberates and judges how he will coordinate these unrelated activities, hence there is "interference and delay." The patient with deconnected hemispheres is not aware of the oppositeness and disconnectedness of the two tasks which are performed by each hemisphere individually, and therefore the patient does not vacillate and proceeds to getting the task done. Ordinary behavior is surprisingly little affected by the operation. But isn't ordinary life what Heidegger calls falling down in average man? This life-style requires only a minimum of questioning, comparing, vacillating, and doubting, and it does not waste time in performing routine daily tasks.

Clearly, Sperry ascribes mental attributes to the cerebral hemispheres. This trend to localize specific psychological functions in distinct areas of the brain rests on the detailed knowledge of deficits of body functions in cases of neurological diseases. These deficits can often be traced back to the destruction of specific neurones and their connections in circumscribed areas of the brain. By studying the performance and the neurological deficits of patients one can pinpoint, for instance, the exact location of the origin of the motor cortico-spinal tracts and of the receiving areas for the somesthetic inputs in the healthy human and in higher primates. This knowledge is schematically represented by four homunculi which stand on their head in the two motor and the two sensory areas of the two hemispheres. The motor homuculus in the precentral gyrus is separated from the sensory homunculus in the postcentral gyrus by the sulcus Rolandi which separates the two gyri. Grinker and Bucy present a drawing of the sensory homunculus (figure 1) that looks like an ugly little monster.[17]

It is a far cry from the homunculus which Wagner, a medieval alchemist, concocts in Goethe's *Faust* "I see a pretty mannikin of delicate shape gesticulating. Come press me quite tenderly to your heart! But not too firmly so that the glass will not shatter."[18]

Goethe's homunculus enclosed in his vial is an island unto himself; and so are the modern neurophysiological homunculi. The deconnected hemispheres work in parallel without mutual communication; they too are islands unto themselves. This is an additional reason why we cannot reconstruct the whole human being out of the separate right and the separate left hemisphere.

Having ferreted the homunculi out of their hiding places in the cerebral hemispheres, we begin to see that we apply a similar compartmentalizing operation on the whole human being. We divide the whole human being into a compartment of mind and a compartment of body; but we actually divide him into a homunculus mind and a homunculus body. Each homunculus ensconces something of the

Figure 1
Sensory-Motor Homunculus

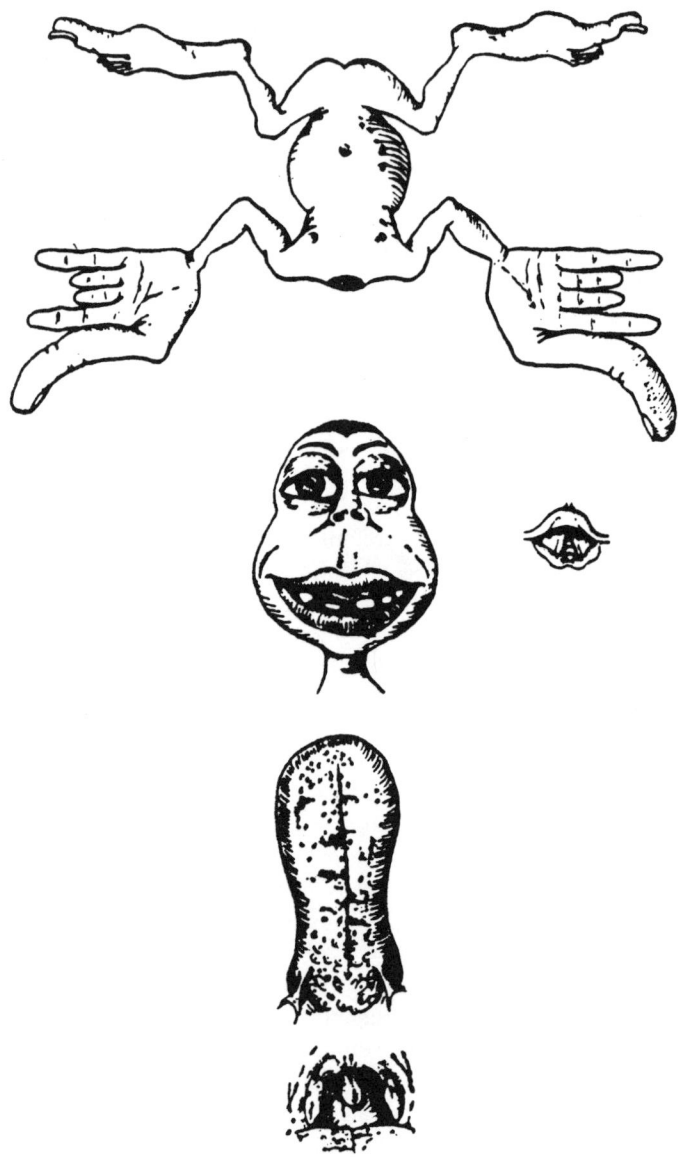

Source: Roy R. Grinker and Paul C. Bucy, *Neurology*, 4th revised edition, 1949, Figure 137, p. 358. Courtesy of Charles C. Thomas, Publisher, Springfield, Illinois.

whole human being, which means that we give bodily attributes to the mind and we give mental attributes to the body. The psychophysical problem gets buried in these two homunculi.

Gilbert Ryle admonishes us as to how we should talk about mentation, but he hardly delineates or defines it. He touches occasionally on the mind-body problem and he criticizes Descartes's exposition.

> Human bodies are in space and are subjected to the mechanical laws which govern all other bodies in space. Bodily processes can be inspected by external observers. But minds are not in space. Only I can take direct cognizance of the state and process of my own mind. But the actual transactions between episodes of the private history and the public history remain mysterious, since by definition they can belong to neither series. It is a necessary feature of what has mental existence that it is in time but not in space. Questions about the relations between a person's body and his mind.... are improper questions. My "mind" does not stand for an organ, not some piece of personal apparatus...he is bodily active and he is mentally active, but he is not being synchronously active in two different places or with different engines. There is the one activity, but it is susceptible of and requiring more than one kind of explanatory description.[19]

Ryle says essentially that we approach the body and its functions which exist in space by extrospection and that we study mind which exists in time by introspection. One comes up with a public history about the body and a private history about mind and thus their union remains a mystery. He demonstrates very clearly that Descartes's explanation of body movements was a mixed, causal-teleological interpretation. Mentation is not causally structured, according to Descartes.

The often recurring critique of the "ghost in the machine theory" of mentation throws a sharp light on Eccles's concept of the self-conscious mind interacting with open modules in the dominant hemisphere. In a diagram Eccles points up a mysterious interaction between the "ghostly," nonspatial, self-conscious mind and the spatial modules. The "ghost in the machine" simile is a most congenial companion to my belaboring the metaphors of the homunculus. Both concepts instill, as it were from the outside, purpose-striving attributes upon mechanical processes. But since Ryle hardly mentions the time problem in this connection, he ends up with a negative statement rather than striving for a specific solution. He simply rejects as improper the question as to how body and mind interact. But he does indicate, in several concrete instances of purposeful movements, that the holism of mind and body is an undeniable fact, such as in the description of the work of a surgeon. We should not describe these activities as if they ran along on two separate tracks, but we should analyze this unified activity with two different kinds of explanatory descriptions.

But Ryle remains silent as to how these methods differ from and correlate to each other. His most challenging criticism of the conventional exposition of the mind-body problem is his denying the status of a scientific method of observation to introspection. He does however give a certificate of legitimacy to retrospection which is, of course, very nearly the same activity as introspection. The contrast of extrospection versus introspection stands or falls with accepting or denying that objective time-space and subjective time-space are useful and logically acceptable forms of experience when we describe human existence. Although Ryle gives due notice to feeling and emotion, he, like Descartes emphasizes the primacy of the intellect. But unlike Descartes he does not see man as an island unto himself.

H. Tristram Engelhardt views the mind-body problem from a different and fresh perspective.[20] He is markedly influenced by Hegel and Husserl, and hence his language makes it sometimes difficult to follow the trend of his discussion, unless one is thoroughly steeped in the systems of thought and the terminology of these two philosophers. However, Engelhardt, a physician as well as a philosopher, frequently balances esoteric sounding concepts against concrete examples of our living in a mind-body relationship in everyday life. He also draws from clinical experience. As I see it, the main thrust of his treatise is that we ought to start out with the unity of mind-body, given in our living as a mind-body in our world. Analytic thinking encounters this unity and dissects it into mind versus body. We are faced with reunifying these two thinglike entities, each with its own causal structure. This unification can be accomplished only on the categorial level, but we continue to vacillate between the unity of and the contrast between these two aspects of reality.

Engelhardt goes on to portray the categorial interconnectedness of the category "mind" with the category "body" most painstakingly. To explain the category "body" we must use the category "mind," while mind can exist only through body. The richer category, mind, must embrace the category body in order to exist as a real entity in the world. The lower category body is a necessary constituent of the higher category mind. The central message which comes through strong and clear is that we ought not to conceptualize mind and body as two things, or thinglike entities, with separate and essentially different causal structuring. If we think of thoughts and things as independent interacting entities we are compelled to invent logical rules for the behaviors of these so-called things. We end up with a rivalry between final and efficient causality, or rather, between causality and purposiveness. After thinkers have made a *res* out of body and a *res* out of mind, separating these two "things" out of the experienced unity of mind with body, then they will be unable to heal the split between mind versus body and body versus mind. Mind and body must be held together categorially as elements of mind-body holism in which we live and of which we are aware in practical daily living. Body is presented as the lower and poorer category and mind is envisaged as the higher and richer category. This valuation of the two categories is not comparable to the conventional contraposition of body-thing over against mind-thing.

Mind enriches body and imparts meaning and reality to body which it does not possess on its own account. Body thus transcends physical reality without being lifted into a ghostly or extramundane world. But mind needs body in order to significantly and concretely be-in-the-world. We cannot understand the category body without at the same time understanding and grasping the category mind, and inversely. This is expressed as identity in difference; this creates the tension between the two categories mind and body.

The tension and the "distance" within the mind-body union flows from the two opposite poles of intentionally and causally organized processes. Hence, the category of the body cannot include the category mind, but mind incarnate must engulf the category body. We cannot see the "mineness" of our body, which is a nexus of physical processes, causally structured, until we become aware that this so-called body-thing is more than a bundle of causally structured processes. The life-support organs which perform compartmentalized functions serve the musculoskeletal apparatus which in turn is innervated by the central nervous system. These are preliminary stages which make embodiment of thought possible. We become aware of their "mineness" if we interpret these causally structured processes as being necessary conditions for the embodiment of mind. We feel that our limbs are "mine" when we see how we perform psychical as well as physical functions with them. The category of body points toward the category of mind and the body thus becomes more than a physical body. It reaches the status of a human body through interidentification with mind, which is intentionally structured.

The poorer category body fills the richer category mind with concrete content, while the category mind imparts meaning and mineness to the category body. We see a hierarchical progression from the life-support systems to the musculoskeletal apparatus to the level of mentation which is the richest domain of significance. Practical mind which creates objective order needs theoretical mind so as to concretely exist in the world. Mind transcends body but it is not immaterial.

The interidentification of the richer with the poorer category, and inversely, may be interpreted as an instance of interpenetration. The category mind elevates the category body to the level of interpenetrating holism of mind-body. Hence, Engelhardt's analysis can be widened in scope by including the concepts of the mental triad and the precious present within its purview. I see one serious oversight in Engelhardt's overall enlightening thesis. It seems that he gives the time problem only secondary importance when he speaks of intentionality and protension, since he fails to point out that projection into the future fringe of a precious present is one of the major if not the major attribute of mentation. He mentions the time problem with the aim of establishing the continuity of mental life, with which viewpoint I take exception.

He places the burden of the continuity of mental life in time upon subconscious processes which are supposed to fill in the temporal gaps which we all experience in our conscious life. If mentation were discontinuous, it would depend on body for its continuity. This continuity is provided by protensions and retentions and not by the continuity of objective time. I find difficulty with the idea that protension could be a subconscious process.

Speaking of "nows," Engelhardt seems to overlook the double meaning of this word, which can function as an umbrella for both the punctiform instant of natural science and for the durational present within a precious present. Retention of past "nows" and protension of future "nows" appear as condensation of memories in the past fringe and projection into the future fringe of a precious present. So far, retention of past nows and protension of future nows are constituents of time consciousness. But when one speaks of any "now," one may refer indiscriminately to both the precious present and to the punctiform instant. I believe that punctiform instant, *nunc stans*, and precious present are insufficiently demarcated by Engelhardt.

The oversight of the importance of the time problem when one tackles the mind-body problem, results in failure to contrapose mechanical movement versus purpose-striving movement, a contrast which is of basic importance in interactions between the two constituents of our existence. This difference is buried in the contrast between efficient cause and final causation. In my opinion "teleological causation" is a self-contradictory concept, which blurs the boundary lines between causality and purposiveness.

The term "teleological" implies that every thing and every living thing that exists, does exist for a purpose. Hence, one runs the risk of investing mechanical movements, which occur in the living body, with purpose-striving attributes, that is, with mentalistic attributes. Teleological causation does not grasp the holism of body and mind, it only blurs the artificial boundary lines between body and mind that have been thought to exist since Descartes conceived of mind as *res cogitans* and the body as *res extensa*.

Engelhardt goes beyond his predecessors in that he does not anchor the "distance" between mind and body and their identity-in-difference on the two distinct and separate forms of time and space. On the contrary, by defining mind as body incarnate and body as conscious matter, he adumbrates the further step which I propose we ought to take. Instead of contraposing time versus space I suggest that we syncretize these two forms into the contrasting forms of objective time-space and subjective time-space. This gives us leeway to combine Engelhardt's categorial analysis with the wider ranging approach to the mind-body problem which I will follow and which is in line with his remark that "we ought to think several things at the same time."

Body as the simpler and the poorer category interrelates with objective time-space, causality, and compartmentalization, which structuring of natural processes compels us to observe them by extrospection. Mind, the more complex and the richer category, interrelates with objective time-space as well as with subjective time-space, with causality as well as with intentionality, which is compartmented but which also interpenetrates, and of which we should become aware by extrospection interwoven with introspection.

There are other reasons why I feel congenial to Engelhardt's approach to the mind-body problem. He mentions the vacillation of the understanding due to the tensions between the categories mind and body, which tensions usher in the search for ontological stability. These are indications of blemishes, fissures, dichotomies, and fractures in our rational undertakings. The frustrations and uncertainties in the mind of the theoretician who seeks to find a solution of the mind-body problem along intellectual lines, color his endeavors with stress and tension, even with emotional turmoil. Insight into the private workshop of the categorial analyst shows that the rational and the emotional facets of human existence cannot be separated. As is the case in all creative and constructive pursuits, suffering is an essential element in the search for a solution of the mind-body problem.

A TENTATIVE SOLUTION

I propose that we approach the mind-body problem by starting out from the whole, psychophysical human being. Let us cease contraposing mind versus body, a view which rests upon splitting time from space. We have to differentiate, and we have to weave together the *forms* of objective time-space and subjective time-space, the *categories* of causality and intentionality, the *methods of observation* of extrospection and of introspection with its ancillaries, the fact that we compartmentalize the body which exists in objective time-space, but that we cannot *compartmentalize* mentation since it is an *interpenetrating* entity in subjective time-space and that body functions are *emotionally neutral* processes, while mentation displays *feeling tones*. Through introspection and its ancillaries we get an intuitive view of man which lays bare our intentionality, our emotionality, and our communicating with our fellow humans.

Neurophysiology, which uses the concepts and the methods of natural science, cannot come up with a description of mind-body holism since it looks upon man as if he were made up out of a compartment "body" and a compartment "mind." Cells and organs within the body and the body as a physiological entity, are indeed compartmentalized elements which interrelate causally. But since mind and body cannot be conceived in this manner, they cannot stand in a cause-effect relation.

A look at mechanical movement and purpose-striving movements will place the concept of holism and the mind-body problem on a more concrete plane. Heartbeat, respiration, intestinal peristalsis, sensory-motor reflexes—all are compartmented cause-effect sequences in objective time-space, which we can lift out of and separate from the general background of body movement and posture. But we cannot isolate purpose-striving movements from the mechanical background movements and postures, because purposive movements always go on in conjunction with mechanical movement. When we perform purpose-striving movements, mechanical movements are occurring also. But mechanical body movements are not always accompanied by purposive movement. Hence, in some instances we are confronted by compartmentalized, mechanical movement, and in other situations we can be aware of the synthesis of purposive movement with mechanical movement. When this happens, we are living concretely; we live into a synthesis of the two forms of time and an integration of causality with intentionality. In order to experience this synthesis we have to follow the combined extrospective and introspective routes so that we can live into the holism of our psychophysical existence.

At least two questions confront us: How can we constitute a whole, notwithstanding our complex spatiotemporal structure? And how can a human being, rent apart by contrasts, blemishes, fissures, dichotomies, and fractures, be a holistic entity? I see two ways in which we can answer these questions: one way, on a rational level, another way on an emotional level. We can show the interrelatedness of the two forms of time-space rationally, by deriving objective time from the precious present by a process of abstraction, and we can derive causality from the mental triad by a similar process (see chapter 10).

On an emotional level, holism is forced upon us by life's vicissitudes, as I will show with concrete examples where body overwhelms mind, and others, in which mind overpowers body. In these instances body-mind holism is a source of suffering, since it is only one of the many faults and breaks in our human existence. In this emotional turmoil we become whole human beings.

This solution leads to the harassing question: Have we traded the mind-body split, which distorts our intuitive feelings of being a whole psychophysical human being, for an equally unresolvable conflict between rationality and emotionality? At this point, the mental triad and the precious present promise a solution, because their interrelation grounds the solidarity of rationality and emotionality. Projection into the future-in-the-fringe of a precious present, is an essential constituent of rational pursuits, and this projection elicits emotional tension within us. On the other hand, emotional tension is the motor which drives rationality onward in the search for solutions of problems. Emotional tension is one of the aspects of mentation which makes us into whole human beings. Living as a whole human being is living as a suffering human being; in that cauldron we create the holism of ourselves as psychophysical entities.

The burning question, so far unanswered, is about the nature of the relationship between brain and mentation, because we can say, for certain, that without the brain there is no mentation, and of course no holism. As we ascend from spinal reflexes to the enormous complexity of the brain, neural functions become more and more complex, more flexible, and less predictable. However, the progression from relatively simple to unimaginably complex, remains within the form of objective time-space and within the category of causality. These exceedingly complex part-functions within the nervous system intercalate with each other, and therefore their future course is hard to predict. According to the *pur sang* neurophysiologist, this unpredictability makes us believe that they are not really causally determined; they only look like purpose-striving events.

This amounts to a denial of the reality of purpose-striving movements. Behind this denial stands the exclusion of introspection and its ancillaries as scientific methods of observation. This denial may trick one into inserting intentionality into the mechanistically conceived neural machinery as an afterthought, as a "ghost in the machine," or as the self-conscious mind. One projects mentalistic attributes upon the structure and function of the brain, and one creates not one homunculus, as Wagner did, but a whole family of these creatures. One runs the risk of giving free rein to one's inclination to identify with cells and organs, especially with the brain and with living organisms. I do not pretend in the least that the following elaboration upon homunculi imagery is a description of how neurophysiologists actually think; it is meant only as a *caveat emptor* against uncritical identification and projection.

The dominant hemisphere is an analytically thinking, mathematically gifted homunculus, who uses the forms of objective time-space, who structures his sensory data with the category of causality, who compartmentalizes these data, and who is interested in their quantitative relations; this homunculus remains aloof emotionally. The minor hemisphere is a synthesizing homunculus who works with holistic, gestalt concepts, within the form of subjective time-space, and who structures his experience with the category of intentionality. Operating mostly with the mental triad in a precious present, this homunculus represents the emotional side of our personality. The corpus callosum is a peacemaking homunculus who sees, with Eccles, that the central nervous system operates in battle formations and who urges the rational dominant hemisphere to interpenetrate with the emotional minor hemisphere, and vice versa, attempting to create a whole out of these two contrasting, compartmented entities. The corpus callosum homunculus would stand in the imagery for a force which drives toward the holism of interpenetrating space-time forms, structuring categories, and especially toward the interpenetration of rationality with emotionality.

I use this imagery which sounds satirical and which looks like a Hieronymus Bosch allegorical painting, to show the pitfalls that we dig for ourselves when we overlook that we have outgrown and suppressed only partially our original uncritical identification and empathy with and projection upon living things, which characterized our thinking during the first decade of our life.

Figure 2
From Antithesis to Synthesis

Forms: Objective time-space	Subjective Time-space
Categories: Causality	Intentionality
Methods of observation:Extrospection	Introspection-empathy-Identification-Projection
Approach toward observed data: Compartmentalization	Interpenetration
Attributes: Emotional Neutrality	Emotionally Charged
Synthesis on abstract, rational level:	Derivation of objective time from precious present
Synthesis on concrete, emotional level:	Living through existential tension, suffering in the mind-body holism

Figure 2 is a visual representation of an initial analytical, dualistic approach, which is developed into two attempts at synthesis. The two forms of time-space are at the top of the two columns, supported by the two structuring categories, which stand on the two methods of observation, which rest on compartmentalization and interpenetration, undergirded by mechanism's and mentation's contraposed attributes of emotional neutrality and emotional charge. The foundations of the entire structure are the two methods of synthesis by rational abstraction and by living concretely and emotionally. The dividing line between the two columns does not mean that there is no interaction between the elements, since synthesis does not happen solely at the bottom of the columns; this process is going on all along the interlocking of the several factors which are inherent in the performance and the experiencing of purpose-striving activities. Synthesis is not just an afterthought that we hit upon when purpose-striving is over. While we are involved in physical and mental purposive pursuits, some or all of the contrasting elements of this experience crisscross, interact, and interpenetrate. This is the case when we live into our holism. The time-space forms, the categories and the methods of observation with which we study mind-body interaction, can no more be separated from each other than mind can be separated from body, nor body from mind. Popper exemplifies the view that one can only separate mind from body if one holds to an animistic view of ourselves. According to "...shamanism...the soul of the shaman may leave the body and may go on a long journey; in the case of the Eskimos, even to the moon."[21]

We cannot get away from the dichotomy mind versus body, but as constituents of our holism mind and body are quite unlike the building blocks of some physical structure or of some chemical substance. To take an almost simplistic example, hydrogen combining with oxygen becomes water, with properties very different from its composing atoms. But if we decompose a water molecule into oxygen and hydrogen, these atoms are identical to what they were before they became part of the water molecule. Atoms, combining to form molecules and molecules decomposing into their constituent atoms, are compartmentalized units which interact but which do not interpenetrate.

Not so body and mind. When, conceptually, you split a whole human being into body and mind, body retains some of the characteristics of mind, and mind retains some of the characteristics of body. These mutual residues are evidence that mind and body interpenetrate. But when, again conceptually, you put mind and body together again, you have something more than the dementalized body and something more than the disembodied mind, making up the whole human being. We are aware of this "something more" as a tension and discrepancy between body and mind. This tension is a sign of the interpenetration of mind with body and body with mind. Interpenetration generates emotionality, existential dread, despair, it makes us suffer, since it is contraposed to compartmentalization. The collaboration of these two opposing processes has to be continuously nurtured and furthered toward our psychophysical holism.

The question how mechanical movement and purpose-striving movements can work together simultaneously remains unsolved. If we start out from mechanical movement and superimpose purposefulness by small increments, we are travelling a dead-end street and we shall not arrive at purpose-striving movements. Positing the whole human being as our point of departure is more promising. By gradually stripping these movements of their purposive attributes and by simplifying their spatiotemporal structure toward the spatiotemporal structure of mechanical movement in objective time-space, we can arrive at an understanding of mechanical movement by a process of abstraction. We proceed from Popper's World *3*, to World *2*, to World *1*; that is, we go from more complex to less complex and gain deeper insight into the interrelations between and the characteristics of both kinds of movement. Ultimately, we are looking for a deeper insight into the relation between mentation and the structure and function of the brain.

When we perform purpose-striving movement, we experience the accompanying mechanical movement as if it were not strictly mechanical. We also experience the purposive movement now being performed, as if it were not strictly purpose-striving. The two contrasting kinds of movement seem to lend some of their attributes to each other, which is something vague and ephemeral and does not really belong to either mode of motion. Interpenetration is, indeed, a mysterious process. Descartes might interpret this mysterious interchange between mechanical and purposive movement as the continuous intervention of creative acts by God. A humanistic interpretation of the interweaving of the two kinds of movement holds that it is man himself who creates the mystery of their mutual interpenetration. The mystery consists of purposive movements instilling some of their temporal-spatial structure into mechanical movement, so that the latter is pulled up out of the realm of existing exclusively in objective time-space and being structured only by the causal category. Mechanical movement then becomes a partner of purposive movement in the realms of both objective and subjective time-space, intentionality, and interpenetration. The two kinds of movement transform and transfigure each other so that both, as we actually experience them, are no longer what they were before, when we had decomposed our body movements into the two contrasting entities by rational analysis.

We can ask a simple, unsophisticated question which touches on the differences between mechanical and purposive movement, as well as on the relation between mentation and the structure and function of the nervous system: "Why, when I wish to move my right index finger, does it move?" If we focus on flexion and extension of the right index finger, then the neurophysiological explanation is on firm ground, because we can interpret these movements as compartmentalized, mechanical motions which proceed in objective time-space and which are structured by the category of causality. Therefore, there seems to be no break in the cause-effect sequence when we state that these movements are initiated by causally structured processes going on in the nervous system. But as soon as we include the intention "I wish" in the question, the psychophysical problem emerges. Moreover, it is very seldom that we flex and extend our right index finger, unless there is some reason to do so. In daily life we perform purposeful movements especially with our hands. But flexion and extension of one of the fingers play usually only a minor part within the global patterns of movements of the other nine fingers, the wrists, elbow joints, shoulder joints, and general body posture, which support the miniature peripheral movements. In daily life we do not compartmentalize our body motions. Unlike flexion and extension of one finger in isolation, purpose-striving movements are not sectioned off in stretches of objective time-space. If compartmentalization of movement does occur in isolation, it excludes interpenetration.

However, the neurophysiologist, since he is a natural scientist, compartmentalizes his subject matter. He studies the brain as if it were isolated from the rest of the body, as a physiologist might study the skeletomuscular apparatus as a compartmentalized organ system. Both study their special field of inquiry with the form of objective time-space, causality, and compartmentalization. Both follow the extrospective method of observation. The physiologist can say without any second thoughts bothering him, that muscles which cause movement of the skeleton cause mechanical motion of the limb. The mechanistic description and explanation of the bone-muscle apparatus satisfies us, but the mechanistic treatment of brain structure and function leaves a major question unanswered. The neurophysiologist asks how brain relates to mentation. Because he must remain within the category of causality, he cannot come up with an answer. He cannot say that causally structured brain process causes mentation, which is structured by intentionality. Does the neurophysiologist have to submit to Wittgenstein's admonition that we have to pass by some mysteries in silence because we cannot talk about them? Does he have to give up on his search for an answer to the question how brain relates to mentation? The answer is in the affirmative if one uses only the form of objective time-space, the category of causality, if one sees brain as compartmented and if one follows the method of observation of extrospection with the exclusion of introspection.

To get out of this impasse we need a wider perspective of methods of observation, of time-space forms and structuring categories. We have to go beyond the strictures of compartmentalization. Over and above these extensions of method, we have to include introspection and its ancillaries as routes of observation.

We reach this impasse as the result of compartmentalizing our subject matter, which leads to the mind versus body split. But this is not the only tool with which we can split mind from body. Another avenue to the mind-body split is an analysis of introspective experience which requires that we perceive ourselves and others as whole human beings, so that we can go beyond causality and compartmentalization. This method of splitting mind from body involves an intuitive, personal approach.

Looking at one's hand, one may wonder, "Look at this hand, it is really mine!" Or, looking at one's purposive movements of hand and fingers, one might ask: "Did I make that movement?" Flexion and extension of a finger can be compartmentalized, but the awareness that my hand is mine, that I am the author of this movement, introduces purpose-striving, "mineness," motivation, into the compartmented mechanical flexion and extension of the finger. We have to go beyond the one-sided approach of neurophysiology. We ought to include attributes of interpenetration, the form of subjective time-space, and the category of intentionality in a study of purpose-striving activities by introspection.

The person who suddenly realizes, "This is my hand," "It is I who made this movement," is flooded with wonderment. This makes manifest his going beyond causality and compartmentalization. He inserts a different universe of interpenetration and intentionality into a compartmented universe which is causally structured. To enter into the combined, causally and intentionally structured world of mind-body holism, we have to follow the extraspective method, integrated with the introspective method of observation. This complex methodology and conceptualization immerses us in body-mind holism and also makes us aware of the mind-body split. The person who asks, quite unscientifically, "Is this my hand?" lifts his hand momentarily out of the holism of his body and mind, but he quickly puts his hand back into the holistic ambience where it belongs. He has had a dim awareness of the mysterious integration of a compartmented part of his body, observed by extrospection, with the interpenetrating holism of the hand with the rest of his body. This holism he observes by introspection. The splitting of mind from body and the reuniting of the two constituents, which we observe occasionally by extrospection combined with introspection, is a process which differs essentially from the conventional splitting of mind from body by compartmentalization.

We can progress toward a Janusian insight into the holism as well as the antagonism between mind and body if we anchor this insight upon everyday experiences. Then we shall become aware of what Engelhardt calls the "identity in difference" between the categories body and mind. The polarity of the two constituents of our existence produces existential tension of varying degrees, from apprehension, to anxiety, to dread. In these emotion-laden experiences, when body overpowers mind, at other times when mind directs body, and then again, when mind and body cooperate harmoniously, we live into our mind-body holism as we overcome the dichotomy of mind versus body. During this healing procedure, we shift from a predominantly introspective approach, when mentation is in the foreground but without ever losing sight of physiological functions, to relying chiefly on extrospection when body structure and function are in the limelight and psychological attributes play a subordinate role. Shifting the mind's eye alternately between the two aspects of our being in the world, we see that they are not separated by an unbridgeable gulf. We can perceive ourselves living, mostly, in subjective time-space or, when we concentrate extrospectively, as living, mostly, in objective time-space. Since we do not live either in the first, or in the second form of time-space, but in a varying combination of the two forms, we are able to grasp the holism of body and mind through a collaborating, teeter-tottering application of both methods of observation. This approach shows us the "identity in difference" of body and mentation.

An example will show that these two methods of observation are not mutually exclusive, since the inward- and the outward-turned approaches are going on in tandem. During an angry outburst against someone, we can observe simultaneously, extrospectively, that our face flushes, our heart races, that muscles tense and tremble. At the same time, we can introspectively feel the depth of our fury, and by empathy and identification we notice the reaction which our outpourings produce in the other person. This complex insight into ourselves and into the others opens a window toward a fresh view of the mind-body relationship. Let us compare four concrete life situations in which the mind-body relationship comes into increasingly sharper focus.

1. You have to go from your place of work to deliver something to a customer. You map out in space the route you will take, and you calculate mentally how much clock time you will need for the trip. The implementation of this plan seems to go on automatically without your thinking about it. During the trip you may be talking to your companion, or if you are alone, thoughts may be running through your head which have nothing to do with your simple mission. It is as if your body "knows" where you have previously directed it to go. But don't be mistaken, you are keeping a watchful inner eye on the body's performance, while other mental and physical activities are going on also.

   The mind-body relationship is revealed in the difference between planning for and execution of the trip. During the planning stage, mentation is in the foreground and you live, mostly, in subjective time-space. During the execution, bodily activities predominate and you live, mostly, in objective time-space. In this routine situation of daily life the mind-body relationship is not: either you live mentally, or you live physically; it is a balancing shift between bodily and mental existence.

2. Many joggers describe a fascinating change in their perceptions of their body movements after they have jogged, maybe, one mile. Then suddenly, the legs go as if by themselves, rapid heart beat and fast breathing do not bother them any longer, and they feel a strange lightness of body and a lifting of spirits. I interpret this series of experiences as follows. First, the mind imposes its goals on the recalcitrant body, which gradually becomes attuned to the task. In part, this is the physiological, homeostatic integration of heart rate, blood flow, rhythm of respiration and oxygen supply, and the dilatation of arterioles in the leg muscles. But the unusual feelings of being transposed onto a different level of existence, which seem similar to feelings of levitation, mean that the jogger suddenly experiences the holism of mind and body beyond the level of physiological integration of body functions.

   Musicians describe similar experiences. Instrumentalists spend a good deal of time each day on exercises which may have very little direct relation to the ultimate goal of music making. Fingers, wrists, shoulders, the entire torso, have to be cajoled and forced to work together in a complex coordination of neurophysiological mechanisms. Suddenly the musician feels and hears that the movements which he has mentally imposed on his body have produced the auditory effect that he has intended. He feels a body lightness and an emotional lift, since he experiences the holism of mind and body in a here-and-now precious present. These feelings are similar to what joggers describe, with this difference—that the goal of joggers is to jog, but the goal of the musician is to communicate his own musical experiences to an audience. This communicative experience raises him far above the private, personal satisfaction of mind-and-body holism which the jogger experiences. When the mystery of the fusion of mind and body happens, when we become an interpenetrating entity, we experience the emotional uplift of omnipotence, happiness, and gratification.

3. In illness, body-and-mind interaction can lead to opposite outcomes since some persons can and some people can't overcome body dominance by heroic effort. The cellist Jacqueline Després is a victim of multiple sclerosis; no longer able to play her instrument, her career is a shambles. She teaches cello seated in a chair or in a wheelchair. She welcomes and enjoys contacts with friends, and her marriage has grown in a different direction. Since she and her pianist husband no longer concertize together, their personal lives are more intertwined, because music making does not consume all of their time together. She has not given in and she has not given up.

4. One of our friends has told me how he has battled with osteoarthritis of both hipjoints for the past twenty years. In his twenties, he went to an amusement park. Of the twelve baseballs thrown by a machine, he hit only two. He thought no more about this disappointing experience until some six months later, when his physician told him after a routine physical examination, including x-rays of the hip joints, that he had beginning osteoarthritis. The doctor said that he might be a wheelchair case within the next fifteen years, but he said to himself: "Goddamn it, a wheelchair is not for me!" Sometime thereafter he tried to lift his left leg over the two-foot-high accordion gate which they had installed to keep his crawling baby safe. He could not get his leg over the gate. Then other symptoms became troublesome. He often woke up during the night with pain and he had to prop his left side up with pillows. Getting in and out of the car was difficult. He learned to put his left sock on from behind by bending his leg in the knee joint. He said: "You have to invent something new all the time." He went to the hospital for special surgery in New York where they replaced the head of the left femur with a plastic cup. They predicted a stay in the hospital of three months, bedrest at home for three months, and a possible return to work after a year. Out of the hospital in about two months, he was back at

work in four months postoperatively, on crutches. He said: "I was in control of it, rather than it controlling me." Thereafter, his right hip joint became less functional and more painful. One year after the first operation another surgeon replaced the head of the right femur and rebuilt the acetabulum with a new metal alloy. That time, he was out of hospital in three weeks and back to work in five weeks, postoperatively, and with minimal pain. At present he has to be careful how he gets in and out of a chair and how he sits. He is most comfortable in a reclining chair which takes the weight off his hip joints.

He concluded, "You have the image of a patient with osteo in a wheelchair. That fear is more debilitating than the osteo." This man's heroic attitude has enabled him, with surgical help and with follow-up physiotherapy, to keep the devastating effects of his illness from ruining his life-style. His determination, imagination, and foresight have prevented the invasion of body mechanisms into mentation from forcing him to live on the level of objective time-space. The triumph of mind over body has kept the holism of his body and mind intact.

From the first to the fourth example, the conquest of mind over body becomes more and more visible and dramatic. In the first situation, one is hardly aware of any particular tension between mentation and physiology, except if one directs his attention forcefully, or out of curiosity, or if one is interested in the mind-body problem per se. In the next situation the discrepancy between what the jogger maps out for himself and what he can accomplish and between the ideal of music making which the musician has in mind and what he actually attains, can lead to contrasting outcomes. If they fail, they may feel harassed, disappointed, or outright depressed or infuriated with their failure. The other possible extreme outcome may be that they feel gratified, comfortable, exhilarated, or even omnipotent. In the third example, we see the heroic effort of a human being who collects the pieces of her shattered psychophysical holism together, creating a new holistic entity with a quite different temporospatial perspective compared to the way she lived in space-time before she became ill. The fourth situation, is a most elevating instance of man's foresight, freedom of choice, and conquest over the recalcitrant body. Anyone who is capable of empathizing with this person, will feel an upsurge of uplifting emotions as one admires his heroic attitude toward his bodily affliction. From example one through four, the temporal-spatial panorama widens and in parallel, emotional undercurrents deepen and become more powerful and meaningful.

Jurrit Bergsma, a psychologist who does liaison work in hospitals between patients and medical staff writes:

The medical specialist is more and more confronted with an inclusive approach, in which the active participation of the patient as a totality comes to the fore.... Medical culture teaches that suffering is unnecessary, because pain can be abolished by technical means.... One has to find a new coherence between objective time and perspective on the future.... On the one hand he grows and ripens while being ill, on the other hand, he is a victim of technology, in which there is hardly any possibility of a human existence and where death might seem more humane than living.[22]

Ideally, holistic medicine, which is the vogue today, should help the patient defend his spatiotemporal structure, which requires the physician and nursing personnel to think in technical terms of the patient as a unity of structure and function in objective time space. They ought also to approach him as a total human being who also lives in the world in subjective time-space.

Our admiration for the heroism of these sick people should not cause us to overlook a problem of natural science: how is the mind-body relationship related to the law of the conservation of energy? Is the amount of energy put out by the physical activities of our bodies (compared to the energy put into our bodies) increased even a little due to our mental activities? If this is the case, no matter how small the difference, theories about mind-body interaction are in trouble. For the natural scientist, the linkage of cause and effect permits no interference from noncausally structured entities. Cause equals effect. Students of the mind-body problem feel as if imprisoned within the deterministic world of natural science. Mind influencing body would go against the symmetry of energy put into the organism and its expenditure of energy.

McDougall brings this problem up several times. He reproduces an elegant diagram which pictorializes the energy exchange between mind and body without coming into conflict with the law of the conservation of energy. In McDougall's model of mind and body interaction, the mind is the cause of a change in the body's motion. He overlooks, however, that any change in a body's state of motion requires expenditure of energy. Hence, McDougall's argument is not convincing.[23]

This is one out of many attempts to solve the problem of energy transfer within the organism. Let us take another look at the billiard game, which one can choose to observe by extrospection exclusively. From this viewpoint, the various phases of energy transfer that are going on on the table, between the impact of the cue on the cue ball, between the player's muscles and the cue, and between his muscles and the neural events that cause these muscles to contract, are all in accord with the law of conservation of energy. But the extrospective method of observation yields a one-sided and restricted view of the real billiard game, because man is being degraded to the status of a mere object in objective time-space, causally structured. It does not grasp man's psychophysical holism as a subject who exists in the two forms of time-space, the two structuring categories, and who is both compartmentalized as well as interpenetrating. However, if one observes the game and the player not only extrospectively, but also by introspection and its ancillaires, then one can conceive of mentation as going beyond causality and compartmentalization, without denying that causality is a costructuring category within man as a holistic entity. Spatially considered, man is only one aspect of the total universe, but this tiny little speck demands that he be observed with a markedly different approach compared to the extrospective method of observation of the natural scientist, who studies the immense world of objects. A one-sided extrospective approach does not collide with the law of conservation of energy but it does fail to grasp man's holism. A one-sided inward-turned attitude does not get the body in its field of vision, and it overlooks the questions about the transfer of energy within the organism. The student of man sees man's body as a compartmentalized unity of structure and function as if the body were an object; adding the introspective route, he gets man into view as a subject. This progression does not do away with causality and determinism, but it goes beyond these structuring categories. Intentionality lifts the body beyond the status of being a mere object. To speak in Engelhardt's terms: mentation enriches the poorer categories of causality and determinism so that they become the richer categories of intentionality and indeterminism.

In terms of the two forms of the billiard game, one can say that the robot game occurs in objective time-space and that it is structured causally, while the player adds the dimensions of subjective time-space and intentionality. This addition does not interfere with or change the transfer of energy between brain, muscles, and cue, but only changes the *form* of movement from mechanical to purpose-striving, so that there is no question of energy being added or subtracted.

We experience the superposition of intentionality upon causality when we perform purpose-striving movements. I conceptualize that this superposition and enrichment comes about due to the activities of the mental triad in a precious present. Does the mental triad contradict the law of conservation of energy? Does it stand before the high court of reason, accused of performing the miracle of escaping from this law? Granted, it may appear to some as if a miracle is going on when we perform purposive movements, but they do not set aside the causal structuring of the mechanical movement which we transform. It is not my intention to present the mental triad as if it were a miracle worker. Mystery, yes; miracle, no. Miracles deny the causal category, mysteries go beyond the causal category but leave it intact.

Then there is the question: Isn't the mental triad another specimen of a homunculus, do not its part functions assume the guise of the total human being? The mental triad is only a one-sided, time-oriented description of mentation, used in order to stress the importance of our becoming aware of our complex spatiotemporal structure. The mental triad stands for interpenetrating experiences and activities, which it helps to structure. This structuring activity changes the form of our bodily movements and does not touch upon the question whether there does or does not occur an energy exchange between mentation and brain.

Leaving the area of theorizing, an analysis of interpersonal communication will bring us to the more solid ground of concrete interaction between two or more whole, psychophysical human beings, whose holism we can leave intact.

INTERPERSONAL COMMUNICATION

Interpersonal communication sets us free from living in isolation in an island existence. It performs this liberating function by a complex process in which the two forms of time-space, the two structuring categories, and the two methods of observation cooperate. The ancillaries of introspection, especially empathy, give us knowledge of and insight into the inner mental states of others. These methods of observation save us from getting entrapped in solipsism, which holds to the conviction that we cannot know what goes on in the minds of our fellow human beings.

Communication proceeds on many levels. The most unstructured and primitive form is fusion-symbiosis between mother and infant. The most concrete and precious kind of communication with which some of us are blessed, is fusion-communication between woman and man. Communication between humans in general is less concrete, more limited, and it is laden with feeling tones of lesser intensity, though two or more communicating people often do partake of a precious present. Communication between man and animals is primitive and limited and askew toward man, since animals do not possess language capacity, so that body language is in the foreground. Identification and empathy must be more and more curtailed, the lower one descends in the evolutionary scale. It is a misnomer to speak of communication between man and insects, since we should not identify with these creatures. These restrictions apply even more to the study of the spatially and causally structured functions of total living organisms, their organs, cells, and the cells' organelles. Biology in the restricted sense should rely solely on extrospection.

We see a gradual progression to increasingly limited modalities of communication which become more and more impoverished. Communication between man and mammals is mostly a sending and receiving of stimuli between two or more bodies, as empathy and identification fade out of the picture. We progress from more concrete to less concrete experiences until we finally reach the abstract world of natural science. It is as if one spirals out from a concrete center toward points farther and farther away from this center onto an abstract level of experience and conceptualization. See figure 3 as an illustration of this theory.

Mother and infant fusion-symbiosis is an instance of primitive communication built on the quasi-fusion of two bodies. This contact starts in utero during approximately the eighth or ninth month of pregnancy, which is enriched during the postnatal care period by mentalistic elements when the baby is still totally dependent upon its mother. The stage of fusion-symbiosis does not open up to the extrospective-introspective approach. The very word *symbiosis* implies that both the infant and the mother partake of each other on an undifferentiated level which is obtuse to the intrusion by an observer. Nevertheless, if a mother is so inclined, she can mentally take a step aside and outside the symbiotic relationship with her infant and she can turn herself into an extrospective-introspective observer of herself and her baby. But she will come up with one-sided and meager results because during the early postnatal period she communicates with her infant mostly on a bodily, physical level. The mother can register her own feelings toward her offspring, but since she must restrain the strong inclination to identify with her baby, she also keeps empathy in narrow bounds. Therefore, we are impeded in developing insight into the fusion-symbiosis relationship. The mother's *tour de force* of examining this relationship does not necessarily prevent her from seeing herself as a whole human being who looks forward into the distant objective future and hopes and trusts that she will see this tiny bit of humanity develop into a mature human being who then will be capable of interpersonal communication with her or his mother.

Figure 3 Communication from Concrete to Abstract

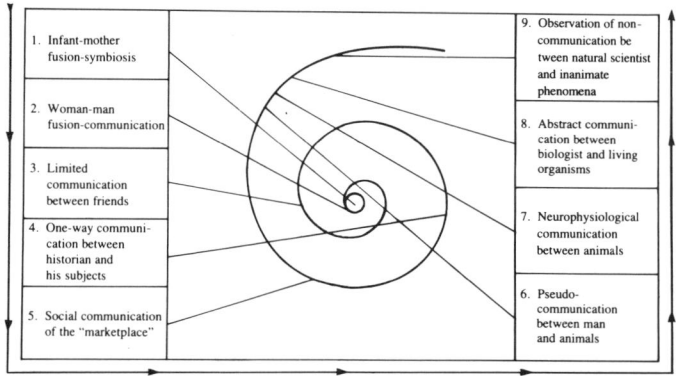

Hindu religion deals extensively with the prenatal phase of our existence. "The Hindu medical tradition, known as Ayurveda,...[states that] by the fourth month, with the formation of the heart, the fetus is considered endowed with consciousness and able to express its desire for things of taste, smell, and so forth, through the longings of the mother....The three strands of ancient ritual, myth, and medicine revolve around a common theme of a highly sentient fetus, aware, reactive, and quite susceptible to extra-uterine influences."[24]

One might interpret this complicated mythology as an attempt to grasp the mysterious fusion-symbiosis which is an obvious fact during the last month or two of pregnancy.

Soon after birth a primitive kind of communication develops between mother and newborn infant which goes beyond mere fusion, although it is still on the preverbal level. The infant wants to be touched, held, rocked and it also becomes interested in things around itself and explores them with hands, feet, and mouth. The first sign of an attempt to escape from the mostly physical phase of communication occurs when the infant starts to smile at its caretakers. Vocalization remains for quite some time on the level of expressing either body discomfort or physical contentment, especially after feeding. But the preponderance of these physical, bodily factors must not cause us to overlook that the young infant's behavior shows evidence of striving into the immediate future, which means that intimations of mentation are already going on in the infant's head.

All humans go through this primary phase of fusion-symbiosis with the mother, which is the obligatory matrix out of which later on the mature modes of communication develop. René Spitz has shown in his studies of babies in a foundling home, who received insufficient mothering care, how important this initial phase is for later physical and mental development. These babies were sickly, they failed to gain normal amounts of weight, they overreacted to infectious illnesses, and they showed a high mortality rate. "A modification in...the ongoing relations with the need-gratifying object...[has] important consequences for future development and maturation....pathogenic environments when extreme, also exert an influence leading to uniformity in the developmental profile over the time in which the pathogenic influence is effective...which results in a uniform deficiency."[25]

Suomi and Harlow have shown with ingenious experiments with infant monkeys who were artificially deprived of normal peer relations, that they not only developed below normal levels in physical respects, but that they lacked various normal behaviors also. Separated and depressed females were unable to submit to copulation, and, if they happened to become impregnated, so to say by accident, they became very poor mothers.

Suomi first deprived infant monkeys from peers, and then, at about age three months, he put them together with slightly older peers. "Peer separations produce clear-cut depressive behaviors.... The present data indicate that monkeys exhibiting behaviors following repeated peer-separations can be returned to age-appropriate social performance through repeated exposure to socially active age-mates."[26]

Is the mother-infant fusion-symbiosis a preview of things to come in later life, when the fulfilling fusion-communication between a mature couple is accomplished? There are vast differences, of course, but a similar mysterious spatio-temporal structure of a shadowy, vaguely demarcated "prototemporal Umwelt" (J. T. Fraser) pervades both relationships.

Now, let us take a big leap forward in time, of two, three decades. Fusion-communication between woman and man is an instance of mind and body holism in its purest, most delicate, and most overwhelming mode. Some women and some men experience this mode of holism during their sexual union, when the physical boundaries of their two bodies fuse, when the man and the woman simultaneously feel sexual sensations in their own body, and when they are also aware by empathy and identification of the sexual feelings of their partner. Concomitantly with physical and mental interpenetration of two bodies and two minds, ego boundaries are lowered. In this burning emotional communication two whole human beings fuse together as if they are at a loss to decide: "Which am I and which is the other?"

While the couple are in the grips of the physical and emotional turmoil of the orgasm, it is impossible for them to register extrospectively and introspectively what is going on within themselves, within the other, and between them. After the union has been consummated, the objectifying attitude may be liberated if one or both partners undertake the rational extrospective and the intuitive introspective, retrospective contemplation of their memories of their holistic experience. Then they will realize that fusion-communication is not just a neurophysiological storm, nor is it merely an interaction between two bodies in objective time-space. Orgasm, if it is a constituent element in a lasting woman and man relationship, based on trust and love, is not the synchronization of body functions in objective time-space. Fusion-communication concretizes and realizes the interpenetration of the two lived-in worlds of man and woman in objective time-space as well as in a precious present in subjective time-space. Their mental triads create the holism of two mind-bodies, fusing with each other and mutually communicating in a precious present.

A real-ideal man and woman relationship does not make the couple immune from the miseries of suffering. Opting for high ideals, they will find that their relationship can be marred from time to time by dichotomies and rent by fractures, which hurt all the more since each sets high goals and expectations for himself and for the other.

In contrast to the seekers of an ideal relationship who are able to hold on to the great expectations with which they entered into this bond, many couples settle for a routine, cooperative relationship in which they find very little of the fusion-communication of a real-ideal bonding. They may be held together by material ties, or legalistic considerations, or by religious convictions, by convention, or by children. With it all, the isolation and the loneliness of the two partners is not resolved. The dichotomies in such a marriage are displayed in mutual resentments and quarrels, disappointment in the other, alternating domination and submission. They may go their separate ways relying on the male versus female division of work areas, or each finding his or her own ways of diversion and friendships. They suffer silently and often hide their conflicts from others, because they find the ideal missing, which they thought they had realized in the early, fervent years of their union, when sexuality was a powerful force binding them together. They are as isolated and lonely now as before the marriage. Some couples resign themselves to the stark reality that they must accept a less elevating existence than they had hoped to find, in what they trusted was going to be a real-ideal marriage, and they may find a common ground in communicating on a friendship level. But the sexual domain may become a more and more body-dominated experience in which mind and body holism falls by the wayside. If mutual disappointment and resentment blaze forth into hatred, then the marriage becomes a living hell in which each makes the other suffer. In modern times, most people sense that this modality of suffering is destructive, that it tears them both down and prevents their growth, so that they may decide to break up the marriage.

Friendships are based mostly on shared common interests and geographical closeness. They reverberate on a less elevated level compared to a real-ideal relationship between woman and man. They do not demand total involvement with the other person, and they are not as all-encompassing as fusion-communication. Most friends talk about what they are doing now in the present, what they plan to do in the near future, and they may share past, common experiences. Men usually communicate on a fairly concrete, physically oriented level, but occasionally one hears that men discuss theological or philosophical problems, more rarely, intimate, personal details of their family life, and even more seldom problems which harass them personally.

Discussions between women in fairly large groups are not very different in character, since they converse about children, housework, clothing, and similar concrete subjects. But there is a subtle difference between interpersonal communication between two or three women and between two or three men. Few men seem to feel the need for, or have found the special person who fulfills the role of a bosom friend, with whom they maintain contact over long periods of time. Many, or even most women, do find and cherish long-lasting and meaningful friendships when they discuss intimate personal concerns and they support each other in such a relationship.

This particular difference between the sexes may well be due to women's role in the bearing and rearing of children. The fusion-symbiosis with an unborn fetus and with her infant during the early weeks of total dependence, is a world that is interdicted to men. Men can only get a secondhand glimpse, as it were, or learn by proxy what his pregnant wife and the caring, nursing mother is feeling. Mothers, empathizing and identifying with their children, extend these intuitive methods of communication to other human beings. Men can learn these subtle attitudes by conscious, rational application, but women know these skills intuitively. This may be one of the reasons why men are often satisfied with concrete, emotionally rather distant relations with other men, while women have a hunger for close and warm friendship with one or two bosom friends.

Making music together can be an experience of friends partaking of a precious present over the detour of entering together into the world of the composer. This highly emotionally charged experience can bring people closer together than the pedestrian type of communication by means of average, daily, or other physical activities. Actors can communicate with their fellow actors and actresses in a precious present through the vehicle of the quasi-real characters whom they portray, who are communicating here-and-now on stage.

We see an even more abstract modality of interpersonal communication between the historian and his human subject matter than the rapport between musicians and actors. The historian enters into a unidirectional, abstract contact with these past lives, in which there is no mind-body involvement. However, the psychological factors of actual communication are very much alive in the historian's mind. He identifies and empathizes with historical personages, and he aims to make them come alive as if they were still whole, psychophysical human beings. While he communicates abstractly and mostly intellectually, he is bound to become emotionally involved in his scientific work, because he deals with, for him, very real rational as well as emotional subjects. He has to be a person with a lively fantasy who "hears" his subjects speak to him in their letters and other writings by "intra-audition." In a minor way, we are all historians of a sort when we read a novel and when we get involved with the adventures, successes, the emotions, and the sufferings of the characters, by identification, empathy, and projection. They become "really alive" for us. The difference between the writer of a novel and the historian is that the former wants to impress upon us mostly the emotionality of his fictitious characters, while the historian struggles to create an image of a whole, psychophysical historical person as a rational emotional person who stands in communication with his social and physical environment.

Turning to the field of medicine and surgery, one can learn a good deal about the mind-body relationship from the work of an anesthetist. His activities are characterized by a shifting back and forth between the introspective approach and interpersonal communication with the patient as a whole psychophysical subject, and on the other hand, his viewing the patient only extrospectively as a functioning body. This alternation between the combined extrospective-introspective and the primarily extrospective observation and finally again the combined approach, correlates with the presence, versus the temporary absence, and points toward the reappearance of a total, mind-and-body whole person.

The anesthetist traverses three phases in his relationship with the patient. First, he meets the patient in the evening before the day of the operation. He tells him what to expect, what he is going to do in the morning, and he gets the trust and the cooperation of the patient, putting him at ease. He identifies with the patient since he can retrospectively be aware of memories of similar feelings of apprehension from his own past, apprehensions that trouble the patient right now. This is the first instance of his communicating with the patient, seen as a whole human being. During the second phase of his work, when he has anesthetized the patient, he checks vital signs, he applies various chemical, electrical, mechanical support measures. During this phase he observes the patient almost exclusively by extrospection, and he sees him as if he manipulates a functioning organism in objective time-space. For the time being, interpersonal communication is suspended since he is obliged for the sake of efficient execution of his duties as an anesthetist, to suppress or even to eliminate any tendency to identify and to empathize with the patient. He has been trained to follow the extrospective route of observation exclusively during this technical phase of his work. During the third phase, when the patient is fully awake again, the anesthetist once more relates to and communicates with the patient as a whole psychophysical being, who exists in objective as well as in subjective time-space. Introspection and its ancillaries are reintroduced into the interaction between doctor and patient.

During phase two, the anesthetist grasps the patient extrospectively as a causally structured unit of compartmentalized part functions, during which phase the patient is no longer a purpose-striving human being. During the first and the third phases, he grasps the holism of the patient's mind and body and he communicates with him in the modes of two complete human beings by introspection, empathy, identification, and projection. This extreme example of, first, the interrelation and cooperation between extrospection with introspection and its ancillaries, next, splitting off introspection so as to be free to follow the extrospective approach, and finally, the reinstatement and interweaving of introspection with extrospection, alerts us to the fact that we veer back and forth between the inward- and the outward-turned attitudes in daily living also, but with very little awareness of these complex and varying synthesizing processes.

Interpersonal communication in our daily life is largely of a physical ilk. For instance, in a supermarket people go about their business and pay very little attention to each other, and the checkout girls are almost noncommunicative. Their attention is locked in on the items that come toward them on the moving platform, the price tags on the items, and which keys they must push on the cash register. They work like robots and seldom talk to the customers. But when they see a friend they shift immediately into a totally different role, and one can hear a lively exchange going on about common past exploits or future plans. After this brief showing of interpersonal communication, they revert back to the robot role, which they play because they deal with objects in objective time-space. Relations with the customers are devoid of empathy and identification, but communication with the friend abruptly rises to the level of objective and subjective time-space, and there is a brief, though superficial, interlude of interrelation between two whole, psychophysical human beings.

Communication below the level of man becomes more incomplete, impoverished, and abstract the lower we descend on the scale of evolution. During this descent we must be ever on the alert, introspectively, to make sure that we restrict identification and empathy with living organisms in keeping with their spatiotemporal structure. Scientists who work in close contact with chimpanzees and gorillas, for instance those who teach them sign language, are at risk of uncritically identifying too extensively with these higher primates, who display so many attributes which are similar to our own.

Interactions between chimpanzees can be strikingly similar to interpersonal communication between humans. The van Lawicks have filmed the entry of a female chimpanzee with her baby into an existing group. She remains seated some distance from the group, then comes nearer, stretches out her hand toward a male chimpanzee, who slowly stretches out his hand, until finally their index fingers touch. Now she has become a member of the group. Any viewer of this scene will be emotionally touched by the near-human behavior of these animals. It is this emotional involvement which is the very reason why we naturally identify and empathize with them. Yet, for objectivity's sake, one should critically examine the question how far one should go in terms of reading interpersonal communication into the animal behavior.

Jane van Lawick-Goodall obviously identifies and empathizes with her "Friends, the Wild Chimpanzees": she cried when she heard Melissa's baby with the broken arm shriek out all the louder, when his mother tried to comfort him by offering him her breast; she shared the pain with the baby and frustration with the mother. All of us who live in close contact with our household pets will sympathize with Jane's reaction, because we experience similar feelings when they get hurt. We identify and empathize with them, hold brief conversations with them, and we bury them when they die.

One of my patients was seriously depressed after her dog died. The dog was extremely attached to her and she was very fond of him. They buried him in their backyard and every time she looked out of her kitchen window the image of her beloved pet, lying there cold and abandoned in his grave, haunted her. The only solace for her grief was for them to move to another house.

As recounted in an earlier chapter, Jane van Lawick-Goodall was very excited when she observed that chimpanzees, sitting on a termite colony, broke off a twig, stripped off the leaves, and used it to fish for termites, which they like to eat. She says that this is the first time that man has observed that these animals fashion a tool, and that therefore we must review our own status as the only *homo faber*. This far-reaching conclusion is in my opinion the result of too close an identification with the chimpanzees.

Let us go down a few steps on the evolutionary ladder and watch what happens when two strange dogs meet. They approach each other slowly with short, stiff little steps, and, if one is close enough, one can see their muscles trembling; piloerection of the neck hair is menacing, no tails are wagging. Then they touch snouts and begin to sniff each other, going around and around, neck hairs lie flat by now, tails begin to wag and playing starts. As we observe this behavior extrospectively, we interpret it as the result of the sudden flooding of the system by excess adrenaline, but we go on to identify with the dogs and we believe that the extrospectively observed behavior is the manifestation of anger and fear which the two dogs are experiencing. But how can we know if the dogs really do feel anger and fear? Is this display no more than a brief neurophysiological storm, which we wrongly interpret as evidence that the animals actually suffer these internal states? Are we justified in going that far in identifying and empathizing with the dogs? If your own dog is involved in this encounter, you will feel afraid for him and you may feel anger against the other dog, and hence it is difficult to keep empathy and identification with your own dog in proper bounds. Then, it is very likely that you will assume a similar attitude toward other animals by extension.

Other observers may feel that it is safer to limit the interpretation of the dogs' behaviors to the first-mentioned physiological explanation, rather than to be convinced on insufficient grounds that the dogs experience fear and anger. These observers would say that each dog sends sensory stimuli of vision, audition, and especially of smell to the other dog which receives them and that this whole process goes on in objective time-space and that it is causally structured. They would say that one should limit oneself to a purely extrospective method of observation and that it is unscientific to assume that mentation, emotionality, and intentionality might play a role in this interplay, which one can explain sufficiently on a neurophysiological, chemical basis. It is almost a matter of taste whether one should include identification and empathy in the extrospective observation of the encounter of the two dogs, or whether one should suppress this style of observation. Therefore, it is difficult to prove or to disprove whether or not the dogs are mechanistic, machinelike creatures, or whether one observes mental activities in these animals. The spatiotemporal structure of their behavior gives a hint toward a solution. Since dogs, like many other animal species, look into the immediate future and since they obviously have a rudimentary memory function, they do bring past and future together in the present, in an uncomplicated and limited fashion. Therefore, we are justified in attributing some modicum of mentation to most animals.

Most scientists will probably agree that one must curtail identification and empathy even more stringently when one observes birds, compared with those who observe subhuman primates and higher mammals. But in ornithology as in the case of the two dogs, there is no cut-and-dried solution to the question as to how far one should go in curtailing or eliminating these intuitive attitudes. Konrad Lorenz says that one of his geese who lost her mate went looking for him for ever-greater distances, emitting a wailful cry, and that she was depressed. One of his ornithologist friends who saw this goose and knew nothing of her history, said, "That goose must have gone through a lot!" Should we accuse these dedicated and astute scientists of carelessly identifying and empathizing with this bird and should we criticize them for projecting human attributes to her? It would be presumptuous for a nonornithologist to assume this stance.

The question behind these unsettled problems, whether a purely extrospective attitude is indicated or whether the combined extrospective-introspective method of observation should be followed, is this: do we have evidence of mind-body interaction in the case of the behavior of animals and of birds? Though it is frustrating, we may have to leave this question as unanswerable on some levels of evolution.

When we contemplate the behavior of insects, the question whether we see mind-body interaction in this realm seems to find an answer in the concept of instinct. We think of the complex nest-building and social organizations of some species as astonishing phenomena on the intellectual level. I take it that instinctual behavior is thought of as causally, rigidly determined behavior. This definition fits the vast majority of insect behaviors, but there are many significant exceptions when an insect seems to break out of such rigid determinism. And if so, does this prove that the insect's "mentation" goes beyond causality? The bewilderment that plagues an entomologist who studies the dance of the honeybee seems evident in a brief report by M. L. Brines.[27]

His description of the dance carries both causal and teleological elements without the author settling for either type of viewpoint. If the newly discovered features in the dance point to the bee evading ambiguity in its message, then we have to ascribe a high degree of intelligence to it. But if we explain this behavior as flowing from the structure and function of its nervous system, then we remain within the causal domain. Some of Brines's language gives one the impression that he projects a level of intelligence onto the honeybee which is almost similar to human intelligent behavior, especially when he speaks of their possessing a dance "language." However, this identification with the insect and excessive projection upon it of our own attributes, is negated or at least kept in check by his statement that "neural wiring" may be responsible for their seemingly purposive behavior. Time and again, we see in the theorizing of biologists that causality and intentionality are not sharply differentiated. Nor does one have hard-and-fast rules on how far one should go in identifying with certain animals and with insects. This bewilderment should not harass biologists in the restricted sense, who study anatomical preparations, cell structure, chemical organ functions in a compartmentalized field. They are restricted to follow the purely extrospective approach, and therefore they should not identify and empathize with their material, since the mind-body problem does not concern them. This is obviously true in the field of natural science. It is of interest to mention this fact only because it has taken a long time of historical development to establish the supreme role of causality and extrospection in natural science, without competition from empathy, identification, and admixtures of intentionality. It has taken natural scientists many centuries to develop from concrete, animistic thinking to abstract, causal thinking. It is as if the modern scientist has to repeat this progression in his own intellectual development from living originally in his animistic childhood world, to ensconce himself in his mature, abstract world of natural science, where no communication takes place between himself and his subject matter, only extrospective observation. One is reminded of the embryological law that ontogenesis has to repeat phylogenesis.

SUMMARY

The conventional approach to the mind-body problem splits the human being into a body-thing in space and a mind-thing in time. From this viewpoint one asks how brain and mentation interact, starting out from the sensory-motor reflex. One progresses toward the increasingly complex, higher levels of the central nervous system and finally to the enormous intricacy of the brain. Purposiveness is thought to be present already in the reflex mechanism. As structure and function of the nervous system become more and more complex on the higher levels, purposiveness gradually attains the status of intentionality. This surreptitious insertion of intentionality into the causally structured mechanisms of the brain is due to a lack of differentiation between causality and intentionality and between compartmentalization and interpenetration. The conventional solution addresses the problem of how it is possible to ascend from the compartmentalized sensory-motor reflex which is causally structured, and which runs forward in objective time-space, to interpenetrating mentation which is intentionally structured and which occurs in subjective time-space. Putting numerous causally structured compartments together does not bring forth purpose-striving mental activities. Brain structure and function in objective time-space, causally structured, cannot be the cause of mentation. The fundamental question how nonspatial mentation can interact with the brain that exists in space, is left unanswered.

Engelhardt offers a solution on the abstract, categorial level; he breaks out of the confines of the conventional solution.

Instead of starting out from the split, mind in time and body in space, our baseline should be the whole, mind-and-body human being. The body is a phenomenon in objective time-space, it is causally structured and compartmentalized and we should observe it by extrospection. Mentation is a process in subjective time-space, which is intentionally structured and interpenetrating and which we should observe by introspection and its ancillaries.

A tentative solution can be achieved by the derivation of objective time-space from subjective time-space and of causality from intentionality. Compartmentalization is a nearly self-evident concept; interpenetration can be portrayed by the mental triad and the precious present. These processes of abstraction satisfy us rationally. An unconventional solution completes this approach, since it follows another path on the emotional level: we heal the dichotomy mind versus body by heroically suffering through the struggle of mind overpowering body and body encumbering mind.

NOTES

1. Charles M. Sherover, *The Human Experience of Time* (New York: New York University Press, 1975), pp. 97, 98.
2. René Descartes, *Philosophical Works*, rendered into English by E. S. Haldane and G.R.T. Ross (New York: Dover Publications, 1955), 2:242.
3. Franklin Fearing, *Reflex Action* (Baltimore: Williams & Wilkins, 1930), p. 20.
4. Harold Höffding, *Histoire De La Philosophie Moderne*, trans. F. Bordier (Paris: Librairie Félix Alcan, 1924), pp. 243-44 (Authors' translation).
5. Stuart F. Spicker, *The Philosophy of the Body: Rejection of Cartesian Dualism*, ed. and with an introduction by Stuart F. Spicker (Chicago: Quadrangle Books, 1970), pp. 10, 11.
6. David Hume, *A Treatise on Human Nature*, ed. L. A. Selby-Brigge (Oxford: At the Clarendon Press, 1965), pp. 35, 36, 178.
7. Hume, *On Human Nature*, p. 60.
8. William McDougall, ed., *Body and Mind* (London: Metheun and Co., 1911).
9. Wilhelm Wundt, *Grundlagen der Physiologischen Psychologie*, 4th rev. ed. (Leipzig: Verlag von Wilhelm Engelmann, 1893) (Author's translation).
10. Henri Bergson, *Matière et Mémoire: Essai sur la Relation du Corps à l'Esprit* (Paris: Librairie Félix Alcan, 1925), p. 194 (Author's translation).
11. Bergson, *Matière et Mémoire*, pp. 7, 273, 247.
12. Gunther S. Stent, "Limits to the Scientific Understanding of Man," *Science* 187 (March 21, 1975): 1057.
13. Karl R. Popper and John C. Eccles, *The Self and Its Brain* (Berlin: Springer Verlag, 1977), p. 38.
14. Ibid., p. 364.
15. Ibid., pp. 329, 335, 332.

16. R. W. Sperry, "Hemisphere Deconnection and Unity in Conscious Awareness," *American Psychologist*, vol. 23, 1968, pp. 726, 727, 724, 732, 724.

17. Roy R. Grinker and Paul C. Bucy, *Neurology*, 4th ed. (Springfield, Ill.: Charles C. Thomas, 1949), p. 358.

18. Goethe, *Faust II* (Leipzig; Utrecht: Pfeil Verlag, G.m.b.H.), p. 373 (Author's translation).

19. Gilbert Ryle, *The Concept of Mind* (London: Hutchinson of London, 1949), pp. 167-68, 50-51.

20. H. Tristram Engelhardt, Jr., *Mind-Body: A Categorial Relation* (The Hague: Martinus Nijhoff, 1973).

21. Popper and Eccles, *The Self and Its Brain*, p. 157.

22. Jurrit Bergsma, *Somatopsychologie*: Op Zoek naar Psychosociale Dimensies van de Geneeskunde [Somatopsychology: Searching for psychosocial dimensions in medicine] (Lochem, Kaatsheuvel, The Netherlands: Uitgeversmaatschappij De Tijdstroom, 1975), pp.13, 149, 180, 183 (Author's translation).

23. McDougall, *Body and Mind*, p. 212.

24. L. D. Hankoff and Ultan Chandram L. Munver, "Prenatal Experience in Hindu Mythology," *New York State Journal of Medicine: History of Medicine*, December 1980, pp. 2008, 2013.

25. René Spitz, *A Genetic Field Theory of Ego Formation: Its Implications for Psychopathology*. The Freud Anniversary Lecture Series, New York Psychoanalytic Institute (New York: International Universities Press, 1959), pp. 34, 70-71.

26. Stephen J. Suomi and Harry F. Harlow, "*Social* Rehabilitation of Induced Depressive Disorders in Monkeys," *American Journal of Psychiatry* 133, no. 11 (November 1976): 1279, 1284.

27. Michael L. Brines, "Bees Have Rules," *Science* 206, no. 4418 (November 2, 1979): 571-73.

8
Faults, Dichotomies, and Fractures

FAULTS AND DICHOTOMIES

Human existence is not a harmoniously interlocking, smoothly operating process. Slight imperfections and blemishes, faults and fissures, contradictory elements, irreconcilable conflicts, dichotomies, and fractures form a kaleidoscopic spectacle which is in some instances admirable and at other occasions frightening. From the viewpoint of time, the dichotomies and the fractures in our existence flow from our complex temporal-spatial structure, in which we constantly flux back and forth within a shifting spectrum of "now" living, mostly, in objective time-space and then again, mostly, in subjective time-space in an uneasy equilibrium. Both forms of time-space intermingle within this spectrum, but we never reach either terminal point in its pure, abstract form.

We can see more clearly against this temporal-spatial background how man's rational as well as emotional nature is bound up with the dichotomies and the fractures. The dichotomies are the lesser problems of living and they are of a more rational variety compared to the fractures. These are the most serious and threatening constituents of our existence and these fractures are heavily laden with emotion. Our attempts to realize values belong under the dichotomies and they do not affect us as searingly as the fractures. The inescapable fracture living-versus-dying affects us throughout our lifetime, and it can unleash the most excruciating emotion which tortures us at times. But most people are not in touch with this emotion-laden foresight, because we humans take various steps to abolish the dread of our personal death.

The other major fracture in our existence, besides dying and death, is loneliness. This state of mind grips us when we are living an island existence, a separate unit unto ourselves without communicating with others. We may experience these feelings during wakeful hours at night, but also while we are in the midst of partaking of the frolicking in a congenial crowd. The feeling of loneliness is experienced mostly in a depressive mood, and, although it does not carry the finality of the dread of death, one senses a distinct kinship between this dread and depression in loneliness.

We must look for meaning in suffering not only in the two most serious fractures, loneliness and dying, but also in the minor blemishes, contrasts, and dichotomies in our existence. This search demands of us that we go through both a rational exercise and that we also involve ourselves as total, emoting, communicating, purpose-striving human beings. Thus we will discern meaning not only in the extreme fractures, but also in the lesser problems of living. In this progression from blemishes to dichotomies to fractures, our emotional reactions increase in depth from slight tension, to anxiety, to panic, to dread. I shall illustrate this increment of feeling tone with a series of concrete experiences from daily life which I take from our involvement with the automobile.

You experience slight tension when you go out of your driveway, as you are wavering a moment between deciding whether to stay safely at home, or to expose yourself to highway traffic and its uncertainties.

You may feel apprehension when you are on unfamiliar roads going to a destination where you have never been before; this slight apprehension may climb in intensity to near panic if the gas tank registers almost empty. But imagine passing the scene of a serious accident; this may cause you to feel anxious or depressed since it makes you aware of the fragility of life and now you may sense distant warning signals of the possibility of your own death.

Panic grips you if your car, on an ice-covered road gets out of control, and there may be the real danger of a collision with oncoming traffic. This happened to me decades ago, and the memory of this near-accident lingers with me as of a nightmarish dream. You are gripped by dread if you are exposed to a head-on collision: now death "stares you in the face" as the possibilities of being or not being, of living or dying, bore in on you. Recently, our son described his going through the threat of such a collision when he was convinced that this was the end; but he did not panic. He stopped his car in his own lane and the other driver veered into the correct lane at the last moment. He evaluates his lucky escape as follows: "After you have faced death nothing else matters, nothing else is important any more." In truth, a reevaluation of values.

In these situations one registers hardly noticeable to hardly bearable emotional tensions. Within this series, rationality and emotionality partake of our internal states in reverse proportion. Options such as choosing between staying home, or being like a lost soul on a strange road tend, to be more intellectually weighed, in contrast to feelings flooding our consciousness when we face the ultimate fracture in a near-fatal accident.

In other situations which are not as mechanistically slanted, one notices similar differences and varying levels of tension states. Before starting to lecture, most people feel a lesser or greater degree of apprehension, moderate involvement in the subject matter, up to enthusiastic alertness. I well remember being up tight and anxious when I first presented my personal work at Yale, because I wondered how the audience would react. In subsequent presentations of related material I may have been somewhat ill at ease, but not frightened or apprehensive. Some people never seem to get over pre-lecture and during-lecture tensions.

One of our professors in Utrecht was an expert in his field but he acted extremely nervous during each of his lectures, though the substance of his talks came across forcibly to us students.

The stage fright of actors and musicians has a variable feeling tone. Arthur Rubinstein, who gave the impression that he loved to come out on stage and give concerts, said that he loved playing for an audience so much that he'd play for free. But in contrast to this euphoric and self-confident attitude there was a picture in *Life* magazine of Arthur suffering from stage fright—warming his hands over a radiator backstage before going on stage. He says: "Fear before each concert is the price I have to pay for my superb life."[1]

Actors too are frightened before going on stage, worried that they may forget a line or that something will go wrong. But, they say, as soon as they are immersed in the character which they are portraying, stage fright vanishes. In this frame of mind they can make corrections or overlook lapses of memory or other mistakes which do happen occasionally even to the most seasoned actresses and actors. A well-known British actor forgot a line in *The Merchant of Venice*, and he made up meaningless syllables and sounds in the identical rhythm and tonal qualities of the correct line, so that only his fellow actors, but no one in the audience, were aware of his lapse of memory which lasted only a few seconds.

Stage fright issues from the obligation to shift suddenly from living mostly in objective time-space to living mostly in subjective time-space. The performer asks himself, "Will I bring off what I have prepared, studied and rehearsed, and will I project it to my audience? Will I be able to transport them from living mostly in objective time-space toward living mostly in subjective time-space?" For the performer the watershed between the two life-styles is localized in a quasi-punctiform instant, that is, when the performance starts.

Most boxers cannot sleep at all during the night before an important bout and they show their extreme tension by pacing restlessly in the dressing room before they go into the ring. Other athletes display similar evidence of insecurity and fright, even if their sport is not quite as realistically threatening and potentially lethal as boxing is. Tennis champions have been known to act out uncontrolled anger, disappointment, and self-criticism right out in the open, before hundreds of spectators. One of these prima donnas fractured his racket after he had missed a shot.

Many students are tense, anxious, and fear-ridden before or during exams, and they may suffer from excessive perspiration, palpitations, tremulousness, and attacks of diarrhea.

In all these situations the interpretation is urged upon us that here are human beings who are insecure, worried, frightened, and even dreading the future in objective and subjective time-space. The extreme instance is that of the boxer, whose brain may be damaged or irreparably traumatized. But, the football player's knee may be injured for life; the tennis player's status in the stratosphere of his profession may be hauled down or ruined; the musician may be ridiculed by critics and competitors; the actor who forgets a few lines may spoil the effect of an entire performance due to rising nervous tension; the lecturer may dread the stares of incredulous or unsympathetic listeners; the student's fear may block his memory and cause him to fail. All these people live into the uncertainty of the objective and the subjective future, each in his own mode of existence and in his special field and circumstances in which he lives between the polarity of rationality and emotionality. Each suffers in his attempts to integrate past and future in a durational present and to synthesize objective time-space with subjective time-space.

Does it make any sense to gather under the title "suffering" the dread of death, the depression in loneliness, the pain of creativity, the tense expectation of the possible fulfillment or of the not-coming-to-pass of an ideal for which you have striven, apprehension about physical trauma or disease, and the slight discomfort of an itch, the pebble in your shoe, a tiny pinprick? A critical analyst may very well object that I confuse suffering with negative affect, sadness, discomfort, pain, impatience, or boredom. It is, of course, utterly irrelevant to compare suffering under the dread of death with suffering from a pinprick. The problem is how to differentiate suffering from the welter of emotional states with various qualities, intensities, and durations, which do accompany suffering. Suffering and emotionality, though interlocked in the realities of daily living, can be teased apart, and they can be placed in contrast to each other by rational analysis. Suffering is not just a feeling, a negative affect. Emotions are always co-present in a process of the mental healing of conflicts and fractures in our existence, but they are not the constitutive, driving forces of suffering.

Should we then reserve the title "suffering" for the high-flying endeavors toward concretizing values, for overcoming the aridity of loneliness, and for the conquest of the dread of death? And do minor attempts to bring blemishes and fissures together, to heal dichotomies, fall outside the domain of suffering? Let us introduce at this point the notions of a higher and a lower modality of suffering, the former referring mostly to the healing of fractures, the latter to the healing of blemishes and dichotomies. Both modes of suffering are healing processes which reconcile and synthesize contradictions, opposite elements of our existence, and therefore both produce unity, synthesis, or holism. The lower kinds of suffering are infused with milder feeling tones, the higher modes are characterized by poignant emotions. This difference flows from the fact that each, in its own sphere, fuses and pacifies contrasts and opposites. The lower modes of suffering are accompanied by emotions of lesser intensity, and they are usually of lesser duration in objective and subjective time-space; the higher modalities are embedded in deeply cutting affects which are usually spread out over extensive periods of objective and subjective time-space.

Suffering, the core of human existence, holds our rent-apart consciousness together. Its omnipresent and necessary coworker is emotionality of various qualities and intensities, depending on what kind of tension-laden, ambivalent challenges purpose-striving human beings delve into. The ultimate source of suffering is the fracture living-versus-dying, being versus not being, which engulfs us with the deepest pain.

J. T. Fraser pointed out to me that common sense equates suffering with bodily pain. One should separate pain in the bodily sense and of a physical nature from the pain of suffering. To show that this equation does not hold true I have contrasted ominous suffering due to the dread of death with the insignificant pain one "suffers" from a pinprick. This seemingly lighthearted contraposition makes clear that we experience mental suffering and physical pain in totally different contexts.

Suffering and pain have in common that we avoid experiencing either one. Also, pain is equated with suffering because we never experience suffering without pain. However, we *can* experience pain without suffering, such as a pinprick. A third reason why we can separate suffering from pain is that the depth of suffering is not necessarily and always coexistent with similar depth of pain. Pain and suffering do not operate quantitatively in parallel, notwithstanding the fact that suffering is always tinged by painful feelings. A patient with osteoarthritis suffers deeply, but his suffering is not congruent with the intensity and the duration of bodily pain which harasses him. To wit, if he adopts the heroic life-style over against his affliction, he can counteract pain and impaired function with specific measures. If he is successful and does not let pain ruin his life-style and if he continues to live a purpose-striving and gratifying life, though on a reduced level, then the pain may be just as excruciating, but he will suffer less. This is so because he continues to project his activities into the future-in-the-fringe of a precious present and in planning for the more distant future he integrates objective time-space with subjective time-space. He does not allow pain to ruin his spatiotemporal structure. He suffers, and thus he safeguards the synthesis of the two forms of time-space. Suffering keeps this structure intact and to that extent suffering is a healing power.

Freud suffered intense pain from sarcoma of the jaw. But he refused to take opiates because he preferred to suffer from the pain and have a clear mind, rather than being prevented by opiates from thinking clearly and constructively. He challenged the bodily pain, and through heroic mental suffering he continued to help his patients overcome their destructive mode of neurotic suffering, and he worked on the construction and reconstruction of his metapsychology, his clinical and other writings, until the very end of his life.

Pain threatens to destroy the temporal-spatial structure of our mind-and-body holism. Besides being a danger signal of something wrong with body structure and function, pain can be a destructive agency in the mental sphere. Suffering on the other hand, is constructive. It is an essential constituent of our spatiotemporal structure. It is a dynamic mental process which heals life's miseries and the depressive feelings which accompany the distortions of our spatiotemporal structure by painful afflictions.

One of the sources of internal conflict is the split within our personality between rationality and emotionality. This split challenges the etymological meaning of the word *individual*, which refers to an entity which cannot be divided. However, in our daily activities we can be aware of our behavior, sometimes being more rational and at other times more emotional. When one of these constituents of our personality overpowers the other, we may experience this dichotomy as painful and we may criticize ourselves severely for allowing this imbalance between rationality and emotionality. But instead of trying to gloss over or even to forget these different periods of contrasting life-styles, we ought to accept this dichotomy in our personality makeup courageously.

Breton seems to take a step aside from the problem of the meaning of suffering when he sees suffering on the one hand as a process which differentiates us from inanimate matter, while on the other hand he sees suffering as the enigma which makes us aware of our frailty and our impermanence.

> Extreme suffering, like death, leaves us confounded. Would suffering and Transcendence then be a part of that reserved realm which has been designated by the name of "unspeakable"?...I do not intend to justify the necessity of suffering. The human experience of suffering is both *passive* and *active*.[2]

Arthur McGill puts his devaluation of suffering in even sharper outline as he vacillates between an opinion of suffering as the giver of life which we have overcome and the conviction that suffering prevents us from growing and developing. This is the attitude of middle-class Americans. McGill believes that the savagery of suffering must be overcome through religious belief. I conclude on the basis of my own experience and by observing the lives of fellow humans, that suffering is not a savage destroyer but a benevolent builder and developer of the human individual.[3]

If we are to become mature human beings, our rationality and our emotionality should be harmoniously interwoven. No field of human endeavor is free from the polarity of these contrasting constituents. This polarity is evidence of dichotomies within human existence, no matter from what angle we view ourselves.

One can detect this polarity very clearly in the world of mathematics which to an outsider presents itself as the paragon of rationality and cohesiveness, since mathematics is based on axioms from which one arrives at conclusions by logical deduction. And yet within the mind of the creative mathematician these rational procedures are charged with an abundance of feeling tone. The great English mathematician, G. H. Hardy, reveals some highly illuminating glances into the working of the mind of the mathematician. He says:

> A mathematician, like a painter or a poet, is a maker of patterns. The mathematician's patterns, like the painter's and the poet's, must be beautiful; the idea, like the colors and the words, must fit together in a harmonious way.... When the world is mad, the mathematician may find in mathe-

matics an incomparable anodyne.... A mathematician may be still competent enough at sixty, but it is useless to expect him to have any original ideas.[4]

What strikes me especially in Hardy's *Mathematician's Apology* is the great emotional impact which pure mathematics has on the mind and the life of a gifted mathematician. He compares his work with that of a painter or a poet, and he comes close to the belief in inspiration in his field, since a mathematician's "creations" are nothing more than descriptions of that which already exists outside of his mind. This remark is all the more surprising because in other sections of his autobiography he describes himself as a nonbeliever in religious metaphysics. He finds *effective succor* against the emotional turmoil within the world in his rational pursuits.

The sad fate of the creative mathematician who is no longer assailed by any original ideas after age forty, threw him into a mood of unshakable depression. He felt useless, and life had lost much of its meaning when he was just one of the competent members of his profession but no longer a trailblazer.

C. P. Snow describes Hardy's depression in detail:

Coming so late [in his forties], this creative urge gave him the feeling, more important to him than to most men, of timeless youth.... It is common to meet great athletes who have gone, as they call it, over the hill, that is the point at which a good many athletes take to drink. Hardy did not take to drink; but he took to something like despair.[5]

Creative people travel the road of dedicated and painful preparation, of heady emotional ebullience when they reach pinnacles of accomplishment, and then they suffer during their descent into the "Flatland" of average people when advancing age undermines their creativity. Most artists suffer great emotional upheaval during their creative activities. Marek compares Beethoven's struggles to get his musical ideas transposed into note script with Gustave Flaubert's battling with the written word.

I am reminded of Flaubert's travail. To write four hundred pages of *Madame Bovary* he wrote four thousand. In his letters one can read about the struggle: "Sometimes, when I find I haven't written a single sentence after scribbling whole pages, I collapse on my couch and lie there dazed in a swamp of despair."[6]

Stuhlmueller senses that a similar struggle goes on in the minds of Israel's prophets.

No one, moreover, suffers more than those gifted people who are called upon to maintain ideals and yet remain tolerant toward the mediocre and overly cautious people....The prophet pays a costly price of sustaining high ideals in a mediocre society.[7]

The suffering of Deutero-Isaiah gives credence to the conviction which I defend in this book that suffering is valuable, that it is meaningful notwithstanding the torture of painful emotion which is suffering's companion.

Truly creative people often descend into the most painful pit of suffering during their acts of concretizing the loftiest ethical, artistic, or scientific ideals.

Compared to the creator's search for the realization of ideals, man as a social, communicating being also encounters many fissures and dichotomies in plain daily living, which are of a different and sometimes even more complex nature. The reasons why we suffer in daily living are more down-to-earth. The differences between children and parents, young and old people, the in-group and the out-group, are ever so many sources of frictions and mutually inflicted suffering. Underneath affectionate feelings and their overt demonstrations between parents and children, there is also much resentment and distancing, each living their own lives, though their roads and minds cross from time to time. The child thinks of the old as slow, weak, unproductive, and losing their mental luster. The in-group feels strong, self-righteous, and does not question its position of power. The out-group feels taken advantage of and demeaned; above all, they feel deprived of a future as they are forced into a corner of doing meaningless, repetitive services for the in-group. The Indian caste system is an extreme example of a fissured society. The dominant role of the male in our Western cultures has similar results in that it separates the sexes and fills females with resentment, while the male begins to feel insecure in his position of artificial power and wonders if it will last.

Blemishes and conflicts in these several groups make the individuals suffer, even if they are the people who wield the power. What kinds of suffering are these states of mind? Some of these modes of suffering are downright destructive, such as those that dissident Russians endure in psychiatric hospitals, or those that Jews and others, persecuted and destroyed by Hitler and his gang of criminals, endured. The suffering of battered children is unspeakable, and they often drag the mental and even the physical scars of violent assaults throughout their adult lives.

Even within these contexts suffering can become constructive, for instance when an out-group sees encouraging positive results from their fight against the unreasonable aspects of the power of the in-group. In some reports coming out of the concentration camps one hears of an occasional person who suffers constructively in this destructive environment. A priest, while working in filth, is challenged and ridiculed by his tormentors, and he exclaims: "God is even here!" Viktor Frankl worked against formidable odds on a scientific manuscript while he was imprisoned in a concentration camp.

He wanted desperately to save the manuscript which contained his life's work; then in a flash of insight he understood that he had to wipe out his whole life that lay behind him in order to survive. Later on he accomplished the Herculean task of reconstructing the lost manuscript on bits and pieces of paper.

> Eventually I began to reconstruct the manuscript which I had lost in the disinfection chamber of Auschwitz, and scribbled the key words in shorthand on tiny scraps of paper.... [At the time of liberation]... I ran back to my hut and collected all my possessions and a few scraps of paper, covered with shorthand.[8]

These are the sagas of courageous people who do not give in and do not give up. I am thinking of a Russian writer who composed poetry and a whole section of a novel in his head, while he was incarcerated in a labor camp. He had these materials so firmly committed to memory, that he was able to transform these compositions to a written text after his release. Some people are able to sustain the integration of their individuality with the aid of less lofty pursuits. Recently I heard of a prisoner during the Vietnam War who was a contractor and builder of homes in civilian life; he kept his mind occupied by building, demolishing, and rebuilding new houses in his imagination.

THE FRACTURES: LONELINESS

Short of the foreknowledge of our personal death, the feeling of loneliness is the most painful fracture in human existence. Many people are no more aware of their feeling of loneliness than they have a grip on their dread of dying or death. An attempt to describe the state of mind of loneliness faces other obstacles as well. We bracket loneliness with being alone, being depressed, isolated, of feeling desolate. It is hard to pry loneliness loose from these mental processes and these various feelings, in particular because loneliness occurs in variable combinations with these closely related data of consciousness. For instance, a deeply depressed person is prone to feeling lonely because he is unable to communicate with others on account of his depression, and he complains bitterly of feeling lonely and alone, even if he is surrounded by people who love him and who are trying to help him.

A person who is alone may not feel isolated or depressed or lonely, depending on how he sees himself in his island existence and what he is thinking and doing. In contrast, we can feel lonely while in the company of others.

Since loneliness is a very private feeling and mental state, it is difficult to express it in the medium of everyday language which is designed to express common experiences, more public feelings, and concrete activities. What one tries to convey with the spoken word is reinforced by the expressiveness of body language. These means are fairly ineffectual when it comes to describing loneliness.

Robert Weiss and coworkers have studied loneliness as a result of crises in many different life situations. They describe how loneliness is felt by recently divorced people, poignantly expressed by some of their clients.

> There is no time in one's life when loneliness ceases to be a threat. There are times when even those whose lives are in every way adequate, nevertheless experience loneliness. One is not only alone but also able to use one's loneliness to recognize with awesome clarity both one's ineradicable separateness from all else but also one's fundamental connectedness.... The lonely person experiences the world as desolate, barren, devoid of others, an empty world, dead, hollow.... The majority frantically seek relief [from] silence, being alone in the world, being dead.... "You don't realize it until you know it, but loneliness is the worst thing you can suffer in life."
> ...Loneliness is not caused by being alone but by being without some definite needed relationship or set of relationships.[9]

Weiss clarifies a most important fact about loneliness: being alone is not the same as loneliness. The fundamental fact about loneliness is that one feels torn between feeling connected as well as being disconnected from self and from others. Some of the clients experience loneliness as similar to leaving and reentering the world of the living. This interpretation of death reverts back to childrens' fear of death as being abandoned by their parents and siblings. Weiss feels, as I do, that loneliness and the dread of death are closely related. But our awareness of loneliness is almost as deeply buried as our dread of dying.

I have chosen to describe loneliness by the indirect route of showing the many different methods by which we evade, suppress, battle, and finally sublimate this feeling. We seldom fight this battle consciously and by intention; most of the time we are unaware of this struggle since we evade feeling lonely by stealth. The indirect route we follow most frequently is that we believe that by evading being alone we can control loneliness. In contrast to this and other evasive maneuvers, we possess two direct ways which attempt to conquer loneliness: the constructs of religions and the reality of a real-ideal marriage. Religions do not take up the battle against loneliness directly, but they provide powerful means to soften its pain. The purpose of a real-ideal marriage is to eliminate loneliness, which the happily married couple often do not even recognize. On the whole, people think that marriage is designed to make a couple happy, which is only one of the additional blessings of this seldom-accomplished goal. The essential meaning of a real-ideal marriage is the relief from loneliness, while suffering is an inevitable constituent especially of a happy marriage.

Evaluating the different compromises and solutions of the feeling of loneliness, one must keep in mind that depression and hopelessness are not themselves sources of the existential dread of loneliness. The oft-quoted poem of Donne who says that man is not an island unto himself, and my insistence that interpersonal communication is basic in the mental makeup of man, both reinforce the conviction that man is ultimately a social being. These statements about human nature are only partially true. Elaborating on Donne's metaphor, man lives sometimes on an island in isolation and by himself, but at other times he wants to escape from his encapsulation and he wants to live on the "continent" immersed in society. In his back-and-forth shifting from island existence to continent existence, man tries to forget his loneliness, which he may experience in either life-style. The struggle against loneliness goes on by various endeavors and on superficial and on deep levels. However, it is hardly the case that the various life-styles which intend to roll back loneliness, are conscious, purposeful challenges to loneliness. We struggle in this arena most of the time unwittingly, as it were willy-nilly.

The ambivalence of loneliness and togetherness can be a source of suffering. While in affable company, we may want to be alone, and, when we are alone, we may hunger for companionship. In a sense, the dread of death and the flight from loneliness are similar: we want to live when death threatens us, and we may want to die when life no longer holds out any hope for us. Suffering fulfills its uplifting task of healing both major fractures in our existence. This healing power runs like a red thread through these attempts at describing loneliness and exposing the countermeasures which we use against this unwelcome state of mind.

A group of relatively successful maneuvers to extinguish the anguish of loneliness carry the stamp of temporal and spatial truncation. Many people seek to challenge the blight of loneliness on a higher plateau of existence in a relation between man and God. Fusion-communication between man and woman in a real-ideal marriage soars above all other attempts in terms of effectiveness of attenuating loneliness. These last two life-styles are more lasting because the whole person is involved in an interpenetrating synthesis of the religious person with God and in fusion-communication between woman and man. The first-mentioned group of methods to cope with loneliness are bogged down in compartmentalization and truncation in time-space.

We experience loneliness as existential dread since we are aware that something is lacking in our existence and this prompts a restless search for one's completion by the other. Fortunate are the few who find this interweaving and interpenetration of two human beings, which is the most basic and final conquest of the misery of loneliness.

A soothing of loneliness in low key may come to pass when people, falling away into average man, generate superficial contacts with others in which curiosity, gossip, and know-it-all (Heidegger) keep conversations going on a kind of stimulus-response style and when interpersonal communication is rather artificial. Workers in industry who make useful objects without aspiring to create anything of enduring beauty or value, by working together and not being alone, can forget their loneliness. Coffee breaks, lunch time, chats about mundane subjects bring about cameraderie between coworkers for brief periods. Eating and drinking together, card games, sports, are compartmented and truncated attempts which go forward mostly, in objective time-space. These social activities reduce separateness in an island existence and they are a welcome escape to the dilution of one's self in a continent existence. These are the best possible strategies that many people can muster against this fracture in our existence. But does sinking away in average man really soften the boring-in pain of loneliness? Is it not a "wild narcosis" as Kierkegaard says?

A lasting, close friendship comes to grips with loneliness on a much higher plane and this relation can keep loneliness in abeyance so that the friendship overcomes its most thorny aspects. A life-long intimate friendship with one or two friends is especially important to women and it is said that this type of relationship is less important to men. Anne Seiden reveals close friendships between women as deeply meaningful forces in many women's lives. They help each other not only in times of crises to which women are specifically prone, but they also attenuate feelings of loneliness since these friendships carry a warm and reliable emotional undertone.

> Research on widowhood has disclosed that a greater number of men have intimate conversations only with their wives, while women are more likely to have a female confidante as well.... The fact that the mental health of single and widowed women is generally better than that of single and widowed men has been attributed in part to the supports available from women's friendships.[10]

Creative activities in which we seek to concretize beauty, truth, or ethical ideals, can counterbalance a state of separateness and feelings of loneliness. Two mathematicians working out problems as a team, unwittingly and without any intention of doing so, battle their feeling of loneliness. It seems that the first time two mathematicians worked as a team was when G. H. Hardy invited an Indian mathematician to join him. This may well be one of the motivations for many mathematicians preferring to work in a group. While they partake of exaltation of the beauty and truth of mathematics, ego boundaries are lowered due to the emotional warmth pervading the group, interpersonal communication is opened up, and feelings of loneliness are reduced. They communicate with each other via the detour of this rational as well as emotional inspiration, which allows them to escape temporarily from their island existence. The elimination of loneliness is only a happenstance, a gratuitous side effect of their immersion in their creative work and it is, of course, not the ultimate goal of their teamwork. Very likely such rapprochement within a group of mathematicians happens especially when they are involved with what G. H. Hardy called real mathematics. Highly gifted theoretical mathematicians create this kind of mathematics, when they find new solutions to old problems or discover entirely new propositions which advance this "mother of all sciences." Hardy calls "dull mathematics" what engineers use and what is being taught in universities. The very terms indicate that the first kind of mathematics stimulates these specialists emotionally, while the second kind, though rationally satisfying, offers only limited emotional gratification. I think that dull mathematics is probably a weak fighter in the battle against loneliness and that it is the real mathematics which is forceful in this battle precisely because of its uplifting feeling tone.

Biologists who work on complex problems often cooperate in fairly large teams so that each member can concentrate in depth on a special aspect of the overall problem. In parallel with their rationally grounded work, their feelings of loneliness retreat into the background, although they too may not even be aware of this temporary release from this burden.

We may be blessed with emotionally uplifting rewards when we partake together of works of art or of music. Two or more people playing music together are brought nearer to each other via the detour of the emotion-laden, shared experience of beauty which brings about a semifusion of the performers and the listeners. These rational and emotional strivings toward interpersonal communication, diminish loneliness during limited periods of time.

During the periods of his creative work, an artist or a scientist may not feel lonely at all as he is enthralled with the concretization of a value which floods him with uplifting emotions. But although these periods of precious presents when the mental triad is extraordinarily active, free the genius from the fear of loneliness, he does not actually eradicate this state of mind and in between the high points of creative activities he sinks back into the valley of average man. During these arid interphases, loneliness is probably more excruciating to him than to most of us who do not partake of such highly exultant moments and accomplishments.

Giacometti's work methods are an unusually stark example of an artist's island existence. He often used his wife as a model, and, after he had worked as the artist all day, he might say to her: "Let me look at you, I haven't seen you all day." As a creative artist he was locked into the island existence of being preoccupied and immersed in forms, masses, colors, and balances, in which his wife fulfilled her role of providing him with the concrete material, to which he gave forms that appeared before his mind's eye. Outside the studio he emerges out of that world and enters into the totally different world of interpersonal, one-to-one fusion-communication with his wife, who is now a whole human being for him, who represents an entirely different and much higher value to him compared to his wife as a model.

Do these two people suffer in their shifting back and forth between the two worlds of creativity and marriage? They veer between two incompatible, contrasting roles, fraught with the dangers of discord, feelings of separateness and loneliness because the women-man fusion-communication is temporarily suspended. Especially his wife may very well fall back into the pit of loneliness, while he has withdrawn from reality into his world of creativity. She may feel envious of him, even angry at him, because she may feel she is being "used," since she cannot actively partake of the creative process going on within him. They both have to find their own precarious, unsettling balance between these two worlds of artistic creativity in which loneliness threatens her, and the creativity of their real-ideal marriage in which they attempt, together, to overcome feelings of loneliness.

Caring for others softens the feeling of loneliness and it has strong elements of valuable ethical deeds. Treating the sick can be charged with deeply gratifying emotions, since it appeals to one's fathering and mothering drives. The relationship between doctor and patient is necessarily one-sided and compartmented, because he is the dominating authority while the patient is the passive, "patient" partner who looks up to the doctor, as if he were standing on a pedestal. When the doctor can really help the patient to get well, he is uplifted by feelings of power and gratification, while the patient's mind is likely to go on a rampage of fantasies of hero worship due to the helper's supposed omnipotence. The tight bonding between these two people in a cooperating relationship can create the illusion that they have overcome their aloneness and therefore their feelings of loneliness. However, this too is a truncated relationship, as it is in the case of the artist and his wife-model. It ends with the recovery of the patient, although both doctor and patient naturally have memories of this highly satisfying period in their lives, but these memories are all that is left from their lifelines crossing each other. If the physician is unsuccessful even after professionally correct intervention in the patient's illness, then the doctor may feel depressed, powerless, self-doubting, alone, and more lonely than before he entered into this particular relationship between helper and helpless.

It is not the usually brief duration per se of a patient and doctor relationship which makes these encounters only limited maneuvers in the battle against loneliness. Long-term psychotherapy and psychoanalysis over a time span of many years, are no more and no less effective in this area. On the one hand, the patient has much more opportunity to spin fantasized tales to himself about a therapist or an analyst, while in reality they know very little about him. The psychiatrist, on the other hand, knows more about the patient's personality and his factual life than anyone else does, and yet he has to build and maintain a professional screen between himself and his patient in the service of the ongoing treatment. He must not enter into a bilateral, person-to-person communication with his patient, as if two people came together to share their life's experiences. This would interfere with therapy simply because the stream of information about the patient's life ought to flow from patient to therapist and not in the inverse direction. Hence, a psychiatrist is just as prone to feelings of separateness from his patient and therefore of loneliness, as the physician who treats somatic illnesses.

A conventional, durable, and successful marriage is a bilateral attack on the feeling of loneliness, when couples attain a compartmentalized unity of two human beings who communicate with each other on several levels. Marriage is an attempt to overarch the psychological and physiological differences between man and woman, which are stubborn sources of loneliness. To heal the dichotomy man versus woman, man seeks to be completed by woman and woman wants to be completed by man. Freud mentions an age-old myth which tells of human beings who were originally a unified male and female structure. But Zeus split them asunder into their constituent parts, and ever since the male part and the female part have been roaming about in their territories, searching to be reunited with their lost counterpart.

In an average, durable marriage multiple concrete factors tend to unify woman, who feels something is missing in her existence, with man, who feels that an essential element is absent from his individuality. These unifying pressures diminish the feeling of loneliness since the awareness of incompleteness is one of the strongest roots of feelings of loneliness. The search for completeness is hampered by imperfections and dichotomies since each partner brings his and her own history, defects, and quirks of personality into the bond. They disappoint or hurt each other from time to time, and they suffer under the load of blemishes that their marriage has to carry. But we do have some power to build bridges between the partners because man can learn about the psychological traits of woman, and woman can begin to understand man's assets and deficits if both develop and apply the subtle attitudes of introspection, empathy, identification, and projection. If these two-way reverberations come naturally to the couple, they can unleash powerful binding forces which stabilize a marriage, just as they enhance other kinds of interpersonal communication.

Most people are not clear about what the goals of marriage really are. One thinks that marriage ought to make two people happy and that crises and periods of hostility and misery should not occur. They should try to forget these low points as quickly as possible and act as if they have not been that serious. Here we come once again upon the rock-bottom conviction that we humans, ideally, ought not to suffer, that suffering is wrong, or is a sign of evil, or is caused by our sinfulness. Therefore, we should strive to eliminate suffering as much as possible. This conviction rejects vehemently any suggestion that suffering is the core of human existence. As a result one is surd to the meaning of suffering and overlooks that suffering is a core element in both an average as well as in a real-ideal marriage.

A serious risk which runs through all types of relationships between the sexes is the fact that the mind-body relationship sets the tone in these modes of intimately living together. In the early stages of courtship the mere natural warmth of closeness between two bodies may take over, and these powerful sensory impressions may create the illusion that the couple have actually found the close physical, mental, and personal rapprochement which they have yearned for. When sexual warmth begins to lose its radiant glow, each partner may feel deceived by nature, by life, and by the other. As time flows along, one may see that aggressive feelings tear the relationship apart, which leaves each member living in his and her own, man-versus-woman, compartmented loneliness.

This failure has come to pass because each person attempted to build a total, ideal, interpenetrating person-to-person union on the limited, sexually slanted foundation of a body-to-body unity to which, in time, each added other compartmentalized elements. They have reached only a limited goal with limited means, and in their disappointment they may give up on building toward a meaningful person-to-person relationship in which sexual gratification is a result of and not the basis of a sound marital relationship. If aggressiveness raises its ugly head, they may decide to break up and they may experience as much suffering from loneliness after as before their marriage. However, if each one can step aside from the ongoing vicissitudes of their lives and if each can take a critical look at his own faults, they may develop deeper insight into their assets, shortcomings, and their expectations. With this wisdom gained they are forewarned against the pitfalls that lie in wait for them in case they should become enthralled with another mate in the future. Unfortunately, many people are incapable of undertaking the subtleties of self-examination and they often commit the same errors again, especially if the wild call of the body overpowers rationality. They suffer destructively, which leaves the disillusioned person even more lonely than before his trials and tribulations.

A marriage between two neurotic people has many features in common with an average marriage, and it is often as durable. These partners too, feel that they are lacking something, and they look to the other to fill in a gap in their personality. Many neurotics complain of a "hollow feeling" somewhere in their body, which one can usually trace back to inadequate mothering and fathering during infancy and early childhood. Going into adolescence the neurotic person may be unable to accept fathering and mothering which the parents may actually try to offer them at this time. Neurotics are likely to keep their distance from meaningful persons in their adult life and this may include the marital relationship. They often say that their partners do not understand them and do not support them sufficiently. What they really expect from their mates is not to fulfill the incompleteness which mature people suffer from, but to heal and fill the hollow feeling which has become part of their life-style. They expect the partner to give them in marriage what they missed getting from their parents in earlier life and to make up for this deficit with its archaic feelings of emptiness. That is, they cast their husband into the role of a father who treats his wife as if she were a little girl, and this husband forces his wife into the role of a mother as if she were now mothering this little boy. Each alternates between exercising an adult, supportive role and assuming an infantile, receiving role.

In a marriage between mature persons, mothering and fathering roles are acted out also, but in a metaphorical sense and on an adult level, and without either partner looking for infantile gratification and without regressing to childhood, or even to infantile behavior. The neurotic mate imposes unfulfillable demands on his or her mate and both suffer from their thwarted attempts to recreate infancy and early childhood. Are they suffering destructively or constructively? It is destructive suffering if they blindly go on squeezing infantile gratification out of the other in an attempt to "regain the lost paradise of childhood." It is constructive suffering if they realize that they are distorting an adult relationship into one of infantile dependency. Often, psychotherapy can help neurotically bonded couples to relinquish this search for a past that is lost forever, which so far has impeded their growing into a mature, mutually satisfying relationship on an adult level.

Two people may combat loneliness relatively successfully in a durable homosexual or in a durable Lesbian relationship.[11] Its durability depends on their capability to keep mutual hostility within bounds, so that it does not overshadow feelings of affection and intimacy. These relationships run the serious risk of neurotic, destructive aggressiveness which breaks many homosexual and Lesbian friendships apart. Many of them drown in *The Well of Loneliness*. Psychiatrists are, on the whole, prejudiced interpreters of these relationships because we see mostly the unsuccessful ones in which the partners have got into serious troubles. If the partners partake of a durable bonding, then they do not seek psychotherapy or psychoanalysis.

A real-ideal marriage is the most direct and the most daring attempt to stave off loneliness, but it is also the most problem-laden and risky undertaking. On the one hand, we place a heavy weight on the male and female human being as being incomplete without the other, but on the other hand one cherishes the idea that each is an independent individual sufficient onto himself and herself. One strives toward a union between a whole complete female and a whole complete male human being. In these times, economical and societal implications of marriage are stressed much less, while the emphasis is laid on the goal of two people seeking happiness due to the relief of their loneliness. Many people enter into marriage with the hope and the faith that they will overcome loneliness by creating a new emergent entity which transcends the limitations of the individual female and the individual male partner. Charles Morgan has described this idealism in a most sensitive vein in *'The Fountain'*.

> Love, friendship, even intimate association between two people has an underlying substance which is distinct from their separate personalities, though it has proceeded from their mingling.... Our love will be ringed about in time having a perfection which, when the ring is dissolved, will change its form but not cease to be.[12]

Morgan believes that a loving couple can create an emergent entity which rises above their existence as two separate human beings and of which each partakes through the intermingling of the two individuals. The delights of sexuality thus become transilluminated by this transcending hypostasis and lift two human beings out and above mere animal existence. He is convinced that this suprapersonal being which these two people have created, as if they were like gods, cannot be destroyed by the death of the couple. Two mature, complete, and separate persons communicate and fuse to become a new interpenetrating entity. This creation can be brought about and become a durable experience if they are capable in reality of underscoring their relationship with introspection, empathy, identification, and projection in order to become aware of the assets and the deficits of each personality. They have to make secure their idealism that they are "being destined for each other" by acknowledging their existence in objective time-space and in subjective time-space. Then they will be able to communicate with each other in years to come and to partake of the fusion of two egos whose boundaries are being lowered. This inward fusion-communication heals the dichotomy male versus female. In the Chinese tradition the fitting together of the yin and the yang principles within a marriage bond is accomplished from without, by parental and societal and other pressures and by the male taking the leading role to which the female submits. In contradistinction, fashion-communication is generated by the two participants from within themselves.

The healing of the female versus male dichotomy is a many-sided process of which suffering is a meaningful element. Morgan makes us feel that a couple who set out to create a supraindividual emergent entity have to suffer during their daring unifying conquest. He makes us aware of the process of woman and man fusion-communication. This power lifted Viktor Frankl up above the brutality of a concentration camp. In a very real, though fantasized experience he partook of fusion-communication with his wife.

> But my mind clung to my wife's image. Imagining it with an uncanny acuteness. Real or not, her look was then more luminous than the sun which was beginning to rise.... But soon my soul found its way back from the prisoner's existence to another world, and I resumed my talk with my loved one: I asked her questions, and she answered; she questioned me in turn and I answered.[13]

Although Frankl knows full well that he reshapes past memories of his wife into here-and-now reality, a subjective reality which is more real than the objective reality of his surroundings, he draws immense power from this fantasized encounter. It actively reinforces his faith that love between woman and man is the highest good to strive for. Frankl was condemned to live in a fractured existence between inhumane, hardly bearable torture in reality and blissful reminiscence in fantasy. The fractures and the dichotomies in his existence in a diabolically contrived environment in which the basest human hatred and murderousness were released did not tear him apart because he had chosen the heroic life-style. He suffered constructively.

In normal, more humanely organized society dichotomies and fractures operate in a more subdued modality, but they come through to us though in less excruciating form and we can take countermeasures to heal some of the fractures. Whether people live in the heartbreaking environment which was forced upon Frankl, or whether they are ordinary, free citizens, most manage to overarch the imperfections in their lives, such as those that exist in a real-ideal marriage.

Dichotomies and fractures are bound to crop up in our striving for the realization of the soaring ideal of fusion-communication in an interpenetrating relation between woman and man. These imperfections cause the most painful suffering that we humans have to endure. Fractures which rent us apart as single individuals, or those which our surrounding world sends our way, are less devastating experiences compared to the realization that faults do appear in a real-ideal marriage. They make two people suffer simultaneously and together as anger, guilt, and doubt about the permanence and the possible failure of their bond pulls them temporarily apart. These dichotomies bring about a mode of suffering which is harder to bear than those which we experience in other life situations. Since for most people an ideal marriage in which two people fuse and communicate is the highest possible value attainable, they are bound to go through the valleys of suffering. Whoever sets out to realize an ideal, a value, a *summum bonum*, has to accept that he or she will suffer. A real-ideal marriage is not designed to eliminate suffering, but to overcome and to heal the fracture of loneliness. We ought to adopt the heroic life-style in the search for this value.

Death is a most threatening source of suffering which confronts a couple who are partaking of a real-ideal marriage and death is more painful within this context than in any other perishable human relationship. At the time of the death of one of the partners, the surviving member not only loses a valuable human being, but the marriage, no longer given as a reality, lies in shambles as a concrete experience. The only values left to the survivor are memories. I think it is a rare person who can find true solace in the fact that while they were together, they did create a suprapersonal hypostasis, which is not destroyed when "the ring of time is dissolved." For many people such transcending idealism might well go up in smoke.

The threat of death and the possible loss of one's mate overhangs us during our entire lifetime, but with advancing age this threat presses in upon us more and more insistently. In our twenties and forties we can keep the thought of the possible death of one of us fairly well within bounds by defensive mechanisms, but after age fifty or sixty this becomes a real concern even if there is no serious illness on the horizon. The naked truth is that a happy marriage does not alleviate the dread of dying and of death. The few fortunate ones who have found the richness of a real-ideal marriage live, like all of us, from crisis to crisis and under the shadow of death. Though loneliness may conquer us during momentary lapses in periods of tumult, the basic accomplishment we have wrested from life is not snatched away from us. But the dread of death and the vacuum of loneliness will engulf us on occasion. We have to put great effort into living on this high level of interpersonal communication.

One might question how the synthesis of objective time-space with subjective time-space can smooth over the concrete feeling of loneliness? The answer is hidden in the several methods of fighting loneliness. One of the more pedestrian ways of fighting loneliness is the sinking away into average man. Superficial though this maneuver is, some measure of spatiotemporal synthesis is going on, for instance in the chats of the workers on a production line who talk about their plans for the future, what they are doing today, and what they did yesterday; they share their precious presents and they organize objective time-space. If one uses higher strategies to fight loneliness, the shared precious presents become more meaningful and charged with brighter feeling tone, the past and future horizons of objective time-space and subjective time-space become more extended, and the synthesis of the two forms of time-space becomes more powerful. Mathematicians working as a team, people playing music together, are instances of these more embracing syntheses.

Religions battle against loneliness successfully within the huge spatiotemporal schemas of various metaphysics. Here we see a synthesis of objective with subjective time-space on the grand scale. On the other hand, the faithful who partake of common prayer, who are together in a place of worship, and who feel the strength and the beauty of repetitive rituals, share the experience of precious presents in the here-and-now. The meaning and the feeling tone of a religious experience in a *nunc stans* are similar to the meaning and the feeling tone which a nonreligious person receives from beautiful things, true propositions, and the doing of ethical deeds in a precious present. The difference between these conceptions is that the nonreligious person places beauty, truth, and ethics at the top of his pyramid of values, while the religious person thinks of these values as subordinated to what is for him the *summum bonum*, the experience of unification with the godhead in a *nunc stans*. The nunc stans is an integral part of the metaphysical theories of religions; the precious present and the mental triad are constitutive elements in the theory about the interpenetrating quality of mentation.

The synthesis of objective time-space with subjective time-space as a baseline for the struggle against loneliness is obvious in the exquisite effectiveness of a real-ideal marriage in relieving this fracture in our existence. The precious present of the first meeting with one's life-long mate is like a watershed in one's life, since fusion-communication is the rarest of interpersonal relationships. Hopes for the realization of this ideal which one has carried in the past, are here-and-now fulfilled, and the strong desire to grow together into more complete human beings infuses the objective future with great expectations. The couple undertake to synthesize objective with subjective time-space in the process of sharing each other's personal histories, plumbing the depths of each other's personalities, carrying out plans in the present which were made in the past, making decisions together toward the future-in-the-fringe and toward the distant, objective future. Helping children to grow up is basically a time-synthesizing activity. Sexual union can be the most precious instance of this synthesis. Smoothing out frictions, distancing from each other and coming closer together again, are processes of mutual growth which may take years and which call forth most painful crises. The heroic life-style will reward the couple who keep their most valued possession alive.

A real-ideal marriage is the most sublime realization of the synthesis of objective time-space with subjective time-space and therefore, suffering is more searing in this mutual bonding than in most other interpersonal relationships.

THE FRACTURE: DYING AND DEATH

Death is the major fracture in our existence. The dread of dying and the fear of death are ever present in the minds of men. In fact, it is this dread which separates man from the animal kingdom. This dread is basic to the human constitution. Therefore, we can see that children at a young age are already plagued by it.

The process of dying is multifaceted, and varies from person to person. If one dies suddenly, without warning symptoms, one is probably not aware of what is happening. Some people report that they did not experience anything like fear or dread after they have been threatened by nearly dying. Others are not afraid of dying themselves, but they are apprehensive about what may happen to the surviving members of their immediate family. We may die, in a metaphorical sense, several times during our lifetime, and phobic patients, unhappily, are tortured by the threat of their imagined, imminent death most of their lives.

It is difficult to think, talk, or write about dying and about death, because we build defenses against this eventuality starting very early in childhood. This became apparent when one learned that doctors, nurses, and other health personnel who work in intensive-care units, or who care for terminally ill patients, literally have to "learn" to get a grip on their personal, repressed dread of their personal death. And a much more unsettling example of this repression is the fact that we try to overlook the possibility of an atomic war as "the unthinkable," since we use the same defences against this horrendous eventuality as we use against the dread of our personal, future death.

The dread of dying is the hidden motivating force behind the creation of cultural artifacts, in which the several defenses against this dread are sublimated. Mythologies generally confuse time with death and death with time, as they deal with this dread in the medium of metaphors and of supernatural forces. (I will enlarge on these topics in chapter 10.)

The premise that animals have no foreknowledge of their future death and that they are unaware of their imminent death in the here-and-now, leads to the conclusion that man is not an animal. Let us see what happens when living organisms other than man are faced with death. These data can be used in defense of the thesis that there is an unleapable chasm between man and animals.

On the lower end of the evolutionary scale, a big amoeba chases a small amoeba until it incorporates the small one. One would be loath to say that the small amoeba flees because it "knows" that it is in danger of being killed. The drones "enthusiastically" follow the queen bee on her "nuptial" flight and the one who impregnates her loses his abdominal stinger and dies. Are the drones "courageously" facing their possible death? And are they aware that, once the workers have turned them out of the hive when the nectar-gathering season is over, they will die? Do the workers "know" that their "lazy," passive hivemates are going to die? It seemed as if von Frisch's discovery of the honey dance of the honey bee released this insect from its instinctual, that is, from its deterministic, causally programmed behavior and therefore other entomologists were initially incredulous. One detail points up the amazing sophistication of this string of behaviors. The frequency of the dance changes in relation to the distance of the cache of nectar from the hive; the dance is repeated more often when the nectar is nearby and drops to a lower frequency when it is farther away. "If the food was between

100 and 200 yards from the hive, the dance was repeated about 10 times in 15 seconds. When it was 1000 meters away, the frequency of the dance dropped between 4 or 5 in every 15 seconds."[14]

Ants too display incredible feats of what looks like intelligent behavior which seem to release them from rigid determinism. However, their lacking the attribute of foresight is evident in a laboratory test.

Konrad Lorenz describes many death-protective behaviors in animals such as the wolf, but he also reports, somewhat to his dismay, that certain pigeons start pecking at each other until some of them are killed. If one takes a rat from the family nest in which it has grown up, where family members live peacefully together and support each other, and if one returns this rat after some time to his original social environment, the other rats at first pay very little attention to him, and then suddenly they will fall upon him in wild rage, with eyes bulging, and they kill him. A tree cut by a beaver often falls on the beaver's paw or tail; the trapped animal scratches and bites everything he can reach, but other beavers of the same colony who pass by do not rush to his aid and he dies eventually. The lioness will seldom allow the lion to play with the cubs because he often attacks them and eats his "own" offspring.

Oblivion of possible death seems to be the rule in the animal kingdom, but the behavior of dolphins seems to be an exception. They will keep a sick or injured mate from drowning by pushing him above the water level. On the other hand, when one goes on a kamikaze mission to destroy a submarine with a charge of explosives tied to his body, it does not seem to occur to the other dolphins to look for him or that they are aware of his disappearance. The characteristic difference between the Japanese kamikaze pilot and the dolphin kamikaze diver is that the human being not only knows that his mission will end with his death, but also that he sublimates this anticipated event, convinced that a hero's death will ensure him heavenly bliss beyond his watery grave.

Goodall and Galdikas have found out to their consternation and disappointment that the higher apes may kill each other and, what is most horrible to Goodall, that they may kidnap and eat each other's babies. A group of chimpanzees from a northern territory killed one male chimpanzee from the southern group in the Gombe reservation. Goodall has observed in detail how a female chimpanzee chased a mother with a baby, seized the infant, killed it, and ate it, sharing chunks of the flesh with its own two offspring. Later on, Goodall saw how these two females met and embraced each other, the aggressive one making no attempt to steal the docile chimp's baby. Goodall is obviously upset about this macabre scene enacted by her beloved friends, and she concludes that this behavior proves that the higher apes are even closer to us humans than she had inferred from their tool-making and tool-using behavior. If these statements about the similarity of simian behavior and human behavior are taken literally, then one has to ascribe an ethical conscience to these apes. But the behavior of the two mothers who embrace each other, Melissa whose baby had been kidnapped and killed by Passion the kidnapper and the devourer of baby's flesh, is most unlike what one would expect of two human mothers. The bereaved human mother would never forget nor forgive what the attacker had done in the distant past. For the present, the past has lost its lugubrious meaning for Melissa. In other words, chimpanzees live mostly in the immediate here-and-now with very little interconnection between distant past and the actual present.

Galdikas was involved, in a rain forest in Borneo, in rescuing and rehabilitating baby orangutans whose mothers had been killed by poachers. She describes how these apes reacted to a dead group mate, and how she found out that one of the older apes in the group had killed several young females. While several orangutans gathered around the dead body and looked at it and touched it, the killer sat far up in a tree, then came down and went through a ritual that Galdikas compares with gestures of some shaman from some obscure tribe of humans. Galdikas had mothered this adolescent killer with intensive, mothering care from infancy, and she is appalled at the result.

> In a crumpled heap of orange hair lay an infant, Doe.... Sugito came slowly down to the ground. His reaction to Doe's body was totally unlike the others. I watched in fascination as he slowly approached. Then, standing on two legs, he raised both arms over his head and brought them down, fluttering in front of him.... Now Sugito was seven and I faced the dreadful consequence of inadvertently raising an orangutan as a human being... an adolescent who was not only incredibly curious, active and tool-making, but one who killed.[15]

Goodall accuses chimpanzees of murder and cannibalism. Galdikas feels guilty because she brought up an orangutan with loving kindness, and he grew up to become a killer of his own species. Aren't these statements evidence of too far-going empathy and identification with these animals and of projecting human attributes onto them? Reticence in respect to this quite natural inclination is indicated because the temporal-spatial horizons of apes is much narrower than those of man. This reticence increases as we descend the evolutionary scale. There are dangers both at the higher and the lower ends: we have to look out for not making insects into *homunculi* and we must not make *homines* out of apes.

Most significant in Goodall's and Galdikas's descriptions of the reactions of apes to death of their group mates is that none seems to attempt to bury the victims. In contrast to the absence of burial practices in animal behavior, archeologists always find some form of caring for the dead body by prehistoric humans.

Max Scheler speaks of the minimal dread of death in primitive peoples.

> The fear of death... was almost totally absent among primitives because of the doctrine of reincarnation in new bodies.... Among the primitive individuals self-consciousness is completely embedded in the community. The individual person and the whole generation sleep and dream, so to say,

without having yet a unique and distinct consciousness of boundaries of their own feelings.... Deeper suffering arises among the civilized because of the responsibility of self, isolation, loneliness, alienation from community and because of worry and anxiety about life.[16]

Huizinga describes how people tried to cope with the fear of death in the Middle Ages.

No other epoch has laid so much stress as the expiring Middle Ages on the thought of death.... An everlasting call of *memento mori* resounds through life.... Denis the Cartusian said: "And when going to bed at night, he should consider how, just as he now lies down himself, soon strange hands will lie his body in the grave."... Nowhere else were all the images tending to evoke the horror of death assembled so strikingly as in the churchyard of the Innocents in Paris.... Living emotion stiffens amid the abused imagery of skeletons and worms.[17]

We must remember, though, that medieval man had to face death much more frequently and much earlier in life than twentieth-century man. It is as if medieval man exposed himself on purpose to an immunization process of a psychological and emotional bent, to be better prepared when the disaster of death did strike him. The "immunizing" effect of the masochistic pleasure felt by people when they visited the churchyard of the Innocents was reinforced by its socializing function, since this chamber of horrors "lay open to the eyes of thousands."

Anthropologists Kunstadter and Gay report on religious rituals of the Lua and the Kpelle's. These primitive peoples live in dread of their deceased ancestors, and they have invented complicated rituals to appease them. It is obvious that this dread is a derivative of the dread of dying and of death, even if it takes the form of combating illness. Their defense against the dread of dying is a displacement of this anxiety onto the ancestors who are still living in close proximity to them. Displacement helps to forget the dread of the ultimate fracture of our existence.

Living through the horror of dying in fantasy is one of the ways in which we suppress and repress this dread. In our own times, don't many of us love Hitchcock's horror movies, and don't the more sophisticated undergo voluntarily the stress of Bergmann's films, which deal so often with dying and death? Here too, we drag ourselves through an unsettling emotional experience in a world of fantasy in order to be better prepared for suffering through the dichotomies and the fractures of real life.

Many studies of child development reveal that children are preoccupied with dying and death at a very early age. Childers and Wimmer studied seventy-five children ranging in ages from four to ten years.

On the question of universality [of death]... 45% of the 4-year-olds... did not know.... The 9-year-olds answered decisively: 100% positive.... The understanding of death as irrevocable shows a greater uncertainty at all levels.... Most of the age groups were clearly not sure of the irreversibility of death.[18]

Stone and Church report what a five-year-old son said about growing old.

Among other things, the child may develop a fear of death.... "I won't ever be able to catch up with you. I won't ever be the same as you at the same time."...Closely tied in with the child's awareness of time in relation to growth and change, is, of course, his awareness of his own and his parents' mortality, again perhaps tempered by a sense of immortality.[19]

Kübler-Ross presents a brief synopsis of the development of thoughts about growing up and dying from age three onward. "Up to age three a child is concerned only about *separation*, later followed by the fear of *mutilation*. He is concerned about the integrity of his body and is threatened by anything that can destroy it.... After the age of five death is often regarded as a man, a bogey-man who comes to take people away."[20]

Bonaparte's observation of a boy of two-and-a-half years, shows that neat schedules of child development in this area are not always true to the facts. This little boy said about his dog, "He mustn't go out in the street by himself. He'd be run over and killed and then we should never see him again."[21]

Dr. Harry Orr told me that his four-and-a-half year old daughter, who knew about the deaths of an old family friend and of her great-grandmother, was surprised and upset when she saw one of her goldfish floating upside down on the surface of the water in the fish tank. The parents explained in appropriate language what had happened to the fish, and then the little girl started to ask questions: "Daddy, are you going to die soon? And may mother die?" Since great-grandmother was of advanced age and had been in poor health and since the child had visited her from time to time, the parents had decided to prepare her for the coming death of the old lady. They read a story to her, which deals in simple language with the problem of death and burial, written for children ages four to eight.[22] It tells how children find a dead bird, and they know it is dead because they feel no heartbeat, and it is cold. They bury it wrapped up in leaves, they sing a song to it and put flowers on the grave, just like the adults do when someone dies. The last picture in the booklet shows the children playing baseball on the right side, and the gravestone with a bunch of flowers on it is on the left side of the picture. An insightful story about young childrens' awareness of death, of the desensitizing effects of burial rites and especially of the opposite feelings of sadness-in-beauty of their singing. They accept the finality of death depicted in the symbol of the gravestone, counterbalanced by life going on as usual, after we have come to grips with the shock of death and after we have put the dread which death instills in us behind us.

Many young children think that the dead person has gone away somewhere and may reappear most any time. This is an uncanny reminder of the rituals of primitive peoples who conjure up the souls of the deceased so that they feel continued contact with them. The denial of the finality of death is widespread throughout human cultures.

A paper by Albert and Jones on bedtime stories confirms, from an unexpected angle, that children are indeed afraid of dying. The transition from being awake to sleep frightens the child, because he wonders if some unforeseen and unexpected disasters might happen while he is not awake. We have seen that the dread of death is at one stage of development transformed into the fear of abandonment. The contact with the adult during the reading of the bedtime story and its expected and carefully prepared termination is designed to give the child the security that, even though he must sleep alone in his own bed, even in his own room, the family will be there if and when he needs them. These interpretations of the story are in harmony with the thesis that the fear of loneliness and the dread of dying are fractures in human existence which already plague the young child.[23] The authors show how the child struggles with the form of continuous, objective time-space and the form of discontinuous, subjective time-space. These insights transform the bedtime story into a device which heals the suffering from loneliness and the dread of dying.

As grown-ups we react quite variously to the dread of dying, as one gleans from Amundsen's report of his encounter with the ice bear and from Barthou's last words when he was assassinated in Marseilles, "Where are my glasses?" Moshe Dayan describes an unusually stoic reaction in a freewheeling interview.[24]

Having almost been killed several times during his lifetime, beginning in adolescence in a fight with Arab youngsters and later during battles with bullets flying round him, Dayan never thought of being hit himself. This is a common conviction of the seasoned soldier. When the end seemed to be upon him here-and-now during a battle in Syria, when a bullet had penetrated his skull and when he was losing consciousness, he accepted the likelihood of his personal death courageously and in his seemingly emotionless, heroic life-style. One might speculate that Dayan had been thoroughly "immunized" against the dread of dying and of death.

Amundsen had a brush with death when an ice bear attacked him and he was waiting for the death blow. He denies that he experienced any elevating panorama of his past life before his mind's eye as so often has been described by people who have almost died, but that he simply asked himself in a matter of fact fashion how many hairpins will be picked up in Regent Street in London on a Monday morning.[25]

Amundsen compares these casual thoughts with experiences that have been described by many people who have almost died and after survival tell the tale of seeing their whole life passing revue in a condensed manner during a few seconds before the mind's eye, and which to many was a most elevating and exhilarating experience.

How one deals with the dread of dying depends to a large extent on one's personality makeup and one's outlook upon life. Dr. Harry Orr told me that his father who suffers from progressively crippling lung emphysema, has been exposed many times to the prospect of dying during the last five years of his life. The old gentleman knew very well that this disease would kill him eventually, but he said, insistently: "I am not afraid of dying as long as I don't have any pain; anyhow, every time I leave the hospital the doctors tell me that I can go fishing again." One night, the respirator stopped functioning, and it took about thirty minutes to correct this emergency. Then he said to his son with a few tears in one eye, "I *am* afraid of dying." But a few hours later, "Did I really say that?" My colleague added that his father always found it difficult to be in touch with his own feelings.

I became particularly aware of the discrepancy between the dread of death and the dread of the process of dying in a conversation with Dr. Hans Syz who, now in his eighty-fourth year, wrote a most sensitive and probing diary fragment in his twenty-first year.

He reveals nightly, recurring, overwhelming anxiety attacks, in which he is disturbed by the nothingness of our being. In conjunction with this ongoing emotional storm he is able to dissect enlightening rational thoughts about living and dying. From insight into life as rigidly determined, he shifts to the dawning realization that by conquering the dread of death we can become aware of the meaning of life. Only if we first come to grips with the fear of death can we go on to develop fully our dormant potentials. Reconciled with the fact that some time in the future we will die, we can adopt a serious as well as a happy outlook on life.[26]

Personally, I think that the most incisive *credo* in Syz's diary is his attributing meaning to each moment. This conviction is in harmony with my attributing intrinsic value to the precious present.

It is fascinating to contrast Hans Syz's thoughts about death in 1916 with his thoughts about death in 1980. As a youth he is convinced that he has overcome the dread of death indirectly, in terms of attacks of anxiety, panic, and dread at the very beginning of the diary. Young Syz's thoughts about death are very similar to my conviction that the dread of death is constitutive of human life and that happiness becomes a reliable state of mind, after we have come to grips with and have conquered the dread of death. Furthermore, I believe that our temporal structure is continuous as well as discontinuous, that thinking and feeling ought to work together in our building and understanding of life and reality. Mostly, one has to go through periods of suffering such as Hans Syz went through in his twenties, in order to experience true happiness and bliss. I interpret Hans's nocturnal and early morning panic attacks as his becoming aware of his own eventual, unavoidable, future death. He experienced this dread of his own death as "nothingness."

During a recent conversation Dr. Syz challenged my thesis that the dread of death is the basic source of normal anxiety and of neurotic anxiety. He defended his diverging opinion on the basis of his own experience after a recent serious coronary attack. After he had been honored at a testimonial dinner, he collapsed on the floor of the dining room. If it had not been for the capable intervention of two medical therapists who happened to be present and who applied cardiac massage and artificial respiration, he might have died right then and there. In retrospect, Dr. Syz does not discover any feelings of dread of dying or of death either before or after his coronary crisis.

I think that his lack of feeling of dread is understandable against the background of his diary and of his medical training as a young man. He did come to grips with the dread of death and conquered it by accepting death as part and parcel of living. Also, being frequently exposed to the tragedy of death during his medical training, he had to learn to overcome the upsurge of this dread frequently during his younger years. In general, doctors have quite active defense mechanisms which assuage the dread of dying and of death. People who are exposed to the spectre of dying and of death only on rare occasions, often possess less active defenses.

Dr. Casey had a most unusual encounter with death. He noticed some blood in his urine, and he underwent a cystoscopic examination under thiopentothal (a barbiturate) anaesthesia and acetylcholine (a muscle relaxant drug) to rule out the possibility of a cancerous growth in the bladder. When he woke up from the short acting anaesthesia and had regained full consciousness, he noticed that he was totally paralyzed. He heard the discussions of surgeons and nurses, he could see the ceiling and parts of his body, and as he tried to put all these experiences together the feeling of terror became unendurable. After some time which seemed endless to him, terror began to decrease and he accepted that the end of his life was near. He tried to contemplate whether the actuality of dying would be more or less terrifying than the panic and terror he had just passed through and by now he felt rather serene. Then, suddenly he saw that he could move his right toe, and he envisaged the possibility that he might not die. In retrospect he believes that when his time to die actually is upon him, he will not be frightened but he will be calm and serene.[27]

Dr. Casey is convinced that the terror, horror, and dread while facing death for several hours, has helped him to overcome the fear of his personal dying some time in the future. The healing power of this devastating experience underscores the rationale of exposing ourselves consciously and artificially to the horror of death in a socialized, formalized, manageable format. Primitive peoples conjure up the dangers of communicating with their ancestors. In the Middle Ages people visited the churchyard of the Innocents, and in our cultures well-organized funeral rites aim to neutralize the fear of the process of dying in reality. Hopefully, dying will be less painful than if we had not been exposed to these artificial encounters with dying and with death.

Dr. Sabon reports briefly on patients who have gone through the "Near Death Experience" during open-heart surgery. He has investigated the experiences of

107 patients who experienced unconsciousness and near death when they had so-called out-of-body experiences. An ongoing duel with the pen within the medical profession can be used for or against one's belief in a life after death.[28]

At this juncture, Heidegger's abstract conceptualization may widen the horizon of our discussions and illuminate our vision. Heidegger describes in a masterful fashion our living toward death and toward the process of dying, though his complicated style makes paraphrasing and translation difficult. "Being-there" (*Dasein*) is a being-toward-death and it constitutes our capacity to become a whole human being. Something that has not yet made itself manifest nevertheless already belongs to being-there. This "not-yet" is both impossible and possible which we experience as the dread of death, which constitutes the nothingness of being-there's existence. Everyday-life clutches on to admitting the opposite meanings of death, its possibility versus its unreachability in order to reduce its power by covering over dying and to make it easier to accept that we are being-thrown-into-death.[29] I gather that Heidegger means to say that death "makes our life into a whole." We can live toward death if we accept the fact of our future death heroically. If we can live heroically at the time when we are a dying person, then we will have the power to make our life into an admirable, uplifting whole, notwithstanding the ultimate spatiotemporal destruction of our mentation and of our body. In view of death bringing about the holism of our terrestrial existence after we have lived a meaningful and purpose-striving life, we can come to terms with the dread of death without denying, suppressing, or ignoring this dread.

Heidegger's theoretical stance can deepen our insight into the practical activities which are going on in an intensive-care unit (ICU). Theory evaluates practice. Practice exhibits the power of theory.

Dr. Campbell describes how he and a senior nurse introduced a group approach toward the problems which face nurses who work on an ICU.[30] Nurses who work in close contact with dying patients and who may strongly empathize with some of them, asked themselves anxiously whether their emotional involvement impaired their professional technical competence. They discussed coping mechanisms, such as help from religion, compartmentalization of their professional from their personal life, recognition of their personal satisfaction from helping others. These factors turned out to be mostly their defenses against anxiety not only about the patient's impending death but also their own personal death anxiety. They learned that this hidden private feeling was in part responsible for generating doubts about their professional proficiency. Support machinery was used not only for obvious technical reasons but also as artificial psychological means to contain the fear of losing patients. The group meetings uncovered these hidden motivations and showed the nurses and doctors that machinery can isolate one, not only from the patient as a person, but it can also stifle the dread of death by imparting to the health professional a feeling of omnipotence over the progress of the patient's illness. The fear of one's own death and concern with the dread of losing patients did diminish during the nurses' progression from student to faculty status.

Overcoming death anxiety is a life-long process. The pitfall and the difficulties inherent in this process of mental growth and becoming of our personality are a source of suffering for all mankind, though it takes on particularly painful proportions in the medical profession. Anselm Strauss has done research in several hospitals, concentrating on the attitudes of medical and nursing staffs toward dying patients. He shows that the conventional attitude is to keep the patient unaware of his impending demise, to which rule he points out many exceptions.[31] The problem of whether one should keep the patient unaware or make him aware of death being near at hand is crucial. If he is kept unaware, one has to constantly invent little white lies and, what's more, unwittingly one controls the patient's style of dying. This may totally distort the patient's outlook upon his own future which, he may think, is still lying wide open before him. Unawareness or closed awareness may raise a serious obstacle between a husband and his wife since the healthy member can no longer share a serious problem with the sick spouse and this life-long style of communication may break down. The changeover from closed to open awareness of the approaching end may cause turmoil and resentment in the patient. On the other hand he may suddenly feel a great relief from now understanding what is going on for him and around him.

Lidz noted such a "conspiracy" to keep a dying patient unaware of her being close to death. A lady of ninety-three was dying from a malignancy, and neither her physician nor her children thought it wise to tell her that the end was near. Lidz soon found out that she was furious with her children who did not want to take her into one of their homes, and she felt they had abandoned her. He told her that she had only a short time to live, that the hospital was the proper place for her to be, and that no one had abandoned or was going to abandon her. The old lady immediately forgot all about her wrath and thought it was foolish of both her doctor and her children to keep the truth from her. As a result of this intervention, harmonious communication with her children was reestablished and she died in peace in the hospital. During these last weeks of her life the whole family had discussed past happy and difficult experiences, present events, and plans for the future, from which reminiscences and anticipations the old lady received great satisfaction and solace.[32] Even in this hospital setting, with the foreknowledge of impending death hovering over this group of concerned people, we see in the old lady's attitude that a human being wants to integrate past with future within the present until the very last breath of terrestrial existence.

Dr. Whittaker allows us to participate in the agony he and his wife went through during the twelve-year lifespan of one of their sons who had congenital heart disease, necessitating three open-heart operations. The first operation was performed when he was thirty-hours old, the second at age six, and the last and final operation at age twelve. The first and the second interventions were beneficial, and all concerned expected even better results from the third operation which unfortunately ended with Derek dying on the operating table because they were unable to get the heart muscle to start contracting again. Both parents discussed Derek's problem quite openly with him, and his father discussed in detail and as far as possible the anatomy and the physiology of the heart with Derek. When the boy asked his mother why he needed a third operation, she told him that he'd be a physical and mental cripple without one and that he would slowly die. At the same time, she held out hope for him for a more normal life.

This tale of woe brings many of our attitudes toward dying and death into the limelight. Dr. Whittaker takes the active, medical-surgical approach which tries to postpone death as long as possible. His solace is therefore on two levels: everything possible has been done, but he believes also that Derek's life had a transcendent purpose, brief as it was, and that his life and death were not meaningless. The mother has seen Derek die in a metaphorical sense "many many times" before her very eyes, but she chastizes herself for not preparing him for the possibility of his death before the third operation. And yet, she did talk to him about a slow death if he did not have the operation.[33]

The parents had good reason to anticipate a favorable outcome of the last operation. But over and above medical opinion, they were uplifted by the deeply ingrained human denial, postponement, and sublimation of the dread of death. Derek was propelled into living the life of a grown-up within a child's body. His battle against the defects of his recalcitrant body prove that he lived a heroic life-style. This history shows that even a child can be a hero.

These examples of prevailing attitudes toward the dying patient show that medical personnel are concerned that if they approach terminal cases as if they were whole human beings with whom one can empathize and sympathize, then their own emotional reactions are bound to interfere with their medical acumen and with their proficiency. Straus, Lidz and many others show that one can approach the dying person as a whole human being who exists in objective time-space as well as in subjective time-space. One empathizes the dread of dying and of death in the patient which is similar to the dread which one experiences within oneself, and hence one can communicate with a dying person, whom one also compartmentalizes and whose bodily functions one manipulates with scientific methods and by technical, mechanistic means.

At this juncture, Heidegger's abstract conceptualization may widen the horizon of our discussions about the dying patient. The personnel which manages the ICU machinery act as an average man (*das Man*). Average man feels the duty to care for, to support, and to change the dying body for the better, with the motivation that death should be postponed as long as possible, at all costs, and they put great personal effort into this goal. They trust that their measures will restore the dying body to a whole, physical as well as mental, human being. They deny, overlook, put off death as long as possible without coming to grips with the two facts that we have been "thrown-into-the-world" and that we have been thrown-toward-death. Therefore, average man sees no problem in his caring for and nurturing this body. This dying body is for them something they encounter, that is on hand, and that they try to change. One can enlarge upon Heidegger's vision of average man and his being fallen away into his caring-for behavior (*Fürsorge*). We can interrelate this life-style with the form of objective time-space, the category of causality, the extrospective method of observation and the compartmentalization of observed phenomena.

Let us retreat from Heidegger's abstract approach and examine how the personnel in the ICU can come to grips with their dread of death. If it is true that these people exist in the life-style of average man, then one would have to assume that they hardly experience the dread of death. According to this one-sided and restricted view, they adopt an attitude toward the dying person which flows from their conceiving of him as a living organism which they encounter. They classify the dying person as an object, not as a subject. For them, the dying person exists in objective time-space, his symptoms are causally structured, and they gather data about these symptoms following the extrospective method of observation. They compartmentalize these observations into quantitatively cohering phenomena, which they can manipulate with the technical means they control. Hence, they have the power to influence these symptoms, which means that they have the power to postpone the patient's moment of death.

In addition to their natural scientific organization of the welter of signs and symptoms which confront them it is obligatory that they also compartmentalize and manipulate their own, inner, emotional reactions in order to be able to do their humanitarian job. Doctors, nurses, and technicians do indeed *have to learn* literally to keep their own dread of death under control by various psychological defense mechanisms. When they learn also, or when they are able to feel intuitively and spontaneously, to suspend these defenses, so that they can empathize with the patient's dread of dying, then they can shift to the more complete interpersonal communication between two human beings. This dualistic extrospective, as well as introspective approach to the patient and oneself, makes one aware of the contrasting duties and functions one has toward the dying patient. Not everyone is able to adopt this Janusian view and this splitness of attitudes.

It occurs to me that I have evaded, so far, to putting down my thoughts and imagery about my own, personal death. This drives home that it is difficult enough to think inwardly about oneself going through the process of dying, let alone talk or write about it. Freud has said that no one can come up with actual, realistic thoughts or imagery about his personal death.

At age twenty-two I attended the funeral of an elderly gentleman who played a rather important role in my life during short vacation periods. During the lowering of the coffin into the grave, I heard by intra-audition the upheaval and the turmoil in the middle section of the funeral march in Beethoven's *Eroica* symphony. I was engulfed by the conviction that even death could not take away a rather idealistic, not to say unrealistic love experience, which later on turned out to be a fata morgana. But the experience of the sublimation of the dread of death through music, at the grave, did jell into the decision to get in contact with the girl.

During one's medical internship witnessing patients die is an unsettling experience, but since others in authority are responsbile for the treatment, one remains only a spectator. Losing one's first patient as a young doctor in practice is a truly traumatic event, but here again, the patient's death is not read as a warning signal toward one's own, future death. Instead, one goes through an entirely practical evaluation of one's treatment of the case and comes up with, "if only" I had recognized danger symptoms earlier, "if only" I had interpreted them differently, "if only" I had done so and so.

In her forties my wife's life was threatened by a serious accident with a broken milk bottle, which caused extensive lacerations of the lower leg and profuse bleeding. I had to rush her to the emergency room, faster than I could summon an ambulance. She nearly fainted several times in the passenger seat, due to the previous blood loss. In later years she had to undergo two rather serious operations. When I heard after those occasions that people remarked about my pale color and worried expression, I became aware, as an aftermath, how apprehensive I had been before and during these crises. I had thought that my confidence in my colleagues was sufficient for me not to have any doubt about a favorable outcome.

The experience of one's own actual dying, or the imagery of such a process are only distant data of consciousness. I gather that this is so because we have repressed the dread of our personally going through this ordeal since early childhood and that most of us are hardly aware of dreading this eventuality. It would seem as if the dread of dying is more of a theoretical, rationally deduced state of mind, rather than an actual experience even if it should be lived through in one's imagination or in dreams. Wayne Oates observes when he contemplates his duties of consoling and supporting dying patients that we are also preoccupied with our own death while we discharge these humanitarian duties. He sees death as the prototype of all other forms of human suffering and therefore he calls death the "parent-suffering." But rationality is not satisfied with this classification since it searches for meaning in this "parent-suffering."[34]

Physical pain, anxiety, loneliness, and the dread of dying are four constituents of human existence which belong together. Physical pain can be a warning signal of impending death, as in a coronary attack. Anxiety, always a concomitant of physical pain, is a derivative of the dread of dying, which is the "parent-suffering" of all other modes of suffering. Loneliness and the process of dying are inseparable but they can also accompany the various modes of dying in a metaphorical sense. Dying is the ultimate fracture in our terrestrial existence, and it is the fundament of our suffering.

NOTES

1. Arthur Rubinstein, "In Paris Hall," *Life*, March 2, 1959, page 94.

2. Stanislas Breton, "Human Suffering and Transcendence" (Paper read at the First Congress on...Suffering, Notre Dame University, Notre Dame, Ind., April 1979), pp. 1, 3, by kind permission of Stauros International, West New York, N.J.

3. Arthur C. McGill, "Our Suffering and the Suffering of the Christ" (Paper read at the First Congress on...Suffering, Notre Dame University, Notre Dame, Ind., April 1979), pp. 8, 27, by kind permission of Stauros International, West New York, N.J.

4. G. H. Hardy, *A Mathematician's Apology* (Cambridge: At the University Press, 1967), pp. 84, 143, 148.

5. C. P. Snow, Foreword to *A Mathematician's Apology*, by G. H. Hardy (Cambridge: At the University Press, 1967), p. 51.

6. George R. Marek, *Beethoven: Biography of a Genius* (New York: Funk & Wagnalls, 1969), p. 160n.

7. Carroll Stuhlmueller, "Biblical Voices of Suffering and Prayer" (Paper read at the First Congress on. . .Suffering, Notre Dame University, Notre Dame, Ind., April 1979), pp. 28, 29, by kind permission of Stauros International, West New York, N.J.

8. Viktor E. Frankl, *Man's Search For Meaning* (New York: Washington Square Press, 1959), pp. 55, 92.

9. Robert S. Weiss, *Loneliness: The Experience of Emotional and Social Isolation* (Cambridge, Mass.: MIT Press, 1973), pp. 228, 11, 127, 184, 17.

10. Anne Seiden, "Overview: Research on the Psychology of Women," *American Journal of Psychiatry* 133, no. 10 (October 1976): 114.

11. Alan P. Bell and Marvin S. Weisberg, *Homosexualities* (New York: Simon and Schuster, 1978).

12. Charles Morgan, *The Fountain* (New York: Alfred A. Knopf, 1932), pp. 207-08, 219.

13. Frankl, *Man's Search For Meaning*, p. 58.

14. N. J. Berrill, *Charade of the Honeybee*, Illustrated Library of the Natural Sciences, ed. Edward M. Weyer, Jr. (New York: Simon and Schuster, 1958), p. 357.

15. Birute M. F. Galdikas, "Living with the Great Orange Apes—Indonesia's Orangutans," *National Geographic* 157, no. 6 (June 1980): 830.

16. Max Scheler, "The Meaning of Suffering," in *Centennial Essays*, ed. Manfred Fringe (The Hague: Martinus Nijhoff, 1974), p. 139.

17. Johan Huizinga, *The Waning of the Middle Ages* (Garden City, N.Y.: Doubleday & Co. 1954), pp. 138, 148, 151.

18. Perry Childers and Mary Wimmer, "*The Concept of Death in Early Childhood*," in *Child Development* (Milwaukee: University of Wisconsin, 1971), pp. 1299-1301.

19. L. Joseph Stone and Joseph Church, *Childhood and Adolescence: A Psychology of the Growing Person* (New York: Random House, 1968), pp. 304, 305.

20. Elizabeth Kübler-Ross, *On Death and Dying* (London: Macmillan & Co. 1969), pp. 157-58.

21. Marie Bonaparte, "Time and the Unconscious," *International Journal of Psychoanalysis* 21 (October 1940): 445.

22. Margaret Wise Brown, *The Dead Bird: Four Children Prepare a Woodland Funeral* (New York: Dell Publishing Co., 1979).

23. Stuart Albert and William Jones, *The Significance and Structure of Children's Bedtime Stories*, in *The Personal Experience of Time*, ed. Bernard S. Gorman and Alden E. Wessman, (New York; London: Plenum Press, 1977), pp. iii-132.

24. Curtis Bill Pepper, " Moshe Dayan: Reflections on a Life of War and Peace," *New York Times Magazine*, Sunday May 4, 1980, p. 39.

25. Roald Amundsen, *My Life as an Explorer* (Garden City, N.Y.: Doubleday Page, 1927), p. 88.

26. Hans Syz, *Vom Sein und Vom Sinn* (Zurich: Editio Academica, 1972), pp. 7, 29, 33, 37. Author's translation.

27. Jesse F. Casey, "A Psychiatrist's Experience of Terror," *Psychiatric Annals* 8 no. 5, (May 1978): 76-80.

28. Michael Sabon, "The Near Death Experience," *Journal of the American Medical Association* 244, no. 1 (July 4, 1980): 29-30.

29. Martin Heidegger, *Sein und Zeit* (Halle a.d.S.: Max Niemeyer Verlag, 1929), pp. 246-51.

30. Thomas W. Campbell, "Death Anxiety on a Coronary Care Unit," *Psychosomatics* 21, no. 2 (February 1980): 131, 135, 136.

31. Anselm L. Strauss, "Awareness of Dying," in *Death And Dying: Current Issues in the Treatment of the Dying Person*, ed. Leonard Pearson (Cleveland: The Press of Case Western Reserve University, 1969), pp. 108-132.

32. Theodore Lidz, *The Person: His Development throughout the Life Cycle* (New York: Basic Books, 1968), pp. 504-5.

33. Linton Whittaker, "Talking about Death with My Son," *American Medical News*, May 23, 1980, pp. 3,4.

34. Wayne E. Oates, "Forms of Grief, Diagnosis, Meaning and Treatment." Paper read at First Congress on...Suffering, Notre Dame, Ind., April 1979, p. 32. (By kind permission of Stauros International, West New York, N.J.)

9
Suffering: The Core of Human Existence

INTRODUCTION

Suffering is the core of the human condition in a world of imperfections, contrasts, conflicts, dichotomies, and fractures. This world impinges upon us, and we manipulate this world within the framework of our dualistic temporal-spatial structure. Rather than ascribing suffering to our sinfulness, I defend the conviction that through suffering we improve imperfections, we synthesize opposites, and we heal dichotomies and fractures. Instead of devaluing suffering as an element in life which we should avoid, I believe that suffering has intrinsic value. Due to insight into our spatiotemporal structure we can apprehend suffering in a combined pessimistic as well as optimistic mood. This Janusian outlook upon suffering runs like an ever-recurring theme through the history of mankind.

Sophocles describes how blinded Oedipus, utterly dependent upon Antigone, arrives in a place which is reserved for the dreaded Furies and how the inhabitants of the city-state of Athens examine him suspiciously. Nevertheless, Oedipus has found some peace of mind, though one would hardly feel that he is a happy man. He says, "Suffering and time,/ Vast time,/ Have been instructors in contentment,/ Which Kingliness teaches too."[1]

Max Scheler discusses suffering mostly from a theological and historical point of view. Buddhism teaches that suffering is a fundamental constituent of being and that it is punishment for sins committed in previous incarnations. Desire and individuality are the sources of suffering during our terrestrial life, and in order to become redeemed from suffering one has to supersede one's individuality in order to enter the blissful state of nirvana. It is required to "step out" of oneself so one can leave the causally structured world behind, which is accomplished by an act of "holy indifference." One meets a similar view of suffering in the Old Testament in which one is told that suffering is punishment for sins which the person, his parents, or the whole community have committed. The suffering of the Jewish people has engendered the messianic hope of release from suffering. Suffering also holds a central place in Christian doctrine, but here it is intermingled with love and occurs as a precondition for one's future redemption from sin. Enduring suffering under the symbol of the cross, the Christian reaches a state of profound bliss.[2]

The temporal perspective of Buddhist doctrine is enormous on account of the importance of the concept of reincarnation which terminates, ideally, in the dissolution of the autonomous individual. This view of suffering compresses the perspective of man's terrestrial existence in objective and subjective space-time to almost infinitesimal, insignificant dimensions. Christianity also points to a way, though in different contexts, in which man becomes released from the yoke of suffering.

Carroll Stuhlmueller finds evidence in the Old Testament that Jehovah suffers together with the Jewish people and that He foreshadows the loving Father of the New Testament. Jerusalem had to pay more than double for her sins, and the voices of prayer and the voices of suffering intermingle. The God of Christianity too has many attributes in common with man. He is benevolent and sacrifices his only son for the redemption of mankind and to relieve man from his load of sin. Sinfulness, suffering, death, and love are inextricably interlocked.[3]

Bowker believes that we can escape from suffering and from the dread of dying into a life hereafter. He extracts support for this belief from the recurring cyclical natural phenomena which take on esoteric form in the ritualistic practices of priests in Central America. As the snake sheds its skin, and goes on living, so does the priest beat the skin loose from a sacrificial human victim, and by stepping into the skin the priest acts out the renewal of life. Bowker intimates the contraposition of human life being pregnant with possibilities of stepping out of its boundaries but also of being hemmed in by powerful restraints.[4]

The point of view of Christianity leaves man's spatiotemporal structure unaltered during his sojourn on earth. Instead we see three constitutive forms of time developing in this short span of objective time: *tempus*, eviternity, and eternity. *Tempus* accommodates man's existence on earth from birth until death. *Tempus* has a beginning and an end. After death man's immortal soul continues to exist in eviternity, hopefully in heavenly bliss. Eviternity begins at death and then lasts forever. Only God exists in eternity which has no beginning and no end. The Christian depends for his release from suffering on the gift of grace from God, who will bless him with this state if the Christian reacts to God's love for him by loving God in return and by having faith in Him. This faith includes the Christian's belief in a metaphysical world which lies beyond man's sensory experience and intellectual understanding of the "real" world.

Joel Gajardo offers a more concrete definition of suffering which fits into the context of the struggle to obtain justice by the oppressed and deprived masses against the wealthy upper classes in South America. He takes exception to the viewpoint that whatever goes on in a human society has been preordained by God and believes that people can no longer accept this view. These oppressed people suffer because they have been deprived of their dignity through human injustice.[5]

McGill examines suffering within the context of interpersonal relationships. In

McGill's paper there may be a slight intimation that man's suffering is related to things he will be doing in the future. He also seems to suggest that some are of the opinion that suffering is a necessary element in the process of personality development. But McGill himself holds, to the contrary, that we isolate ourselves behind a wall in order to avoid suffering. McGill's complex paper interprets suffering in a negative vein, as something that comes upon us from the outside by dint of the demonic power of evil and he sees it as preventing the creation of selfhood.[6]

The Russian conception of suffering also interprets it as a social rather than as a private process, since it is embodied in the national belief and value system. Russians have a mystic faith in their destiny as a nation and believe their country will fulfill a leadership role among nations, so that rivalry and materialism will be eliminated. They see their victory over Hitler as a miracle which testifies concretely to the reality of this destiny.[7] The Russian viewpoint of suffering places it in a large temporal-spatial schema which overarches long periods of the nation's history. Therefore, the suffering of the individual is drowned, as it were, in the immense sea of suffering of the entire nation. This interpretation is clearly an attempt to break loose from the constraints of our narrow, temporal-spatial existence and to give meaning to personal suffering within a wider context.

Kierkegaard sees suffering as a struggle which is going on in the individual human being as an expression of creating unity within the contrasting elements which constitute man's existence. Man's dualistic structure flows from the determinants of his existence, which are emotional tension, despair, and suffering.[8] He places man in time and in eternity. Man is neither finite nor is he infinite, but he strives toward synthesizing these two temporal aspects of his existence by striving toward becoming concrete. Man realizes himself as an authentic individual within this polarity of opposites which causes him to despair. Kierkegaard reveals the interrelations between our spatiotemporal structure, the emotional turmoil of despair, and the constructive power of suffering.

Cassirer expresses similar convictions about suffering as a constitutive element in our existence. Man is not a homogeneous phenomenon since he is a mixture of many contrasting elements. In contraposition to the rules of traditional logic, which is based on the law of contradiction and which applies only to things and processes which are free from contradiction, man is struggling to establish a dynamic equilibrium between opposite forces which keep him under emotional tension. These contrary powers cannot be made to live in peaceful coexistence although they are not mutually exclusive. But man has to suffer through the process of forcing them into a interdependent relationship. Cassirer lays stress on the fact that we live into our future although we are also influenced by and conscious of our past. But since we live primarily and mostly into the future we are harassed by doubt and fear.[9]

This brief enumeration of how some other people interpret human suffering shows that they have one feature in common, namely that they all stress the importance of the time problem in this area, which Cassirer brings out more clearly than most others.

It is my goal to construct an evaluation of suffering which is based explicitly on our spatiotemporal structure and to show that the modes and the varying depths of suffering depend on how we live into the future in objective and in subjective time-space.

Suffering implies for most thinkers that a human being is in the pit of despair, wringing his hands in utter hopelessness. This is true in certain extreme conditions, but we also experience milder states of sadness, or feeling blue, we anticipate disagreeable tasks awaiting us, or we may feel tense due to the general uncertainties of living. Expressed in terms of feelings, we traverse a spectrum of slight tension, anxiety, fear, and dread, depending on the situations which confront us in the outside world or conflicts of varying severity which threaten us from within ourselves. One may question: at what point in this spectrum do we actually suffer?

The levels of emotional tension and the physical concomitants which accompany and in part express our inner emotional states, do not help us to determine the dividing area between emotional discomfort and suffering. Physical symptoms of feeling tones vary widely from person to person; some people are quick to tremble, perspire, or to have cardiac palpitations, and others experience and display such symptoms to a minimal degree. This means that extrospection does not enhance our insight into our own emotional states or those of others, and this is one reason why this method of observation cannot clarify the meaning of suffering. We suffer in many different modes which we cannot compartmentalize, and therefore we cannot measure them. The quality and the physiognomy of the imperfections, conflicts, opposites, dichotomies, and fractures determine these modes, not their quantity or their numerical strength. The quality of these modes is in part dependent upon their temporal structure, that is, whether we are concerned mostly about the future-in-the-fringe of a precious present, or whether we worry about events which may develop in the distant, objective future. Beyond our tension, anxiety, or dread toward the future-in-the-fringe we are also aware of possibilities in the further, distant, objective future which may beckon us encouragingly, or which may menace and threaten us. The combination of tension toward our personal future and toward the objective, communal future cause us to worry, fear, panic, or dread. It is impossible to render a quantitative description of the gamut of these emotions because of their dualistic spatiotemporal structure. If one recognizes objective time-space as the only form of time-space, then one might conclude, following the extrospective method of observation, that there is a step-by-step, quantitative increment in our emotional tensions which we interpret as suffering at the maximum terminal point of the spectrum. From the viewpoint of subjective time-space and by the introspective route, the several varieties of suffering fuse into each other as they display different qualities and physiognomies, rather than a mere increase in quantity and intensity.

This we can clarify by differentiating between suffering due to physical illness versus suffering from emotional or mental illness. Contraposing these two modes of suffering does not culminate in two disconnected modes of suffering, since they constantly interpenetrate each other. But since they are seldom in equipoise with one another, one mode may be in the foreground, the other being less noticeable.

SUFFERING IN PHYSICAL ILLNESS

We may suffer in the physical mode due to circumstances in our environment, or due to abnormal processes within our body. People suffer from excessive heat or cold, from hunger, natural disasters, the convulsions of war or revolution. In these situations the threat of death is located outside ourselves in the environment, in which we can include our body. However, the physiognomy of this mode of suffering shows that there is more involved than just physical discomfort, pain, or damage to our body. These calamities tear a society apart, and therefore the individual is suddenly an island unto himself though he may be surrounded by other humans who also suffer. His immediate family and his friends may be lost, and he misses their emotional and societal support. This loss may hit him more deeply than hunger or cold, or damage to his body. This mode of suffering can be destructive, rather than constructive, since many people will give in and give up under extreme duress.

Physical illness unbalances the mind-body relation and it may threaten to destroy the body. The awareness of one's possible, imminent death overlays bodily pain and physical suffering. In the following brief case histories one can see the variable psychological reactions to physical illness.

> Sudden, severe backache is a common affliction (*Hexenschuss*: shot by a witch). The patient is frozen in a fixed position which makes him feel totally helpless and also quite hopeless. This pain can be excruciating, and, as is always the case if one senses bodily pain, the patient with back pain may worry if this means that severe damage has been done to his vertebrae or to his spinal cord. But beyond suffering from pain in and concern about his body, he may feel very anxious as to whether he will ever be released from this fixed position. He wonders if he will be able to get up to respond to the urgent calls from his bladder and fantasizes that he may regress to the incontinence of babyhood. Beyond his concern about the here-and-now, he also questions if he will be able to dress himself and to go to work and if not, how his illness may affect his distant future. The insistent danger signals arising in his body tie him down to an image of his distant past in infancy and childhood when he was totally dependent on others. These memories interfere with his projecting into the future-in-the-fringe and his contemplating the distant, objective future. Two processes going on in unison lock the patient into the form of objective time-space. The malfunction of his body blocks the operations of the mental triad in a precious

present. Superimposed upon suffering bodily pain, he suffers from overwhelming anxiety about the future in objective time-space. Frozen in objective time-space, he cannot perform purpose-striving movements in subjective time-space. This destroys, temporarily, his spatiotemporal structure since physical malfunctions intrude excessively into subjective time-space and therefore his mentation is disturbed also.

An elderly farmer sustained a fracture of the neck of the femur which caused him to suffer from severe pain and prevented him from standing and walking. The consulting orthopedic surgeon decided not to put a Smith-Petersen nail in the neck of the femur, for reasons which escape me now, and he put the patient in an extensive cast. The pain soon disappeared, but, convinced that he would never be able to tend to his farm again, the patient sank into a severe depression. I encouraged him to look forward to the time when the fracture would be healed, the cast removed, and then we would start with physiotherapy. In due time I instructed his son to erect a contraption over the bed with ropes and pulleys, so that his father could exercise his leg and hip joint at first passively and then gradually actively. After many months he was back on his farm. The initial pain due to the fracture was quickly overshadowed by his mental imagery of being a helpless cripple living in a "fractured" life-style. The vision of his spatiotemporal life-style in ruins was far more devastating to him than the bodily pain and the many physical discomforts he suffered. Suffering on the mental level in subjective time-space weighed him down much more than suffering in the bodily realm in objective time-space. To quote once more my friend with osteoarthritis: "The image of a patient with osteo in a wheel chair is much more frightening than the osteo."

The pain due to occlusion of a coronary artery is tremendously frightening, because it literally scares the sufferer to death. This is aptly described in the term *angina pectoris*, which is derived from the latin *angere*, to suffocate. This pain immediately releases the dread of one's personal death in the here-and-now, because pain paralyzes the defense mechanisms against the dread of death as well as the sublimation of this dread. Pain, dread, oxygen deficit, prevent the patient from performing mechanical movements in objective time-space as well as purpose-striving movements in subjective time-space. In fact, the better one is able to refrain from all movement, except deep breathing, the better are one's chances to survive. Most victims of this frightening condition immediately take countermeasures, call for help, and get to a hospital. One of my colleagues who made the diagnosis himself that he was having a recurring coronary attack, called the ambulance and admitted himself to the intensive care unit.

At first the acute pain and dread of dying destroys the patient's future-in-the-fringe, but, if he does not give in and does not give up, he fights to restore his spatiotemporal structure which was almost destroyed initially by the dread of dying. While in the intensive care unit he has to be resigned to live, mostly, on the level of objective time-space. In that unit the chemical, electrical, and mechanical equipment of modern medicine reduces him temporarily to living as a causally structured physiological specimen, functioning in objective time-space, in which surroundings purpose-striving movements are reduced to a minimum. If medical treatment is successful, he is gradually released out of this truncated existence, first by disconnecting him from the complicated support machinery, which allows him increasing freedom to perform purpose-striving movements and later, ambulation in measured increments. Finally, his complete temporospatial structure is restored. However, many recovered postcoronary patients have to change and rebuild their life-style by curtailing both kinds of movements, which means that they have to readjust their way of living in both objective and in subjective time-space. This is no small task. It demands a great deal of self-discipline, imagination in finding new solutions, dropping old habits and hobbies, and finding new ones. It is of great importance to learn to control one's emotional reactions. Not all recovered patients are capable of a different and reduced life-style. Compare the readjustment of the doctor who admitted himself to the hospital and who is now back in practice on a sharply reduced schedule, with a retired physician who was furious after several mild coronary attacks because he could no longer go hunting in Maine, and he complained bitterly that he could not even lift his gun to practice target shooting in his backyard; he died in his summer home after a lobster feast.

Dr. Sheehan reveals suffering from bodily stress in a totally different context in a combined physical and emotional realm of experience. A practicing cardiologist and an ethusiastic marathon runner, he speaks from first hand experience about the states of consciousness which a runner traverses.[10] He describes one of the remarkable ways in which we humans first get in touch with our personal dread of death and then overcome or neutralize or conquer this dread. The athlete forces himself to live to the limits of the physiological capacities of his body during which struggle he makes the recalcitrant body do what it does not seem to be able to do. He feels pain in many muscle groups, his heart and respiration are pushed to the breaking point, and he feels utterly miserable. Then, if he is able to persist during a period of time that he knows from previous training periods, there suddenly arises within him a feeling of power over his body, of legs going by themselves, of all organs working in unison. A few moments ago he felt as if he were killing himself, now he senses the holism of mind and body. This is an uncanny feeling of lift and happiness. During the brief interval of time, which Dr. Sheehan calls an instant, when the athlete's struggle turns into a feeling of omnipotence, past, present and future become as one. He overcomes hardship within a present interval within which past extending behind him and future protruding ahead of him interpenetrate. That is, the athlete lives in a precious present and he experiences the time-binding power of the mental triad. Since the athlete pushes himself to the very limits of possibilities of being he lives through many little deaths. For him, the initiate, death is not threatening any longer.

Paraphrasing Dr. Sheehan's words, I would say that the athlete dies many times in a metaphorical sense which is one of the many maneuvers by which we sublimate the dread of death. Dr. Sheehan describes vividly and concretely how the athlete experiences the interpenetration of past, present, and future in a precious present in his all-out efforts, enduring the pain and then the release from the dread of death. He shows that the precious present need not always be a blissful and happy state of mind which one traverses, since the athlete suffers agonies both on the physical and the mental levels. The precious present in conjunction with the mental triad synthesize the future fringe and the past fringe in a durational present, and they constitute the agents which bring forth the dread of death on the one hand and bind this dread on the other. This experience of the athlete is precious because it carries forward with it into the objective future, after his efforts have been successful, feelings of omnipotence, subjugation of the recalcitrant body, self-confidence and self-worth. The athlete grows in these experiences, and he becomes more of a whole person, having suffered through the ordeal of dying in the metaphorical sense, and he is a different person than he was before he forced himself to undergo this experience.

Einstein calls the separation between past, present, and future an illusion. I presume that Einstein overlooked the difference between compartmented objective time and the interpenetrating essence of the precious present. The synthesis of the three aspects of subjective time on which the emotionally charged experiences of the athlete are grounded is far from illusory; on the contrary, it is rock-bottom reality.

SUFFERING IN EMOTIONAL AND MENTAL ILLNESS

Suffering on the emotional and on the mental levels cannot be distinguished sharply from suffering due to physical illness. There are physical elements in neurotic and in psychotic illnesses, and there are mental and emotional elements in physical illnesses. One sees in physical illness that the distortion of the patient's spatiotemporal structure brings the dread of death to the fore. A phenomenological analysis of emotional and mental illness illustrates even more clearly the correlation between this dread and the patient's abnormal temporal structure. We are faced with two questions: How do neurotic patients and psychotic patients live into the future-in-the-fringe and how do normal people live into their private future? And also, how do psychiatric patients cope with the dread of their personal death and how do normal people deal with this dread? Three brief vignettes of patients in long-term psychotherapy exemplify three different modes of coping with this dread. I give them fictitious names for obvious reasons.

Miss Alice, now in her seventies, suffered from cancer phobia, travel phobia, and excessive dependence on her physician. Her mother was a depressive, dirt-obsessed person who mismanaged Alice's toilet training. Alice is an only child. Notwithstanding her neurotic impairments she lives a very active social life, and before retirement she enjoyed deep satisfaction from her profession. However, she had to check several times a day to see if there was any rectal bleeding, and her travel phobia centered on the possibility that toilet facilities might be inadequate or even absent wherever she wanted to go. Psychotherapy revealed to her that her mother's punitive superego was still active in her own restrictive superego. She gained insight into the connection between her mother's phobias about poverty and her own reluctance to spend any unnecessary money on herself. She learned that she experienced these ancient warnings as if they were still active in the here-and-now. Sex and dirt having been equated by her mother, she was severely sexually repressed, and she was deeply ashamed when she told me about short, fantasized erotic attachments to adolescent boys. Finally, in her sixties, she allowed herself to enjoy a superficial hug-and-kiss relationship with a man her own age without feeling guilty. When psychotherapy began to relinquish her from her pathogenic past, her toilet habits were no longer tension-ridden rituals and she stopped checking for rectal bleeding. That is, she began to master her dread of death in the guise of her cancer phobia. Travel phobia improved and she has visited friends in Africa. Now, she sees her physician only when necessary.

Mrs. Bernice, in her forties, is the product of a broken home; her mother abandoned the family for another mate. Her father was a nonfathering individual who looked down upon women as an inferior breed of human beings, and he was authoritarian in the management of his daughter and two sons. There was a long succession of housekeepers, none of whom functioned as a surrogate mother. Bernice is married to an alcoholic and she fears that her two sons may follow the same route. She worries about money although her husband holds on to a good job since he never drinks during working hours. She suffers from periods of depression, and at one time she needed hospitalization for this condition during one month. Psychotherapy allowed her to express and to face her intense hatred against her husband, whose worsening alcoholism caused her to feel trapped and hopeless. She said: "What is there in the future for me?" As she gained insight into the roots of her hostility which go back to her hostility against father and brothers, she also took some realistic steps to conquer her depressive moods. She is holding down a secretarial job quite effectively. Finally she begins to give herself credit for doing a good job as a mother. Recently she mustered the courage to face her husband with the choice, "Either go to a drying-out place or I'll divorce you." He heeded her warning and he has been able to control his destructive alcoholism to some extent.

Mr. Charles, in his thirties, is single and a professional person. His father is a "burned out," chronic schizophrenic patient and his mother is a passive personality; his sister has a spotty work record and she is overweight. Charles has harbored murderous fantasies throughout his life. In his teens he had an imaginary military arsenal in the basement and he blew up whole sections of his hometown. In his twenties he had vivid, bloody fantasies of killing a superior. In his thirties he imagined he drowned certain women who were in high social positions in their bathtubs. Usually he escapes because he has also fantasies of supernatural powers of omnipotence, but often he was thrown into jail and the guards beat him up and threatened to kill him. He has suffered from several depressed episodes. In psychotherapy he shared these fantasies with another person for the first time in his life. Once he told me about a fantasy of coming into my office with a big stick, slamming it on the table and yelling: "Do something about it!" Since I did not flinch or berate him for this verbal expression of aggression against me, the flow of reports on his fantasies was not interrupted. Gradually the fantasies became less murderous in character, became less frequent, and have finally stopped. After many years in therapy he revealed almost by accident, as "a slip of the tongue" as he said himself, why he thinks that girls will reject him: he thinks he is ugly, which he proves to himself several times a day by looking at his profile in a mirror. This fixation on his body image goes back to age twelve when his classmates nicknamed him "Dumbo with the big ears." After many discussions about his profile, which at first I had to practically force upon him, I began to suggest that he try to look at his profile not during a split second, but to try to lengthen the duration of this ordeal little by little. Now he looks less frequently and some times for a few seconds. But he still thinks his profile is ugly.

These three people have in common a similar distortion of their temporal structure. Their living in the present is dominated and darkened by memories of past experiences, which they relive as if they were still the actual, lived-in present. Therefore, projection into the future-in-the-fringe and moving ahead into the objective future are truncated. Their temporal structure is faulty both from the side of their past and their future horizon is also severely limited. The mental triad cannot function freely in this maimed temporal ambiance.

Their life-style is related to their temporal structure. We see a fierce struggle in these three people against the dread of death. Miss Alice concretizes this dread in her cancer phobia and her inability to travel freely where she wants to go. Mrs. Bernice, worn down by the gradual dissolution of her marriage, harbors death wishes against her husband, and she expresses these destructive trends by saying to him, "I wish you were dead!" and at the same time she has a dread of being alone and abandoned. Mr. Charles kills people in his fantasies and the jail keepers threaten to kill him. Their distorted temporal structure and their being tortured by the dread of death are interrelated. Therefore, I see the task of the psychoanalist and of the psychotherapist in terms of helping their patients to rebuild their temporal structure. We strive to release them from being dominated by their past, so that they can gain freedom to move into their future. We lead our patients toward placing themselves squarely within their personal present. By this maneuver they learn to face their personal future with new attitudes and from a new perspective. Liberated from repetitively reliving their past, they become less anxious when they attempt to move into the future-in-the-fringe of a precious present and into the objective, distant future. Early childhood experiences and archaic memories come to the surface during the therapeutic hour, when they are at liberty to give vent to primitive anxiety and destructive drives here and now in the reassuring presence of the therapist. Catharsis defuses some of the intensity of this anxiety and the patient dares to move into the future. Released from the weight of their past, they become future oriented. The mental triad can now fulfill its functions in a precious present.

Being future oriented, they begin to cope with the dread of death within their liberated, spatiotemporally, enlarged life-style. The neurotic patient expects punishment for his infantile aggressive drives, which is meted out to him by his infantile superego. First, he is punished by neurotic symptoms for sins which he committed in a dim, distant past. Underlying these symptoms, he is tortured by the dread of death, the ultimate punishment. When he begins to see that his superego stands for the all-powerful punishing parents of long ago, then the patient begins to strip himself of this paralyzing influence. He develops a constructive, healthy, ethical conscience, which does not punish him for so-called sins, but it urges him to face the fact of his ultimate, future death as a completion of his life, not as punishment. Now he is capable of coping with the dread of death by the means which are open to all of us, especially by sublimation, exactly because he dares to move into the future-in-the-fringe of a precious present and into the distant, objective future.

Dr. James Gorney describes in vivid detail the extraordinary, storm-filled sessions with a borderline patient whom he calls Sarah. The central, constitutive role of the dread of death, both in the patient and in the psychoanalyst leaps dramatically to the fore in his revealing article. Sarah suffered one psychological trauma after another in her childhood and during her teens. Her mother died in cachexis due to cancer; a stepsister committed suicide; so did Sarah's own sister; her well-meaning and supportive stepmother became less available to Sarah after her ex-husband committed suicide. Sarah was well educated and gifted artistically and musically and she was quite articulate during treatment sessions. However, at the end of each hour she invariably stormed out of the room and slammed the door behind her with a loud bang.

After approximately seventeen months a drastic change came about in the direction of the analysis during a tumultuous scene. Sarah suddenly cried out in a spine-chilling scream that the office was like a tomb for her and she rushed to the door. Dr. Gorney says:

> Without pausing for thought, I leapt up, bounded across the room and just beat her to the door, effectively blocking her exit with my body and said: "I want you here with me."... Tears poured out of her in astonishing profusion. Through it all she blurted out the words, "This office is a tomb." At that instant I looked squarely at her, never having been in such close physical proximity. As I looked at her emaciated, contorted face I had a sudden hideous vision of a skull, . . . I said: "This office has become a tomb for both of us. We need to escape from it to find each other." ... Walking side by side the conversation was lively and animated.... After that meeting she warmly grasped my hand.... Sarah began to gain weight and once again became energetic. Nine months later she was able to leave the hospital.[11]

This crucial intervention was the watershed during which brief interval of time Sarah's destructive suffering was turned around toward constructive suffering, since she had come to grips with her dread of death, and therefore she became able to begin to master this dread. Dr. Gorney postulates, "It might be that most analysts are characterologically devoted to mastering their own latent sense of ontological despair, as well as to modify through interpretations their patients' protestations that life is indeed 'meaningless'."[12]

I see the dread of death as the central theme in this case history. This immutable fact of life is the most severe cause of human suffering and resides in the center of the core of human existence. The dread of death assumed many forms in Sarah. She is frightened to death by suicide, but she is "too dispirited to make any real plans." Nevertheless, she dies in a metaphorical sense by almost starving herself "to death." For her, the fusion-symbiosis in a transference relationship with her psychoanalyst stands for a replica of the early, ancient, fusion-symbiosis between her and her mother. She fears further loss of ego boundaries in that transference, since her ego boundaries are shaky enough already.

Dr. Gorney sees his anorectic, cachectic patient "dying before my very eyes," and at one point he has the "hideous vision of a skull," while the office represents death to him, in the image of a tomb. Altogether this paper is an extremely touching exposure of the core of human existence, which is our ontological despair, grounded in our dread of dying and of death.

Not as dramatically overt as Sarah's dread of dying and not as disturbing to the therapist, is this same dread in phobic patients who hide this dread behind a variety of symptoms. Driving phobics give us an opportunity to compare the inner mental states of patients, with the minor fearfulness which tantalizes many of us in certain traffic conditions. The driving phobics are tortured by crippling fears that they may cause an accident if they should lose control of their car, for instance by depressing the accelerator instead of the brake pedal. If they come into psychotherapy they will say initially that they fear they might kill a pedestrian or a child, but later on they describe in great detail gruesome fantasies in which they themselves are mauled or killed. A variant of this fantasy is that they may be so confused and tense that they may drive on and on, not knowing where they are going and they die somewhere all alone and forgotten by everyone.

Driving phobics experience their personal death, in fantasy, in the here-and-now, repetitively and in almost the same format. This is destructive suffering, and it has no relationship to dying in the metaphorical sense which is constructive suffering. When driving phobics pass the scene of a lethal accident they will think, just like the others, "Not yet, not I"; they too deny and postpone the possibility of their personal death. But their fantasies of actually dying "now" are quite another matter. While they are in the grips of the phobia, their future-in-the-fringe is so frightening that their mental triad is stopped dead in its tracks.

The dread of death is handled in a very different manner in the normal life-style. One accepts that death cuts off projection into the future-in-the-fringe, destroys the precious present, and terminates functioning of the mental triad. Death also "returns our body to dust." Death is a double spatiotemporal truncation of our mundane existence, and it is utterly final. But man is loath to accept this finality, and he creates various countermeasures against the dread of death. Besides the mental defenses such as denial or postponing, sublimation is the most effective and constructive defense against this core dread in our existence, against the "parent-suffering." In the normal life-style the dread of dying does not paralyze the functioning of the mental triad in a precious present which, therefore, can work together with these defenses. But the future-in-the-fringe is so threatening to a driving phobic that the cooperation between the mental triad and sublimation is cut asunder.

In a hospital one often sees the sad sight of decompensated, severely depressed people in whom sublimation does not work during the depth of their illness. The faith of the religious person is powerless, the rational thinking of the philosopher is ineffectual, the artist's emotional involvement in his creations does nothing to touch the ravages of the temporarily fractured life-style. In these severe conditions sublimation and the mental triad are not coordinated in a precious present and they cannot function as an interpenetrating whole.

SUFFERING IN A FRACTURED EXISTENCE

In the foregoing examples one sees that human suffering can heal conflicts, dichotomies, and even the fractures in our human existence. In contrast, I shall relate a personal observation of the behavior and the verbal exclamations of a near-psychotic nature of a suffering person who was at the time incapable of creating unity in diversity through the process of suffering.

When I was about sixteen years of age, Free Domisse, authoress of *Krankzinnigen* ("Mentally Ill People") was visiting us. While I accompanied my sister who was singing one of Bach's *geistliche Lieder* (religious songs) "O Jesulein süss, o Jesulein mild ...", Free was looking at a reproduction of Michelangelo's *The Last Judgment*. She suddenly cried out: "If only you could believe that!"—meaning the words in Bach's song—and she stormed out of the house. We went after her, knowing that she had been hospitalized several times for mental breakdowns. I happened to find her, tried to calm her down, and brought her back to our house. In her autobiography she describes several episodes of aggressive outbursts and nagging doubts in relation to religious problems.

At the time of the crisis which I have described, the religious and the artistic sides of life were presented to Free at the same time through auditory channels by Bach's music and by visual channels by the work of Michelangelo. This simultaneous contrast was too much for her to bear internally. Bach's song flows along without a ripple, but Michelangelo depicts the judging Jesus condemning sinners to hell, whose suffering he makes visible in almost gruesome detail. The fracture between the ascending, ecstatically happy, saved ones and the descending, tortured, and hopeless condemned ones, will disturb any thoughtful contemplator of this mural. The peaceful music and the turmoil in the painting rent the sick person's mind apart, which she acted out to the consternation of all who were there, witnessing how these gaps and unresolvable contradictions flooded her mind. Her autobiography ends on a much happier note, since she was able to find a modicum of synthesis of her personality in later years, when she did heal some of the contrasts, dichotomies, and fractures in her existence, with the help of her farsighted psychiatrist to whom she dedicated her book.

The dread of death which Free acted out in disturbed psychotic behavior is felt in a subdued emotional climate by normal people also. Religious metaphysicians attempt to heal the fracture living-versus-dying, but they get caught up in the fracture between some souls living happily in heaven, while other souls suffer a living death in hell. The normal person may accept this solution rationally, but he too does not find internal peace in the face of the threat of death, since he also is upset by Michelangelo's dramatic exposure of this ultimate fracture.

Kübler-Ross has encountered failure of sublimation of the dread of death in many religious patients. Patients who accept the standards of their religion may find some solace from these doctrines but not much more than nonbelievers find in other supports. Those with true, intrinsic religious beliefs and true atheists found most comfort in their battle with conflict and fear.[13]

Breton struggles with the idea of the dread of death on a conceptual level, and he seems undecided whether it is an unmitigated evil or whether it contains some hidden value. Death returns us to the empire of inert matter. This trend of events seems absurd, and suffering becomes inexplicable from this point of view. Reason will not accept such an uncomfortable destiny which is contraposed to the normal course of life. We have evolved from being reflex automata to rational creatures who carefully plan our defenses against life's vicissitudes. But this explanation of suffering is too simplistic for those people who see suffering as consubstantial with life itself.[14] I see in Breton's contemplation of suffering an uneasy and insecure shifting back and forth between the idea that suffering may be meaningful and the conviction that suffering brought into the world by evil forces is like a foreign body which mars the perfection of life. He does not accept that suffering is the core of the human existence.

My intuition leads me to believe that Chopin did accept this fact. He knew that he was dying from tuberculosis and expressed his sadness about his losing touch with music. He could not think toward the future but was drawn back toward his past and remembered his mother and his sisters. But the world of music and how the Polish people sing was fast slipping away. He wrote two short compositions and tried to perfect some other recent compositions, but he was unable to put his mind to the vigorous task of really creative composition.[15] One is impressed by Chopin's heroically going through the pain of the process of slowly dying. His grappling with his impending death brings to mind the clinical and the educational work of Dr. Kübler-Ross with patients dying from malignant diseases. According to her experience, dying patients pass through several stages before they grasp the final state of acceptance. They have to separate themselves step by step from their physical and societal environment, first from casual acquaintances, then from close friends, and at last from their truly beloved ones. Monumental though this task is, it rewards some patients with renewed closeness to those who are deeply meaningful in their lives and instead of alienation, they may deepen and strengthen these bonds as a result of their communal suffering in the face of the inevitable end. Patients fear dying because they are convinced that this process is bound to bring about their helplessness and their utter isolation.[16] Here she puts her finger exactly on the double origin of the dread of death. In dying we have to give up communication with our loved ones, which leaves us utterly alone and lonely no matter how deeply emotional our leave-taking may have been. In death we do not only face the destruction of our temporalspatial holism, but we also have to descend into the pit of loneliness. This is the dual core of our human existence. Others such as Dyan and Amundsen report no such excruciating mental ordeal when they faced the imminent possibility of death.

The following case history by Kuhn and Bradnan is a pertinent illustration of how some people cope with their dread of dying. Their patient returned to the hospital four years after the removal of a sarcoma, so that the surgeon could perform a biopsy of a swollen lymph gland in her neck. Instead of being worried about the outcome of this test, she was depressed because she had severe pain in the left side of her chest and abdomen. She would talk only about the pain, but the psychiatrists found out that she had become more and more concerned about her death. Finally shedding tears profusely she said that if she died, her eleven-year-old son would be in the care of her alcoholic husband, and after this discussion of her realistic worries about the future the depressed mood vanished. The semistructured psychiatric consultation made it possible for her to contemplate her own and her son's future, to make plans for his future, and therefore she could shed the veil of irrelevant somatic symptoms behind which she had hidden her dread of dying.[17]

Suffering, the Core of Human Existence 237

We see various ways of coping with the dread of death within the context of burial rituals. Only man performs these rituals, because only man is aware of his impending future death. Antigone heroically disobeyed the orders of the king, who left her brother's corpse unburied as punishment for his fighting with the king. Antigone's dread of what would happen to her brother's soul was stronger than her sense of duty to the king and the fear of punishment that she was sure to incur. It was better to be punished by a mere mortal than to suffer under the wrath of the gods.[18]

Rodin has concretized the torment of the burghers of Calais and showed how six people cope with the dread of dying, each in his own style. Some radiate an inner sense of power against the dreaded event, which power they have gained after a period of desperate suffering. Others are suffering until the last moments of their life. Their suffering for the ethical ideal of giving up their personal life to save hundreds of other lives from the massacre gives them strength to continue on their march toward death. Rodin has often depicted suffering in his sculptures, but here he shows the subtle variation in which each individual comes to grips with the dread of dying and how each in his own way is able to sublimate this dread; therefore they can carry out their humanitarian mission.

> Eustache de Saint-Pierre...does not hesitate, he does not fear. He advances steadily, his eyes turned inward, upon his soul....he is the first who has offered himself to be part of the group of notables whose death must save their fellow citizens from the massacre. The burgher who has the key in his hands stiffens his whole body to find the strength to support the inevitable humiliation. A man...marches almost too fast, he seeks to shorten the time which separates him from the execution. A burgher who is clutching his skull in his hands, abandons himself to violent desperation. A fifth notable passes his hands in front of his eyes as if to dispel a frightening nightmare. The face of a sixth burgher...is contracted by terrible anxiety. But his companions are marching, he rejoins them.[19]

Literature abounds with instances in which an author comes to grips with the problem of dying. Galsworthy's *Forsyth Saga* begins with the death of an idealistic young architect who loves beauty, and roaming around in a London fog, defeated and hopeless, he is run over by the traffic. *Indian Summer of a Forsyte* relates the death of a young soldier from dysentery in the Boer War; a most touching and saddening interlude. When an old Forsyth is buried, the funeral cortège goes forward in measured steps in the richer sections of London, but the horses are made to break into a trot in the poorer sections. A realistic illustration of the socializing value of burial rituals. In the end, Soames, the central figure in the entire work, dies after he has tried to save his collection of paintings during a fire. Concern about death goes like a constantly recurring theme through the entire opus, as Galsworthy shows us how different people cope with this dread.

Thomas Mann's novel, *The Magic Mountain*, is a *Zeitroman*, a novel about time which depicts how the large cast of characters live in time. Hermann Weigand details in his monograph how patients distort their temporal structure during a long sojourn in a tuberculosis sanatorium.

> Chapter III gives a exhaustive account of Hans Castorp's impressions of the first day of his sojourn in the sanatorium. Chapter IV covers the first three weeks, the period originally allotted for the visit. Chapter V takes Hans Castorp through the seventh month of his hermetic existence, climaxed by the night of the Mardi Gras. Chapter VI consumes time at an even faster rate, carrying the story to the point of Joachim's death, two years and four months after Hans Castorp's arrival at Davos. Chapter VII, finally accounts for the remaining four-and-a-half years of the total span.
> ...There is nothing accidental about this apportionment. It is governed by a law of perspective.[20]

Joachim, Hans' nephew, warns his visitor from the outset that the indwellers of the sanatorium do not think in hours, days, weeks but in "monthlets" and "yearlets." Then he tells Hans that the dead patients from a less luxurious sanatorium below the "Berghof" are transported to the "Flatland" by bobsled but that horse-drawn vehicles do this service more properly for the fatalities of the "Berghof." They arrive in the dining room for supper in stitches of laughter, tears streaming down their cheeks, denying in this behavior a philosopher's definition of laughter, "I laugh so that I may not weep." They try to forget, postpone, and deny their dread of death. At the end of the book, Hans, who is now an infantry man at the beginning of World War I, is swallowed up and lost sight of in the masses of his marching regiment. It is as if Thomas Mann does not want to face the possibility of Hans's death in the trenches, a character with whom he has intimately identified during the creation of his masterwork. One might interpret the message of *The Magic Mountain* as the sad tale of our being "thrown-into-the-world" and toward death and how the patients in the sanatorium use distortions of the temporal perspective as a means to keep their dread of dying and of death under control. From beginning to end, the dread of death lurks behind the patients' preoccupation with time. We see here the common identification of death with time and time with death.

SUFFERING AND CONSCIENCE

Most human beings listen to the voice of their conscience. This capacity to judge ourselves and our deeds makes us painfully aware of the blemishes, imperfections, and the dichotomies in our existence. Conventionally, we think of our conscience as a predominantly nay-saying activity, which places us before black-or-white, either/or decisions and choices between good or evil, reward or punishment. We listen either to our imperfect inner voice, or to those of wise deities, or to the perfect voice of God, according to religious interpretations of conscience. A short summary of this wide-ranging problem exposes various philosophical conceptions. It is often thought that ethics develops after man begins to take stock of his actions, which by some is attributed to man's innate moral sense. Others hold that ethical imperatives of conduct are a result of people weighing the results of past actions. The age-old search for absolute standards of morality may be grounded on the belief in religious absolutes. The ultimate guidance in moral decisions for those who defend the doctrine of emotivism is the subjective desires of the individual.

> Throughout the history of philosophy men have sought an absolute criterion of ethics. Frequently moral codes have been based on religious absolutes. Among modern ethical theories are instrumentalism for which morality lies within the individual...; emotivism wherein ethical considerations are merely expressions of subjective desires of the individual; and intuitionism which postulates the immediate awareness of the morally good.[21]

Some of these definitions of conscience are derived from a view of human nature which emphasizes our past; others imply man's future-striving tendencies without explicitly saying so. According to most religious-ethical theories, conscience reminds us of sins which we have committed in our recent past and loads us down with sins which our forefathers have committed in an exceedingly remote past. Conscience is regarded as a past-determined function which sprouts forth out of original sin.

Freud conceived the superego in a past-tending context; it is generated by authoritative and punitive parental figures who exert powerful influence during infancy and the child's early developmental period. These early experiences are the formative roots from which the ethical conscience of the adult person grows.

I miss in these conceptions about conscience any explicit reference to our temporal-spatial structure. I propose that we build a concept of conscience which is anchored upon the present and which looks out on the personal and the objective future. In conjunction with the mental triad operating in a precious present, conscience loses its threatening, negating, and tyrannical characteristics since it is not a backward-looking power, but a constructive and forward-looking activity. Without being punitive, it urges us to be self-critical, so as to learn from our past mistakes. This ethical conscience does not assume the role of a mythical or historical person who holds up a mirror which reflects our past misdeeds, which makes us feel guilty and which guilt makes us suffer. A past-oriented conscience devaluates the meaning of suffering, because it accuses us of past sinfulness in which we have played no active part. The superego tortures us with revived memories of archaic death wishes against parents and siblings. It makes no sense that we suffer as an adult as a result of these drives which we have long since outgrown.

A time-oriented concept of conscience separates sin from suffering. We suffer because we fail to live up to our own rules of conduct, or because we have failed to concretize an ideal which we have in mind. It is this synthesis of the imperfections, fissures, and dichotomies in our existence which inevitably requires that we suffer. Hence, suffering, under the guidance of our conscience, has intrinsic value. Conscience is a creative participant in the synthesis of objective time-space with subjective time-space. A present-oriented conscience which is future-attuned does not overlook what we have done or not done in our past. It evaluates the imperfections in our personal history, and it encourages us to take counter-measures against repeating these mistakes. It urges us to prepare ourselves more carefully for a future undertaking in which we attempt to concretize values. In order to be capable of listening to the voice of our conscience we must synthesize the two forms of time-space. Only then can we internalize conscience' constructive admonitions and execute them in creative activities.

We suffer pangs of conscience when we do something that revolts against our life-style and our personality, but we do not rake over our past transgressions for the sake of punishing ourselves. An ethical conscience does not give us the latitude to do whatever we wish to do at the moment, since it sets guidelines and markers beyond which we should not and ought not go. Conscience holds hedonism in check. It is a severe taskmaster. We suffer when we act against our own rules of conduct, or when we fail to concretize an ideal on which we had set our sights.

Conscience is an internalized process that goes on between different levels of our personality, but it is not a struggle between our present self and archaic, infantile memory-images. From the viewpoint of our temporal-spatial structure our conscience struggles with us so that we may become more worthy, more value-oriented, better-integrated human beings, with an eye to the objective and the subjective future. When the internal voice of our conscience disapproves of our thoughts, our feelings, our actions, then we are at war within ourselves and we suffer. If we fail to reach a goal, we do suffer, but not because our conscience heaps punishment upon us for this failure; in fact, it may approve highly of our intentions even though we have failed to reach an ideal. After we have made an honest effort, our conscience usually approves of our actions. Not reaching an ideal is hardly the same thing as sinning in the religious sense. The internal voice of conscience makes us more acutely aware of the splits which divide our person. Conscience itself is a divided function, since it is split into a speaker and a listener who conduct an inner dialogue with each other. Here we see that intra-audition and conscience are correlated in close mutual cooperation. Because this inner dialogue is concerned with matters which are important to our whole personality, it is as if a struggle is going on between these two effectors of conscience, the one criticizing, the other defending.

Putting this theorizing in more concrete form, I refer to Allport's Watcher on the hilltop and the Oarsman on the river. In this imagery, it is as if a dialogue is going on between the watcher, who fulfills the role of conscience, and the oarsman, who is the doer. If we fuse these two imaginary people into one-and-the-same person, then we see that the oarsman has split himself in the doer who is being watched by his conscience. The dialogue between the watcher and the doer may sound something like this, "You are lucky to have gotten away with the risk of challenging the forces of the turbulent rapids. Next time you may not be so lucky. So now you are in calm waters; but look out, much more serious danger may lurk ahead, we don't know this river. Let's get some information from people who do know it, or let us buy a detailed map." The doer may listen scornfully to the watcher and he may say, "I have gone through many dangerous rapids and this one was just child's play; why do you make such a fuss?" "You don't know how frightened I was for your safety, watching you from the hilltop; I saw all the dangers ahead that you couldn't see. I don't want to go through that turmoil again." Hopefully the oarsman listens to the voice of his conscience and makes preparations for possible disasters during the rest of his trip down the river.

This elaboration is not meant as a frivolous parody on Allport's thoughtful discussion of determinism versus free will. This contraposing of the watcher who personifies our conscience over against the oarsman who stands for the doer within us actually does go on in our head. We can and do split ourselves into a conscience and a doer. Conscience eggs us on to coordinate our activities better than we did before and to plan more carefully for the future in objective time-space. This allegorical representation of conscience looks artificial because I have compartmentalized the functions of watcher and doer into two separate human beings, whereas these functions, in reality, interpenetrate with one another in one and the same person. Our conscience directs, steers, and encourages our activities by means of a verbally expressed, quite concrete dialogue which we hold within ourselves.

The tensions between the watcher and the doer on the river are minor compared to some life situations where ethical problems are in the foreground. But even in the case of the oarsman facing the rapids, there is an ethical factor. Should he fail, then bystanders such as the watcher on the hilltop may rush to his aid with great danger to themselves. Apart from such an eventuality, which the watcher anticipates in a state of apprehension, is one justified to put one's life on the line where there is really no compelling reason for such recklessness? Is it ethically responsible to go into the boxing ring and run the risk of brain damage and even death, while inflicting these same miseries on one's opponent? One throws the value of one's spatiotemporal structure to the winds. Religious people would say that one ought not to destroy one's body, which belongs to God. A humanist would hold that it is unethical to destroy the holism of one's mind and body because it has intrinsic value.

The voice of conscience can affect us in many different ways. It can give us a mild tug of encouragement or of discouragement, it can be a propelling force to go forward with an activity against heavy odds and it can tear us apart when it makes us aware of irreconcilable life situations. Most of these tragic experiences are hidden from public view in the average turmoil of daily life. Occasionally one is allowed to get a glimpse of a person in public life who is caught up in the painful trap of unsolvable, ambivalent feelings and contradictory motivations. Such an exceptional confrontation took place when Pope Paul VI, after the Maundy Thursday foot-washing ritual, took to task the priests who were leaving the priesthood and who chose the married life instead. The pope compared their behavior with Judas's betrayal of Jesus and chastised them for being driven by "vile earthly reasons." His voice was shaking by the emotional turmoil he underwent and it was as if he wanted to give expression to his inner feelings by speaking in the first person singular. "I cannot think of that tragic drama without associating it in my mind as a bishop and a pastor, with thoughts of abandonment, of the flight of so many brethern in the priesthood."[22]

Here we can sympathize with a man who is torn between his urge to love his fellow human beings, even if they do sin, and his duty as the supreme shepherd of his church, which constrains him to chastise those who fly in the face of one of the tenets of the church, which is the celibacy of its priests.

In ordinary daily life the stings of conscience may plague us in less overt format as parents, because of the contradictory options which are open to us who are concerned how best to bring up our children. We have to shift back and forth between permissiveness and authority, reward and punishment, giving and withholding love. These major and minor crises go on in the sanctity of the home and they are hardly noticed by others, outside of the immediate family. Usually these pangs of conscience are rather low-key and just nagging. But apart from its low emotional intensity, this questioning oneself if one did the right thing during a certain confrontation with one's child, often goes on for years. What is most distressing is that one does not find a final and satisfactory answer to one's doubting if one did the right or the wrong thing. It has been said that in complicated cases the relatively simple prescriptions of conscience tend to prove quite inadequate, which remark is pertinent to interactions between parents and children. There are no hard-and-fast rules how one should bring up children. Rules change from generation to generation and even from decade to decade. Some parents may accuse themselves of having been too strict, while others may regret having been too lenient. Both groups of parents will say as an aftermath, "If only I had known then, what I know now." Thus, conscience plagues us all and makes us suffer.

Although conscience is a future-attuned function of our personality, there is no question that it makes us suffer also because we do look back over our past. But conscience does not make us stop at this point since it also urges us to look ahead into our future and into the communal future. It is the attempt to synthesize past and future within our present which makes us suffer. Hence, suffering is the core of our existence. Conscience urges and reinforces us to undertake this synthesis. Our ethical conscience works in consonance with the mental triad in a precious present in the complex process of integrating the two forms of time-space, the two categories of causality and intentionality and interpenetration with compartmentalization.

NOTES

1. Sophocles, *Oedipus Rex, Oedipus at Colonnus, Antigone: The Oedipus Cycle,* An English Version, by Dudley Fitz and Robert Fitzgerald (New York: Harcourt, Brace, & Co., 1949), p. 82.
2. Max Scheler, "The Meaning of Suffering," in *Centennial Essays,* ed. Manfred S. Frings (The Hague: Martinus Nijhoff, 1974), pp. 121-63.
3. Carroll Stuhlmueller, "Biblical Voices of Suffering and Prayer" (Paper read at the First Congress on Human Suffering, Notre Dame University, Notre Dame, Ind., April 1979), p. 18, by kind permission of Stauros International, West New York, N.J.
4. John Bowker, "Suffering as a Problem for Religion" (Paper read at the First Congress on Human Suffering, Notre Dame University, Notre Dame, Ind., April 1979), pp. 16, 18, by kind permission of Stauros International, West New York, N.J.
5. Joel V. Gajardo, "Suffering Coming from the Struggle against Suffering" (Paper read at the First Congress on...Suffering, Notre Dame University, Notre Dame, Ind., April 1979), pp. 2, 11, by kind permission of Stauros International, West New York, N.J.
6. Arthur C. McGill, "Our Suffering and the Suffering of the Christ" (Paper read at the First Congress on Human Suffering, Notre Dame University, Notre Dame, Ind., April 1979), pp. 13, 25, 21, by kind permission of Stauros International, West New York, N.J.
7. Olga Carlisle, "Reviving The Myths of Holy Russia," *New York Times Magazine,* September 16, 1979, p. 48.
8. Kresten Nordentoft, *Kierkegaard's Psychology,* trans. Bruce H. Krimmse (Pittsburgh: Duquesne University Press, 1972), pp. 137, 129.
9. Ernst Cassirer, *An Essay on Man: An Introduction to a Philosophy of Human Culture* (New Haven: Yale University Press, 1944), pp. 10, 228, 53.
10. George Sheehan, "Athlete and Death: Easier To Live, Easier To Die," *New York Times,* November 23, 1975, Sports Section, p. 2.
11. James E. Gorney, "The Negative Therapeutic Interaction," *Contemporary Psychoanalysis* 15 no. 2 (1979): pp. 320-21.
12. Ibid., pp. 333-34.
13. Elizabeth Kübler-Ross, *On Death and Dying* (London: Macmillan & Co., 1969), p. 237.
14. Stanislas Breton, "Human Suffering and Transcendence" (Paper read at the First Congress on Human Suffering, Notre Dame University, Notre Dame, Ind., April 1979), pp. 4, 8, 10, by kind permission of Stauros International, West New York, N.J.
15. William Murdoch, *Chopin: His Life* (New York: Macmillan Co., 1935), p. 373.
16. Kübler-Ross, *On Death and Dying,* pp. 494-95.
17. Clifford C. Kuhn and William A. Bradnan, "Pain as a Substitute for the Fear of Death," *Psychosomatics* 20, no. 7 (July 1979): 494-95.
18. Sophocles, *Antigone,* p. 221.
19. Auguste Rodin, *L'Art* (Paris: Bernard Grasset, 1924), pp. 102-13 (Author's translation).
20. Hermann J. Weigand, *Thomas Mann's Novel* "Der Zauberberg" (New York: Appleton-Century, 1933), pp. 14, 15.
21. *New Columbia Encyclopedia,* 1975, s.v. "Ethics"
22. Paul Hofmann, "Pope Paul Compares Defecting Priests to Judas," *New York Times,* April 9, 1971.

10
Healing through Suffering in Four Modes

INTRODUCTION

To assert that we do suffer is one thing; to plead for its value as a healing agent and to describe and explain how suffering fulfills this function is quite another task. Suffering heals blemishes, dichotomies, and fractures by making them interpenetrate in subjective time-space. Our interpenetrating spatiotemporal structure makes it possible for us to heal fissures and fractures in our existence. It is as if suffering holds in one hand the faults, the contrasts, the fractures in life and in the other hand its healing power, since it makes fractures partake of healing and healing to intermingle with fractures. While we suffer we are aware and we experience that dichotomies and fractures are being healed by our suffering.

We have to go beyond causation and compartmentalization to describe suffering's healing qualities. This step needs the cooperation of the concept of interpenetration which sets us free from encapsulation in causality. To secure its freedom to make contrasting aspects in our life interpenetrate, suffering uses the mental triad functioning in a precious present to fulfill its healing activities. Interpenetration can be effective in the healing process on the condition that one accepts also the sense of the constrasts between several other building blocks of our existence: objective versus subjective time-space, causation versus purpose-striving, extrospection versus introspection and its ancillaries, compartmentalization as the counterpart of interpenetration, emotionally neutral phenomena versus emotionally loaded experiences, and holism.

How do imperfections, suffering, and healing correlate? It is not as, if, first, we note that fractures exist which separate the real from the ideal aspects of our individuality and that, next, these blemishes and fractures cause us to suffer and, subsequently, these fractures are healed by suffering with the result that, lastly, we find peace and happiness. Healing does not operate within the before-and-after succession in objective time-space and it is not a causally structured power which is compartmented. But neither is our drive to concretize ideals the *cause* of our suffering. All we can do is to make manifest how these three components in our existence—the imperfections, our suffering, and its healing power—interrelate.

Since our self and the state of affairs which it encounters are made to interpenetrate, we lift our fissured, dichotomized, fractured existence onto the levels of holism and values. We suffer because we enter willingly and heroically into contrasting and imperfect states of affairs which we encounter, and we go to work toward healing and synthesizing them.

The mutual interaction between the healing process and the state of suffering is not comparable to the process of wound-healing, which is causally structured, going on in objective time-space and which is compartmented. The example of a surgical operation will do to make this difference clear. The surgeon makes straight incisions in the skin and the underlying soft tissues and finally in the fascial layer so that he can see, isolate, and excise a diseased organ. The last stage of this procedure is to close the wound he has made initially. He brings the separated surfaces of the fascia, of the subcutaneous tissues, and finally of the skin together and secures them by numerous sutures. During wound-healing, newly formed cells and blood vessels grow from opposite sides and restitute the unification of the separated tissues so that the surgeon can remove the superficial sutures in due time. These newly formed tissues interweave within the compartmentalized opposite sides of the originally separated wound surfaces, but these cellular components are themselves compartmented and they do not interpenetrate.

This example of a surgically, neatly incised wound which was meticulously restored to its original coalescence demonstrates that wound-healing can restore the organism to its original structure and it can function essentially in the same way as it did before its continuity was surgically interrupted. Since a similar process of healing can occur in an accidentally inflicted wound which was not surgically treated and since even then the scar, though visible, may not impair the function of the body, we ascribe value to the wound-healing process. However, we may overlook in our admiration for the intricacies of the wound-healing process that this is a derived value. Wound-healing is one of the necessary conditions under which a living organism survives as a causally structured unity in objective time-space. But a causally structured entity is without intrinsic value because it is incapable of purpose-striving in subjective time-space and it is not a holistic entity. The wound, whatever its origin, interrupts temporarily the psychophysical holism of an individual who is striving to concretize values and his wound may make it for the time being impossible for him to continue with his work. But after wound-healing has restored his organism to almost the same causally structured unity as it was before he had been afflicted by this wound, the individual can resume his value-striving pursuits. Therefore the healing process borrows as it were its becoming valuable from the now holistic human being, because wound-healing has ensured the possibility of his survival and certainly the intactness of his body. The scarred person can seek again to concretize values. But wound-healing is intrinsically without value. We attribute value to wound-healing because we believe, rightfully, that human life is valuable, and we deduce, wrongly, that wound-healing has intrinsic value because it aids in the survival of a valuable human being.

I contrapose the physiological process of wound-healing over against the mental process of healing through suffering, because this contrast shows that healing through suffering is of a different nature compared to wound-healing. This contrast is based on the different temporal-spatial structures of the two kinds of healing processes. Wound-healing is causally structured, it happens in objective time-space, and it restores a decompartmentalized organism to its original compartmented structure and function. Healing through suffering is purposively structured, it goes on in subjective time-space, and it is an interpenetrating activity. Suffering leads us toward internalizing the fractured situation and we insert ourselves into it. Hence, after we have suffered through a certain painful, fissured state of affairs we are not the same person as we were before we healed this fissure. Through the process of suffering and healing we have grown mentally and personally, and we have developed into a more valuable and a more value-directed individual. The physiological process of wound-healing, on the other hand restores us to almost the same organismic, causally structured unity of structure and function.

It may appear to some that the following incisive question may disturb the entire argument: if suffering does heal, then why does it not heal us from suffering? Suffering does not eliminate or annihilate the imperfections in our existence, but it does make it possible for us to live heroically with these faults and dichotomies. Suffering also opens additional channels of succour. While it executes its healing function, suffering enables us to interweave faults and fissures with religious and cultural activities and with treasures which make the blemishes and fractures in our existence more bearable. In conjunction with suffering these religious and cultural pursuits reduce the formidable and threatening characteristics of life's contrasts.

We are apt to turn away from and to devaluate suffering since we do seek peace, harmony, stability, constancy, happiness, and ultimately we strive for immortality, either in a metaphorical sense or through the belief in an eternal afterlife of the immortal soul. Therefore we may overlook suffering's intrinsic value which is its healing power.

We do not enter into a state of continuous bliss through suffering. It does not do away with the inconsistencies of human life, it does not accomplish a sort of homeostasis of a psychological, mental, rational-emotional nature. But suffering does relieve us from teeter-tottering between reality and ideality by directing these opposite constituents of our existence to interpenetrate in subjective time-space by intervention of the mental triad in a precious present. Happiness, though captured by means of the healing power of suffering, is a temporary state of mind. In the face of new problems we have to traverse yet another healing process brought about by suffering which is adequate to overcome this new problem. Every repeated period of suffering and the relief which its healing effect brings to us, gives us increased strength toward future, upcoming periods of suffering. This is a step-by-step, discontinuous process of individual, personal growth as we live from crisis to crisis. A "lesson" from each crisis becomes "nested" in the next crisis, which is the way suffering heals imperfections by making future and past experiences interpenetrate within the present. Suffering's healing power prevents our being torn apart by this saccadic stop-and-go process of psychological growth and becoming. In fact, without suffering psychological, mental, personal growth cannot become a reality.

In the following sections I shall describe the healing function of suffering by proceeding from a more abstract, rationally slanted viewpoint toward a more emotionally, intuitively oriented viewpoint. In the final section this approach tends to be the most emotionally grounded as it is based on personal experiences. The viewpoint of this last section is of a different physiognomy compared to the foregoing sections because retrospection, introspection, and intra-audition are superimposed upon one's observational arsenal of extrospection, empathy identification, and projection. Therefore, we can compose a more profound insight into suffering's healing capacity than if we rely solely on the experience of our suffering fellow humans. From this healing capacity it follows that suffering has meaning.

HEALING IN ABSTRACTO

I shall aim to unify the dichotomies objective time-space versus subjective time-space and causation versus purposiveness. This is a rational maneuver which overarches and unifies partially the dualistic view of human existence. We need not leave the contrasts between the two forms of time-space and the two structuring categories of causality and intentionality standing as if they were irreconcilable adversaries, since we can approximate them to each other by a process of rational derivation. I shall show that we can derive objective time-space from subjective time-space, and causality from the mental triad.

Objective time interrelates with the precious present by compressing and magnifying the constituents of the precious present in several stages. The durational aspect of the precious present, which is its central core, is reduced to the durationless punctiform instant. This is the pivotal concept within the form of objective time, the Aristotelian 'now', the *nunc stans*. It is the 'now' over which philosophers, theologians, and, lately, psychologists battle within themselves and among each other. In most languages this concept is encapsulated in monosyllabic words, such as the Greek *ny, nyn*, the Latin *nunc*, the English 'now', the Dutch '*nu*', the German '*nun*' and '*jetzt*'. The French '*maintenant*' seems to be an exception, although the word '*or*' can be used to indicate a brief moment. Other exceptions are the Russian '*teper*' and '*seychas*' and the Hebrew '*atta*'.

The concept of the punctiform instant has confused our thinking about time because this nontemporal entity is often regarded to be the "now" in which we live, act, and emote and in which we communicate with each other. The religious connotations of the *nunc stans* support this overvaluation of the punctiform instant. Stuhlmuller says that God's word is so powerful that his acts can happen at the same moment in which they were prophesied. In such a 'now', an 'atta', prayer can transform us so that we become what we have long wished to become.[1]

Since these elevating religious experiences as well as purely psychological activities go on in durational time, we must discard the punctiform instant as the form of time in which we live as purpose-striving total human beings.

In contrast to the compression of the durational core of the precious present to zero dimension, the past fringe is enlarged to become the objective past, which recedes indefinitely backward into the past of objective time. Enlarged to these huge proportions, the past fringe is shorn of its attributes of limitation, discontinuity, and heterogeneity. Since the objective past is homogeneous, continuous, and without meaning for an individual human being, it cannot be the repository of our personal memories. The enormous magnitude of the objective past is dissimilar to the past fringe of a precious present, which is sometimes short, while in other instances it is relatively far extended into our personal and into our communal past.

But there is an even more basic difference between the objective and the subjective past. The mental triad condenses only smidgens of past information in the past fringe, which memories are stored here and there in our memory bank and these items are discontinuous in time. Added to this discontinuity of our personal past and the use we make of it in condensation, some of our memories are of overriding importance in some acts, while other memories are only somewhat supportive of the new emergent which the mental triad is forming and getting ready for projection into the future fringe. These complex, heterogeneous, and discontinuous attributes and patterns which are characteristic of condensed memories, are absent from the concept of the objective past which is conceived of as an undifferentiated, homogeneous medium.

On the other hand, the future fringe is extended and becomes the objective future which proceeds into the unlimited, homogeneous, continuous future. Unlike the future-in-the-fringe, which has personal, vital meaning for us personally, the objective future, as such, is without meaning for us. However, the objective future can become meaningfully structured, once our activities and our planning imbue the objective future with some of the aspects of projection into the future fringe of a precious present.

The before and after is the fundamental attribute of objective time. It is created by the forward movement of a punctiform instant from the objective past toward the objective future. This is the function of the *nunc fluens*. The functions of the punctiform instant are not only to separate the objective past from the objective future, but it also moves along in the before and after in objective time as the past grows and the future diminishes in magnitude. Thus we construct the concept of linear, unidirectional, compartmentalized objective time.

The punctiform instant performs yet another, exactly opposite function in the process of creating objective time. The still-standing punctiform instant dissects the continuous forward flow of time into compartment intervals of objective time. An interval is a sharply delimited, homogeneous temporal construct which does not possess the attribute of interpenetration. Therefore we can represent objective time with spatial schemas and thus measure it with clocks.

These manipulations of compression and enlargement which we operate upon the precious present construct homogeneous, compartmented, measurable objective time out of the heterogeneous, interpenetrating, not quantifiable precious present. Objective time is the derived and the poorer form of time, the precious present is the original or primary and the richer form of time. The differences between the two forms of time come into clearer perspective since we designate the precious present as the temporal milieu in which the mental triad operates, while we conceive objective time as the appropriate temporal ambience of causality. These manipulations do not reduce the form of objective time to a dependent status or to an extension of the precious present. We appoint each form to its appropriate function: one belongs to our inner subjective world, the other to the outer objective sector of reality.

For us Western people, living in industrialized societies, it is common sense to think of objective time as the only necessary and usable form of time. This conviction is based on the fact that social, industrial, and professional activities are unthinkable and unmanageable without clocks and calendars. We are slaves to these instruments of time measurement, which enclose objective time in spatial compartments.

Westerners who live in the Near East, or the Far East, or in Africa, may be in for a shock, because their hosts and friends in these countries take a very casual attitude toward objective time and they pay little attention to appointments which are made in advance in specific time slots. One may be invited for a Friday night supper and find no one at home when one arrives at the appointed time. If one finally gets together with one's Eastern friends, say, on the following Monday, they are happy to see you, make no apologies, and blithely invite you for some other day in that week. Should one express disappointment with the missed supper engagement they may say, "What difference does that make? We are together now!"

Because the clock and the calendar have locked us into the concept of objective time it takes a special effort to become familiar with the concept of subjective time and the precious present. Classical metaphors depicting time in philosophical terms and in myths, add to the difficulty of grasping the concept of subjective time. Heraclitus' representation of time as a streaming river which we see flowing past us, coming out of the past and moving into the future, has had and still has a powerful influence on our thinking about time.

However, this is not the only ancient Greek imagery about time. The three Moirai suggest a conception of time as a triplicate entity: Clotho, Lachesis, and Atropos spin, measure, and cut off the thread of life. They have ultimate power over the lives of men and even of the gods.[2] The most powerful Moira is Atropos who decides about living or dying. Hence the dread of death is pictorialized and concretized in these three awesome women.

We see an even more sinister threesome holding sway over the fate of men in

Norse mythology. Urth, the most gloomy character of the three women stands for death's doom and for the Past, while Vertandi and Skuld seem to represent Present and Future. Skuld brings all the ills to men. The Nordic saga comes a little closer to differentiating between the three temporal facets of human life, as if the Norse had an inkling of what we now would call the concept of subjective time. Both in the Greek and in the Nordic mythological representations of time we see the ever-recurring fusion of time with death. "The Norse Urth, was conceived in a gloomy light, making her name often equivalent to death doom; later two others were added making the Norse trio of Urth, Vertandi, and Skuld, or Past, Present, and Future, in England represented by the weird sisters of Macbeth. Two give the blessings, the third the ills of life."[3]

One can derive support for the attempt to see objective time as a derivative of subjective time by taking a brief look at the development of our temporal structure from infancy through childhood to adolescence. One can readily teach young children of four or five years how to read a clock, but they have hardly any feel for any notion of objective time. Jean Piaget has shown in his meticulous studies of the intellectual development of children that this capacity does not develop until age eight or ten. He reports that young children have very concrete ideas about what it means to become an adult. Aging means becoming taller and moving more slowly; however, these are not continuous processes but a compartmentalized going from stage to stage. In between these stages time stands still. Bright children who know the date of their birthday accurately, who already know that we divide the day into twenty-four hours and the hour into sixty minutes, often think that when they become older and will be as large as their parents, then they will also catch up with them in the number of years of age. "Aging is not a perpetual and continuous process, but rather a process tending toward certain states; time ceases to flow once these states are attained; age is equated with size."[4]

Paul Fraisse maintains that as the child learns to use clocks he becomes aware of the homogeneity of time, but he too has observed that a conventional conception of time does not develop until age seven or eight. "The ability to measure time is apprehended gradually...which the child of 7 or 8 acquires through concrete operations....[this capacity] is apparent before his actual conception of time in general."[5]

The child is wrapped up in the immediate, lived-in present with hardly a concern for what has been before and what will come later. The objective distant future is for him dim and uncharted because it has very little meaning.

During adolescence one is very concerned about his future, about what he is going to do in the objective distant future three, four or ten years from now. Especially graduation from high school is for many young people a threatening time, preoccupied as they are with what they are going to make of themselves. As a young child, the student was living in the comfortable enclosure of the actual present, but he is now overshadowed by the formidable temporal dimension of objective time which awaits him in adulthood.

From childhood, through adolescence, to adulthood, we see that our temporal structure develops from living concretely in a precious present toward our living more and more in abstract, objective time. In parallel with the precious present being relegated to the background, adult life loses much of the brilliant feeling tones of childhood existence.

We can derive objective space from subjective space by a process of abstraction which is similar to that which we have applied to subjective time. Subjective space, or personal, lived-in space, displays attributes of upward versus downward, forward versus backward, right versus left, which make subjective space meaningful to a human being who lives in this space. The distinction between upward versus downward has the several meanings of being lifted upward toward heaven, versus being dragged downward into the grave; of overcoming the force of gravity by jumping, dancing, especially the ballet, acrobatics, flying; space travel during which one lives temporarily in a spatial environment where g equals zero, is the most extreme instance in which man overcomes gravity. On the other hand, when we are tired, the heft of our body may force us to sit down or to lie down. Moving forward versus moving backward has specific meaning for us, as we can see in many metaphors of common language which play upon these terms.

Right versus left are embedded in our neurological organization, hence this constrast is easily overlooked as an important factor which structures our lived-in space. If a right-handed person tries to do some manipulation with his left hand, he suddenly becomes aware of the agility of his right hand and the awkwardness of his left hand. Merleau Ponty reminds us that soothsayers interpret right as the wellspring of lucky events while left spells out failure or even disaster. He experiences how his right hand is the skillful manipulator while his left hand is "all thumbs."[6] A most penetrating observation of Ponty reveals that right and left have specific and very contrasting meanings within our behavioral patterns, but that right and left are important determinants of direction and structure of subjective space.

Right-handed people structure their subjective space in exactly opposite directions compared to left-handed people. Some bright Jewish children who learn left-to-right Latin script in public school and the right-to-left Hebraic script in temple school, may end up at the bottom level of both school systems. One shakes hands right-handedly, one raises his right hand if one has to swear in court. The legal and ethical worlds *right* and *wrong* indicate that we attach a higher value to our right hand than to our left hand. The slang expressions "righty" and "lefty" split the 10 percent left-handed people in the population from the "correct" right-handed majority.

We construct the concept of objective space out of the concept of subjective space by stripping subjective, personal, lived-in space of its meaningful psychological attributes. We create homogeneous, compartmented, measurable, objective space by truncating heterogeneous, interpenetrating, not-compartmentalizable and not-measurable subjective space. Subjective space is the primary, original, richer, meaningful form of space, while objective space is the derived, poorer, meaningless form of space.

We compartmentalize objective space with geometric points, straight or curved lines. We measure it with a measuring rod and we apply the number series. Descartes reached a crucial point of abstraction of subjective space by the invention of his three-dimensional coordinate system, which facilitates the elimination of the meaningful attributes of personal space by abstraction. Objective space, measured within the coordinate system, numbers, and mathematical equations, has facilitated natural scientists to build a mechanistic conception of the universe.

The derivation of objective time from the precious present and of objective space from subjective space, underlines the fact that time and space measurement rest on similar principles. We compartmentalize time and space with the dual, spatiotemporal concept of the geometric point-punctiform instant, in which time as well as movement come to a standstill. And this is exactly the reason why we can express objective time in terms of spatial parameters. Bergson said that we spatialize time.

Time and space measurement are essential operations within the mechanistic conception of the world which surrounds us. The overreliance on this outlook upon what is outside of ourselves, including the extrospective observation of our body, leads us to believe that objective time and objective space are the most important forms of the phenomena which we observe, and therefore the forms of subjective time and of subjective space have remained in the background until fairly recent times. We ought to relinquish our thinking from the dominance of objective time-space since it is the derived and the poorer form, which cannot accommodate the mental triad and its constitutive characteristic of interpenetration.

The process of derivation by abstraction only bridges over the dichotomy objective versus subjective time-space, but it does not eliminate this dichotomy by any means. I am not driving toward resolving these two forms into a third, ultimate form of time-space. Within a dualistic interpretation of human terrestrial existence we have to accept that we exist on two levels of integration in both forms of time-space. Objects find their proper milieu as they exist in the form of objective time-space, while human subjects need "breathing space" for their creation of new emergents in subjective time-space. Within this perspective the geometric point-punctiform instant functions as the kingpin in time-space measurement; the precious present provides the milieu in which the mental triad can function.

In parallel with the derivation of objective time-space from subjective time-space, we can derive the causal principle from the mental triad. Chains or networks of causally related phenomena come down from the unlimited objective past and they impinge upon chains or networks of future effects in a geometric point-punctiform instant; these effects, in turn become causes of subsequent future effects. The "history" of these causes recedes into the unlimited objective past; the resulting effects reach forward into the unlimited objective future. Even as the punctiform instant separates the objective past from the objective future, so does the geometric point-punctiform instant separate past causal chains from future chains of effects.

The concept of past cause is constructed out of our memory bank and out of the heterogeneous, discontinuous memories which are being condensed by the mental triad in the past fringe of a precious present. One eliminates the psychological, personally meaningful attributes from condensed memories and one constructs continuously forward-flowing homogeneous chains of causes in objective time.

One restructures projection into the future fringe of a precious present into chains or networks of effects, which stretch forward into the objective future. One removes the specific, personal, valuable aspects of projecting into the future fringe in order to construct chains or networks of impersonal, homogeneous effects which will come to pass in the far distant, objective future.

In contrast to the mental triad, the causal principle presents a dyadic temporal structure and it fits into the dyadic structure of objective time. The causal principle structures compartmented events which accommodate to the symmetrical structure of objective past and objective future, and hence, the causal principle can be expressed with mathematical precision in the causal equation.

The development of the conception of causality during later childhood supports my intention to derive causality from the mental triad in a vein similar to the derivation of objective time from the precious present. Until age seven or eight and sometimes not until an even later age, the child is not aware of the before and after relationship in events which he is observing. In fact, he believes that they are temporally reversible.

Piaget did some ingenious experiments to show this lack of temporal and causal integration. He shows children two connected flasks, separated by a cockstop. The upper flask is filled with a colored fluid, the cockstop is opened and closed from time to time, and he asks the children to draw what they see, which they do quite accurately. Then he takes the drawings and cuts the upper flask from the lower flask with a pair of scissors; he scrambles the halves of the two drawings and stacks them up in this random fashion. Then he asks the children to arrange the half drawings in the order of the events which they have observed a few minutes ago. He finds that the younger children fit any upper half they happen to lay their hands on with any lower half, without paying attention to the decreasing of the fluid level in the upper flask and the increasing level in the lower flask. Until the age of six, seven, or eight, they have no conception of the flow of the fluid in the before-and-after of objective time, nor of the causal connection between what happens in the two flasks. In his concrete, semistructured world the child is aware of only the vaguest temporal-spatial markings and the concept of causality is foreign to his thinking up to age seven, eight, or beyond.[7]

The child lives in an animistic world in which he himself and everything around him strives for goals. As he moves into his teens, the forms of objective time and space and the causal principle begin to have meaning. The teleologically organized, animistic world of childhood is gradually supplanted by a world which is predominantly causally structured and which proceeds in objective time-space. The more concrete world of childhood develops gradually into the more abstract world of adolescence. This shift from concrete toward abstract happens as the concretely functioning mental triad in a precious present relinquishes its earlier predominance to the more abstract causal principle and as the more abstract objective time-space gains dominance over the more concrete subjective time-space. We adults elevate the causal principle to its weighty importance in the mechanistic universe of inanimate objects which surrounds us. We create this world out of the animistic world of our childhood by means of the process of abstraction which transforms the mental triad into the causal principle and the precious present into objective time-space. But this process of abstraction does not release us from the concrete world of purpose-striving in subjective time-space. We are intentionally, as well as causally, structured beings who live in objective, as well as in subjective, time-space. The result is that we live in a world which is sundered by contradictions, dichotomies, and fractures. For this reason we adults regard the child's world as if it is blissfully happy, because the child does not suffer as yet from the dissonances and the blemishes of our adult world, in his semistructured world of childhood. A child hardly thinks about the objective future, nor does his distant past weigh him down. Because he is wrapped up in the immediate, lived-in present, his existence is richer than ours in one respect: he experiences everything that happens within him and outside himself with acute intensity, curiosity, and interest. The child lives in the highly, emotionally charged world where the mental triad holds sway and in a fresh precious present, while adult, practical daily life can be, and often is, emotionally impoverished.

We usually evaluate the world which occurs in objective time-space and which is causally structured as the real, fundamental, primary world. This world is the base upon which we can build our fragile, discontinuous, temporary, mundane existence as an ephemeral superstructure. This interpretation of our existence gives us a firm footing in the causally structured world in objective time-space, and it counterbalances the insecurity which the discontinuous mental triad functioning only occasionally in a precious present engenders within us. However, at the same time that we gain a degree of security within the ambience of continuous objective time-space and the predictability of causal events, we lose the insight that the mental triad and the precious present are the wellsprings of our emotional life. In a world image which is causally structured and which exists in objective time-space, man is reduced to an exclusively rational being due to the oversight in such a world image of our emotionality.

A conventional extrospective "outlook" upon the world and an unconventional introspective "inlook" into ourselves, offer alternative interpretations of these two sectors of reality. I propose that our private world and our interpersonal communication with our fellow human beings are the primary source material of our knowledge not only of ourselves and the others, but also of our surrounding world. It may seem a risky undertaking to assign this limited sector of the world, this insignificant speck of a human being and his peers, proceeding in subjective and objective time-space, structured by intentionality and causality, as the point of origin for the development of a more embracing view of ourselves and of the world. One may feel safer if one starts out with the world in objective time-space, structured by the causal principle, and from there on takes a closer look at ourselves. I feel justified to take this risk since it is my goal to illuminate the fissures, the contradictions, and the fractures in our existence from a fresh viewpoint. Starting out from man as a purpose-striving, time-organizing being, we can derive a mechanistic interpretation of our body functions and of our surrounding world by a process of abstraction. Therefore, we can heal the dichotomies objective time-space versus subjective time-space and causality versus intentionality by a process of rational abstraction. Healing *in abstracto* overarches some of the dichotomies which plague our existence.

It is questionable whether one can completely separate causality from intentionality. One has the insistent, insuperable feeling that an element of dynamism pervades the concept of causality. When we say that c causes e, we accept this as a statement of a causal relationship only if there is evidence that e has followed c *of necessity*. We do not interpret the mere succession of events as a causal relationship; the classical example is that we do not think that day *causes* night because night always follows on day. But once we realize that it is the rotation of the earth which causes night to follow day, then we conceive this succession as a causal relationship because now there is an element of dynamism locked into this succession. One might say that the rotation of the earth "makes" night follow day "of necessity." The source of this dynamic attribute within the causal principle is an occasion of philosophical dispute.

The concept of a contemporaneous cause-effect relation (see chapter 5, note 14) flows from pursuing the process of deriving causation from the mental triad to the limit. If one accepts that this process of abstraction is valid, then one can undergird a critique of the concept of instantaneous causation. It rests on the assumption that totally rigid bodies do exist, but this class of bodies is the abstract construct of mathematicians and physicists.

If one pushes the process of deriving causation from the mental triad to excess, then one eliminates the interval of objective time-space between cause and effect since one reduces it to a punctiform instant. We have seen that no movement can occur in a punctiform instant.

If one does not go that far and if one retains this time interval between cause and effect, however minuscule, as a remnant of the durational aspect of a precious present, then one retains also the element of necessity within causation as a remnant of the dynamism of the mental triad. Thus, by keeping abstraction within boundaries one secures a basis for the rock-bottom conviction that the progression from cause to effect carries a note of necessity. The derivation of the causal principle from the mental triad supports our intuitive conviction that causality is a dynamic principle. One must realize that this process of abstraction need not be pursued to the very end of eliminating the time interval between cause and effect. It is a never-quite-completed process. Hence, the interaction between cause and effect retains some of the intentional, protensive characteristics of the mental triad. The conviction that an effect follows its cause *of necessity*, flows from the future-attuned activities of the mental triad, in which the act directs that certain memories condensed in the past fringe *shall be* projected into the future-in-the-fringe of a precious present.

The process of abstraction which transforms the concept of the mental triad into the causal principle does not aim to establish some form of monism; it leaves dualism standing on its own. One reduces a triadic relationship to a dyadic relationship, which is not the same as merely numerically reducing an entity which consists of three elements to a similar entity which contains only two elements. The crucial point of the transformation is that an interpenetrating triadic entity is abstractly reconstructed into a dyadic compartmentalized entity. This abstraction does not eliminate dualism: the two forms of time-space and the two structuring categories retain their specialized applicability to interpenetrating and to compartmented phenomena. They remain contraposed to each other.

The basic reason for this contrast is that intentionality and subjective time-space include emotionality, whereas causality and objective time-space exclude emotionality. And, interpenetration is one of the foundations of emotionality, while compartmentalization eliminates emotionality. Therefore, subjective time-space and intentionality are the richer form and category, objective time-space and causality are the poorer form and category. Healing these dichotomies *in abstracto* brings the dualistic nature of forms and categories into clearer view. Although this process may look elegant from an intellectual viewpoint, still, we are left with the deterministic life-style which is based on objective time-space and causality and the indeterministic life-style which moves in the ambience of subjective time-space and intentionality. In brief, we live between the two opposite poles of causality and intentionality. Rational abstraction does not touch the hustle and bustle of life, and we have to come out of our ivory tower from time to time, as Hume intimated, and move around in the practical world of society. Nevertheless, a theoretical understanding of some of the dichotomies in our existence can help us to cope better with the problems and the contrasts of daily living.

Moreover, healing of these particular dichotomies *in abstracto* is not a purely rational, emotionally detached process. To the contrary, it is a mixed, rational as well as emotional activity which contains elements of suffering. The restlessness which pervades this approach stems from the back-and-forth shuttle in real life between the two forms of time-space, the two structuring categories, between extrospection and introspection, compartmentalization and interpenetration, and a more abstract view toward the world and a more concrete view into ourselves. We are apt to think that the concrete and the abstract worlds can be separated from each other, since we assume that the child lives in a concrete world and that the adult lives in an abstract world. But we never quite outgrow the world of the child from which we have originated, nor can we return to the so-called lost paradise of childhood. Derivation *in abstracto* does not give us a resting place in our balancing of these two worlds over against each other.

Rational clarification gives us a clearer view of the emotional side of life's dichotomies and fractures. Healing *in abstracto* is a first step toward healing *in concreto*, which requires of us that we live in the heroic life-style. Rationality and emotionality each come up with their limited view of man and of our surrounding world. Therefore one should strive to enhance and complete rational understanding with intuitive insight. We ought not to separate rationality, which is the systematic doubter and whose father is Descartes, from emotionality, the enthusiastic seeker of solutions.

HEALING IN CONCRETO

Healing dichotomies *in abstracto* does not relieve us from living on a concrete level, shifting back and forth between living, mostly, in objective time-space causally structured, and living, mostly, in subjective time-space and driven by intentionality. The dichotomies between the deterministic and the indeterministic life-styles do not impose either/or alternatives upon us; to the contrary, the two styles coexist. Sometimes we live close to objective time-space, and our actions and thoughts are well nigh causally determined and emotionality plays a minor role. On the other hand, we can live on the other side of the temporal spectrum, mostly, in subjective time-space in the world of freedom of choice, decision making, and creativity, when deep feelings grip us. In this restless shifting from determinism to indeterminism, living from crisis to crisis, the emotional turmoil of existential dread may overpower us and suffering plagues us. Now we seem close to realizing a synthesis of the two life-styles, but in the next moment, or the following day, or the next year, these goals seem to escape us. First we are joyfully happy, next we feel disappointed or even hopeless. *Himmelhoch jauchzend, Zum Tode betrübt* ("Praising the heavens, Sad unto death"). The word *suffering* carries with it a sense of doom, of glum resignation and hopelessness. Suffering can also be an unsettled state of mind when one is at a loss where one is situated on the spectrum between objective and subjective time-space. While we suffer we may feel neither a depressive mood nor despair. The depth, the quality, the physiognomy of suffering relates to what kind of fissure or fracture faces us. Some are blemishes in our existence we simply wish were not there but we can "take them on the chin." Others are threatening, and we may live in dread of the fractures. After one makes a decision one wishes one had not made, one suffers, but this irritation with oneself may gradually heal over as one makes the best of the mistake by taking countermeasures, or even by plain forgetting. Sometimes, as the captain of the Andrea Doria did, we severely criticize ourselves for making wrong decisions and self-criticism may plague us for years after. The loss of a loved one fractures our existence and this fracture never quite heals, although extended suffering may usher in a state of peace of mind.

Daily living forces a tug of war upon us between opposing sets of motivations and feelings. A doctor goes out on a house call in the middle of the night, though he is bone-tired from a day's work; he battles the body and mental fatigue because the call of duty is stronger than the opposite signals from his body, which tend to pull him into the deterministic life-style. A musician who already feels the pangs of stage fright before a concert, discovers that he has a fairly high fever; saturated with aspirin or with antibiotics he goes through with his concert because his ideal of music-making and communicating its beauty to others keeps the demands from his body in the background. An actor goes on stage and plays a comic role, while his private life is in upheaval; he is playing comedy while he is living tragedy. These three people suffer, each on a different level, between fatigue and duty, fever and performing, comedy versus tragedy. Their suffering brings about a synthesis of the contrasting and irreconcilable elements of their lives. While we suffer, we unify the fissures and the fractures in our existence.

When the deterministic and the indeterministic sides of life disturb us in less dramatic modes than is the case in the foregoing examples, healing of the dichotomies can go on almost by stealth. You want to put together some delicate machinery and it won't work; you fall short of what you set out to do and you feel annoyed, inefficient, and clumsy, but you would hesitate to use the word *suffering* to describe this mildly negative state of mind. But even this subtle tension is evidence of a blemish, a hidden dichotomy or fissure in our existence. When we take care of daily, recurring, matter-of-fact tasks, we do not step out of a deterministic world and enter into an indeterministic world, and then return to the first world after a while. During these pedestrian activities we do not feel the anguish and the emotional turmoil of being torn between two modalities of existence, but even though we may be bored, we do get a feeling of mild satisfaction even from routinely taking care of things (*das Besorgen*—Heidegger). Even in these situations we hear a whisper of the voice of our conscience, which would criticize us slightly if we did not do the expected or self-imposed daily tasks and which leaves us in peace if we stick to our duties.

Our conscience praises our accomplishments and it criticizes our failures. It approves of our moving beyond the deterministic life-style toward integrating this style with the indeterministic life-style. It harasses us when we are content to remain passively enclosed in the deterministic life-style (*das Verfallen in das Man*). We listen to the voice of our conscience which is one of the inner sources of suffering. In the varying life situations to which we are exposed, our conscience may sound a mild warning, it may disapprove and nag, it may torture us with outright condemnation. Conscience plays an important role in determining the shallowness, the severity, or the depth of our suffering. Together with suffering it is a potent force toward healing blemishes, contradictions, dichotomies, and fractures in our existence.

HEALING THROUGH MEDICINE

The work of the physician can be interpreted as a fight against death. Stated more humbly, we try to delay the advent of death. We are and should be proud of our technical ability to eradicate certain diseases such as malaria by getting rid of anopheles mosquitoes and even severe cases of tuberculosis can be treated with combinations of antibiotics and other chemical and surgical means. We have progressed far in advance of the state of medicine when Calmette, speaking in Utrecht in 1930 called tuberculosis *Ce fléau de l'humanité* ("this scourge of humanity"). Sometimes we can postpone death in an otherwise rapidly fatal disease. For instance, inoperable, metastisizing cancer can remain under control for many years with chemotherapy and/or radiation therapy in a fair number of cases. Although these heroic measures may necessitate frequent blood transfusions, while many patients suffer from distressing side effects, many of them can live a life of good and satisfying qualities with minimal pain. The progress in the medical and surgical treatment of heart disease is nothing short of miraculous. Coronary-bypass operations can come close to curative results and they can make life bearable and relatively active for long periods of time. The intensive care unit keeps many patients alive who came into the hospital near death.

On the other hand, many patients are not as fortunate as the successfully treated ones, and they continue to suffer from angina pectoris even after operation and also under the heavy load of multiple medications. They are subjected to an unconstructive, nonsynthesizing mode of suffering. This quality of life can pull the patients down toward the level of a structural-functional living organism existing in objective time-space, which may prevent them from pursuing goals, ideals, or creative activities in subjective time-space, in which they were involved before they became ill. Unfortunately, they cannot rebuild their previous spatiotemporal structure. Physicians strive to reach the perfect goal of releasing the patient from the mode of destructive suffering, so that they can resume their life as total, purpose-striving, human beings. But even if we can release some patients from existing as a physiological specimen, so that they can restore a normal life-style, this does not mean that we can eliminate all suffering. We strive to change them from living in a deterministic life-style in which they suffer unconstructively, to an indeterministic life-style in which we humans suffer constructively.

The goals of psychiatry and of psychoanalysis are similar to those of medicine and surgery. Neurotic and psychotic human beings suffer in a destructive nonsynthesizing mode. Their mentation is largely determined by their past experiences and by their preoccupation with dying in frightening fantasies, which prevent them from constructively living into the objective and into the subjective future. Psychoanalysis, psychotherapy, and the administration of appropriate medications, are multifaceted attempts to release these patients from the shadow of their emotional or their mental illness. Our goal is to help these people to grow and to develop into more complete, harmoniously organized human beings who are no longer imprisoned by the memories and the reliving of their early life experiences and by the dread of dying in the here-and-now. We enable them to move forward into the future in an active and gratifying life-style where neurotic or psychotic suffering will be minimal. If they can heal, with our help and support, the fractures in their neurotic or their psychotic life-style, then they will live close to normal lives, since destructive suffering has been overcome. But this release does not guarantee that they are now immune to suffering in their more normal life-style. The psychoanalyst and the psychotherapist share the frustration of medical and surgical colleagues who do not reach the ideal goal of restitution of their patients to perfect health in all cases. Our common goals are to set human beings free from physical, emotional, or mental illnesses so that they can synthesize reality with ideals and values, body with mind, living and dying and, ultimately, to overcome the dread of death. Sometimes we can help patients who are close to death to accept this fact courageously, stoically, and heroically. This is where the function of the physician and the clergyman overlap. Our letting the patient feel that we share with him the foreknowledge of the nearing end of his life, may mellow the patient's fear of his death. The physician can enter into this duplicate role of the one who seeks to delay the entrance of death and yet who sees his attempts to delay death failing. He can only fulfill this duty if he has come to grips with his own dread of his own eventual death. Identifying and empathizing with the patient's plight, he can partake of meaningful communication with the patient.

HEALING THROUGH CULTURE

Healing of imperfections, dichotomies, and fractures in our existence through culture has many aspects. Man's most daring and his loftiest endeavors partake of these modes of healing. Religions, philosophy, the arts, music, even plain everyday work and also humor have a function in fulfilling these goals. The strongest motivations which urge us on in these efforts and which energize them, are our never-ending fight against the blight of loneliness and our dread of dying and of death.

RELIGIONS

Religions are the most persistent and efficient cultural modes dealing with human suffering, because they offer solace against the state of loneliness and relief from the dread of death. Religions draw this power from metaphysical belief systems which are built around huge temporal-spatial perspectives and the concept of the immortal soul. Bowker offers a brief resume of various religions. Religions regard suffering as a fundamental aspect of human life and they describe this fact in various contexts. For Buddhism, *dukkha* is a transient integrating power which reveals truth in the world of appearance. Muslims believe that everything around us and within us is created by God's omnipresent will and that suffering is an instrumental constituent of His omnipotence because suffering is an *ayat*, or sign, of God. In Zoroastrianism the opposites of good and evil result in our suffering, while Hinduism perceives the unification of opposites through the only reality which is *brahman*. Chinese religious thinkers seek to restore the original harmony which once reigned over the universe, while Christians believe that the power of evil has already been annulled. Buddhism promises to eliminate suffering through the way of asceticism which overcomes "thirst" and desire, allowing a person to withdraw from "determinations" which define the world.[8]

Eastern religions think within the temporal perspective of huge cosmic cycles, while Christianity sees the history of the world as beginning with the creative act of Jehovah, progressing in a linear development toward Jesus' birth, crucifixion, and resurrection, and terminating in the Day of Judgment. Both the cyclical and the linear visualizations reduce man's terrestrial life to a brief period of objective time of such narrow confines as to approach zero duration. In the huge spatio-temporal schemas of Zoroastrianism and Hinduism, man loses his original state of bliss and he has to fight to regain this desirable state of being. Zoroastrianism is essentially the saga of the war between Good and Evil, with the ultimate victory of the Good; how the bad people who perish in hell fit into this cosmic imagery is left to one's imagination.[9] The history of the universe is divided into four periods of three thousand years each, in which the battle is fought between Ahura Mazda and Ahriman. This warfare goes on for nine thousand years, during which humans help either the good or the evil hosts. They are rewarded or punished for their conduct when they cross the bridge of the Separator. During the fourth period, Sayoshant arrives who raises the dead from their graves and decides who will go to heaven and who will remain in hell, but now Good reigns forever.

The conviction that the past was a happier time than the present, stands out clearly in Hindu cosmologies which divide the history of the universe into four ages, covering a total of more than four million years. Each following age is shorter, darker, and less perfect than the preceding age and the fourth and final age is darkest of all and is terminated by the destruction of the world. In this enormous cyclical time of unimaginable dimension, the innumerable incarnations of human life go forward on many different levels.

The ancient Greeks on the other hand, seem to lean more toward a linear temporal schema, as in Hesiod's cosmology. The *golden age* was ruled by Cronus and the elder gods, and mankind suffered from no illnesses and enjoyed perfect happiness. When Zeus took over with the younger gods during the *silver age*, man lost his innocence, became irreverent and prideful, wherefore Zeus annihilated him. During the *bronze age* people were excessively homicidal and destroyed themselves. During the *heroic age* the battles for Troy and Thebes were fought. Hesiod lived in the *iron age* among degenerate mankind burdened by toil and miserably selfish.[10] A very depressive, fatalistic image of human history.

The blissful, paradisian life of Epimetheus and Pandora exudes the ever-recurring theme of original happiness lost. They were living happily together until Pandora opened the box given to her by Zeus, letting loose all the infirmities and diseases from which man suffers. A measure of balance between present misery and a more optimistic outlook upon the future was restored when she opened the box again and Hope emerged.

Genesis tells the story of how the happy, paradisian existence of Adam and Eve was terminated because Eve, seduced by the Evil One, ate fruit from the tree of the knowledge of good and evil—an act expressly forbidden by the Lord God. Adam also ate the fruit, thereby sinning, that is, disobeying. Mankind has suffered and is still suffering, due to this Original Sin, described in folklore dating back thousands of years.

A variation on the same theme, according to Cassirer, is that man's possession of the *lingua Adamica* ("language of Adam," that is, a universal tongue) during a Golden Age of happiness was lost when man tried to build the tower of Babel and therefore was punished through polyglotism for his sinfulness in trying to equal Jehovah's power.[11]

Relying on the field of anthropology, Bowker brings out that the myth of Paradise Lost is still very much alive among the South Sea islanders of the Gilbert islands. They believe that the supreme ruler of gods and men gave the people two palm trees, one where the men were to live and the other for the women. When they saw each other for the first time during a meeting convened by the ruler they were so intrigued that the men joined the women during the absence of the ruler. Instead of being eternally youthful they soon showed signs of aging, and when the ruler returned he expelled them from his realm, throwing plagues after them. He had allowed them to choose which tree to take with them, and they had chosen the tree of the women, which was the tree of death.[12] Again, here is the concept of suffering and dying as punishment for sins committed.

However, Stuhlmueller reminds us that not all religious metaphysical systems idealize the past. Israel's religion was imbued with a faith-compulsion toward the future, in contrast to other Near Eastern religions which revel in their Golden Ages.[13]

Huizinga calls to our attention how ill-founded our belief is that the Middle Ages were a long period of peaceful happiness and emotional stability. He shows how hazardous it is to draw conclusions about the spirit of an age based upon contemplating its visual arts. He claims that a person living in the nineteenth century would depict the Middle Ages as grim and dark, while a contemporary person might admire this epoch as one of pure and naive beauty and profound mystic faith. Historians on the other hand are in touch with the harsh realities of the miseries of human life during that age.[14]

Suffering as punishment for sin runs like a constant refrain through ancient mythologies. Aeschylus describes how Prometheus was punished by Zeus because he thwarted Zeus's design. Zeus, who was disgusted with man's sinfulness, decided to let man perish and die. Prometheus gave men fire so that they could learn the arts. He taught them how to build sun-lit houses so they could emerge from their dark cave dwellings. He gave them the understanding of numbers so they could develop science. He delivered them from deathly illness by showing how to mix herbal remedies. He gave them whatever skills mortals have mastered. He delivered them from the dread of death and gave them foresight and reason, the prerogative of man, proud *homo sapiens*.

> Blessings that on man / I lavished, have involved me in this fate, / And that in a hollow fennel-stalk / I sought and stored and stole the fount of fire, /

When men all arts have learned, a potent help....From the thoughts of death I freed the minds of man....They lived, like small ants shaken with a breath, / In sunless caves and burrowing buried life. / I taught them Number, first of sciences. / I first put harness on dumb, patient beasts, / Obedient to the yoke; and with their bodies / that they might lighten men of heavy toil. / ...This first and foremost: did a man fall sick, / Deliverance was there none, or 'twixt the teeth, or smeared, or drunken; but for the very lack / Of healing drugs they wasted, till I, / Showed them to mix each virtuous remedy, / Wherewith they shielded them now from all diseases. / ...Nay, take the whole truth briefly: in a word, / All skill that mortals have, Prometheus gave.[15]

Prometheus had to suffer because he relieved mankind of suffering. This askew view of suffering relieves man of this burden, but it is loaded onto Prometheus who acted out of love for man. Behind this saga too stands the belief that sin is the reason why we suffer.

The three friends who come to visit and console Job, mourning in sackcloth and ashes, end up interrogating him, propelled as they are by the obstinate belief that Job is being punished by Jehovah for sins he must have committed. They accuse him of craftiness with his windy self-defense and say that he turns his spirit against God. The three "friends" allege that no man born of a woman can be righteous and that Job's own words condemn him.[16]

Religions are not only preoccupied with the distant past, they also look into the distant future beyond the grave, that is to the limitless eternal future. But since religious metaphysics thinks in terms of the huge dimensions of the limitless past and the limitless future, man's terrestrial life becomes compressed into minuscule, insignificant dimensions. Li Tai Po's poetry, rendered in exquisite German by Hans Bethge, speaks to all mankind. It extolls the earth's perduring for a long time. But it bewails man's limited privilege to enjoy earthly enjoyments for less than one hundred years.[17]

Religions have given mankind powerful and meaningful sustenance throughout the ages. Viktor Frankl gives testimony to the strengthening impact of his religious experience in the devastating enclosure of a concentration camp. The prisoners found most unlikely places for their prayer services where they expressed and experienced religious faith in depth and vigorously.[18]

Kunstatter heard the Lua say, "If we didn't have a Swamang, we should live like the apes in the jungle.[19] The Swamang is not only their secular headman but also their religious leader.

The Hindu who becomes integrated with the *dukkha*, the Mohammedan who is guided by the *ayat* of Allah, the Christian who lives "the good life in Jesus," believe that they have gone beyond being merely human. A religious person may agree that aesthetic accomplishments, true propositions, ethical deeds, do indeed loftily express man's humanity but for him they do not reach up into the sphere of eternal values. Truth, ethics, and aesthetics ought to be lifted up above the realms of objective and subjective time-space to the sphere of eternity in order that man may partake of the divine. The Divine Being draws man upwards and onwards above human terrestrial existence toward an eternal existence beyond death. In the mutual embrace of godhead and man, man becomes immortal and he is also liberated from the prison of loneliness. Only then, the religious person believes, does he partake of the world of values. For instance, it is not enough when two people fulfill each other in the fusion-communication of a real-ideal marriage; even this elevated mode of interpersonal union ought to be sanctified by religious ceremony, and it should be inspired by their religious faith so that they can rise above their man-made hypostasis. Then the couple can partake of the transcendental value of their union. In brief: according to religious belief man cannot become truly human by his own efforts.

But in order to open himself up to accepting the offer from above, the individual human being must surrender himself to the loving God in an act of faith. The religious person takes this blind leap of faith because he thinks in terms of two different kinds of knowledge. We accumulate one kind of knowledge through a rational process of logical deduction and we establish thereby true propositions. But over and above this man-based kind of knowledge, the religious believer is convinced of the possibility of inspirational knowledge which does not tolerate doubting critique and which is surd to logical argument. Inspirational knowledge completes the rationally grounded world image by emotionally fulfilling religious metaphysics, belief, and ritual. Knowledge received through inspiration by the godhead presupposes not only a leap of faith by man, but also the belief in the understructure of religious metaphysics from which such a leap can be undertaken. This act of faith involves the suspension of systematic doubting as the method of critical thinking.

A humanist holds closer to the givens of terrestrial experience. He too goes beyond critical doubting and logical analysis which produce true propositions and which satisfy, mostly, our rationality. He recognizes a process of intuitive insight going on within himself: this I try to describe with the concepts of the mental triad and the precious present. Intuition-grounded insight lifts us up above the causally structured world in objective time-space. To gain this insight presupposes that we have faith in introspection, empathy, identification, and projection as valid methods of observation. But both logic and intuition are man-based. Knowledge acquired by logical thought processes and knowledge acquired by intuitive insight both strive in the same direction from reality toward ideality. The religious person goes beyond man in his blind leap of faith. A humanist feels secure in his faith in intuition and in man's capabilities; he does not feel the need to go beyond man.

The religious person may object that from this limited view of man's temporal-spatial existence, suffering has no meaning. To which a humanist may answer that suffering does have intrinsic meaning even if death terminates our existence. The religious person partakes of the *mysterium tremendum* of his unification with the godhead. The humanist may settle for the *mysterium rei* of the mental triad functioning in a precious present. Each strives in his own way to heal the two major fractures in our existence.

I see a meeting ground for these two visions about man's life on earth in the concepts of the *nunc stans* and the precious present. In both forms of time people experience emotionally deeply charged, and powerful transporting thoughts, feelings, and activities, which have lasting value since they give them direction throughout their lives. Subjectively and phenomenologically speaking, the experiences in a *nunc stans* and in a precious present are of the same nature, but rational analysis in retrospect comes up with differently oriented interpretations. The religious person believes that he experiences eternity (eviternity) in a *nunc stans*, but the humanist is satisfied with the powerful transport of a creative act in a precious present, embedded in limited subjective time-space and which has abiding value for himself and others beyond the duration in a precious present, in the objective future. A humanistic interpretation of human life tends to develop a future-oriented view, based on here-and-now actual experiences in both forms of time-space, which view derives its power from faith in the creative essence of mentation.

Both religion and humanism evaluate suffering as the core of human existence. Religions believe that suffering can be healed only if man interpenetrates with the Eternal Divine, which unification relieves man from heavy hearted loneliness and conquers death. The humanist strives to heal his suffering within the confines of objective and subjective time-space and by driving toward an interpenetrating deterministic as well as indeterministic life-style.

PHILOSOPHY

Philosophers have searched for various solutions to the problems of suffering and death on an intellectual level. Socrates held a long dialogue about the negation or the affirmation of life after death, while his own personal death was ominously near. He rejected the escape route held open for him by his friends, which would have been all too easy for him to accept. But the voice of his conscience, which he often described as an inner warning or disapproving voice, forbade him to take the coward's way out. Plato describes a stoic Socrates, who was up to the last moment in contact with his environment, reminding his friends that he owed a cock to Aesculapius. Notwithstanding this coolly objective stance, Socrates must have suffered deeply as he had to leave his friends and his beloved Athens, which was treating him shamefully. His suffering unified these dichotomies and one is convinced that the life of this heroic man was not fractured by death.

Cicero says that he wrote the *Tusculan Disputations* as a relief from grief over his daughter's death. After this short introductory remark, he goes on to quote and enlarge upon what other philosophers have thought about death. He adds a few original thoughts of his own. He does not tell us if his writings and involvement with the philosophical thoughts of his day helped him to get over the grief and suffering over the death of his daughter. He warns his readers not to expect any ultimate and true statements about dying and death. He becomes involved in the various belief systems as to what happens to the immortal soul after death.[20]

I submit that philosophers have not dealt with the process of dying and the fact of death in any great depth, until modern times. They left this dark subject to their theological confreres to worry about. On the whole, philosophers have been preoccupied with man's rationality and relegated our emotionality to a secondary plane of interest. Moreover, most philosophers used to live in their proverbial ivory towers and, being loners leading an island existence, they did not think very profoundly about interpersonal communication. Kierkegaard has altered this scenario radically, and we have seen that Heidegger widened the scope of Kierkegaard's thoughts by showing that while living toward death we create our lives into a whole.

Kierkegaard says that we make decisions in immediately experienced time during which we disclose our individuality. These subjective moments change the spectre of death because death makes a difference for life.

> In the subjective thinker's immediate experience of time, the future has priority. This future generates uncertainty and anxiety. The subjective thinker, when he penetrates into the core of his subjectivity, thus finds the uncertainty of life itself. Death is one of the most ethically significant uncertainties of life.... In the subjective moments of his engaged existence the subjective thinker discloses his existence as qualified by individuality, becoming, time and death.[21]

Heidegger points out that dying is an exquisitely personal experience, since it is "mine" and it never belongs to the others, the "they." "They" flee away from death, "they" have always more time and "they" do not understand the being-toward-the-end. Death is covered up by time which passes away "in itself."[22]

Eugène Minkowski, who was trained as a physician, brings the relatedness of time with death in even clearer perspective.

> The experiential present is a most complex affair which is not simply 'specious present.... Death, which is the negation of life, is necessary for meaning, significance and direction: its impending denial of continuity enables us to live the living present which we essentially structure in terms of a thrust to a future which may be indefinite but which is definitely finite.[23]

Death gives meaning to life and the living present, from which death robs continuity. Expressed in my terminology, the discontinuous precious present with its short future fringe is the haven in which we experience anxiety about our future death.

Sherover defends the turn of thought that a feeling of finitude generates awareness of our future death.

> The fear of death has greatly influenced the logical analysis which philosophers have given to the problem of time. The desire to survive death and to live eternally, in the sense of unlimited time, a desire obviously incompatible with physical facts, has thus led to a conception in which eternal life is

not life in time but in a different reality. In order to escape the "passing away," a timeless reality was invented.[24]

The flow of time represents a superhuman force which we cannot escape because it is irreversible. Hence, time stands for death. Philosophers and theologians have created an abstract timeless reality.

Leaving the problem of life after death aside, Kastenbaum discusses some of the psychological defenses against death anxiety in young and in old people. "If futurity becomes less salient, so does death. One does not have to deny death, but its meaning is altered. Replacing significant experiences in a time-free realm affords protection from the steady march 'from here to eternity'."[25]

Some young people find solace by imposing strict chronological terms on their lives and they try to forget the immutable flow of time. Some find solace by compartmentalizing time in minute intervals which they presume may last forever. If one glosses over that we live into the future, death becomes less formidable. Older people may "replay" their past experiences and thus become released from less meaningful, present, trivial activities.

J. T. Fraser points out that there is an essential difference between man and animals, specifically between our and their reactions in the face of imminent dying and death. Only imminent death fills animals with terror. Only man experiences diffused anxiety when he anticipates his distant death. Or, as I have said, the temporal structure of animals may preclude their being aware of their death in a distant future. Fraser challenges the idea that we need the concept of the immortal soul with which religions come to grips with the dread of death. Our freedom to act and choose gives us the courage to face the dread of death. Courage in the face of death need not be based on immortality of the soul but on the awareness of our self as an indestructible being.

> I wish to consider now temporality in the phenomenon of suffering....this presentness is not an ignorance of "yestermorrow," but a demand for the reintegration of the mental present in terms of the most comprehensive level of temporality in man, which we have been calling nootemporality.
> ...Let us revert to the notion that in "my death" the observer and the observed regress together; this would imply that, before the food-for-worms state is reached, the Umwelts devolve back to the eotemporal and the prototemporal.[26]

Fraser sees suffering in relation to our complex spatiotemporal structure. He places man and his world against a five-leveled, temporal perspective of simplest to most complex temporalities into which he inserts the lower, the higher, and the highest strata of the evolutionary process, all the way from inorganic matter to creative man. Fraser makes here a big leap forward in pointing out that the evolutionary process has been, so far, described in terms of the increasing spatial complexity of living organisms. He has added insight that, *pari passu* with increasing spatial structure and function, purpose-striving becomes more and more complex. Therefore the temporal structure of higher-level organisms also becomes more and more complex. The terminal point is man's capacity to integrate Fraser's five levels of temporality which he designates as a process of "nesting." I should prefer to say that only man is capable of synthesizing objective time-space with subjective time-space. Only man suffers, because only he is able to synthesize the dichotomies in his existence. Fraser interprets pain and suffering in somatic disease and due to accidents as a reduction of the complexity of our temporal structure, because past and future are 'now' not important, only the present. Therefore the highest level of nootemporality is lost, which the patient and those who care for him attempt to restore. When we are dying and when death has overtaken us we regress even further into the eotemporal and the prototemporal Umwelts toward the "food-for-worms" stage.

Fraser is most sympathetic to the idea of grounding human suffering in man's spatiotemporal structure. Tension, anxiety, dread stem from this "nesting" configuration. Tension toward the subjective future-in-the-fringe and the distant future in objective time is reinforced by the apocalyptic view of the end of the world which is characteristic of many religious cosmologies. Fraser reminds us that time has been feared as destructive and welcomed as constructive, that being and becoming presuppose both the unpredictability of the subjective future and the lawfulness of the world insofar as it is causally structured and as it exists in objective time-space.

These thinkers demonstrate that we live into the future in objective time as well as into the future in subjective time. Existential philosophy, psychiatry, and psychology open up a new approach to the problem of suffering, loneliness, and death from the viewpoint of time. This interdisciplinary method grasps man as a rational as well as emotional being who communicates with his fellow humans. Thus we become released from philosophy's one-sided view of man as *homo sapiens* who lives an island existence.

Philosophers overlooked the problem of subjective time because they concentrated on objective time. Therefore they failed to develop a theory of emotionality until Kierkegaard placed the problem of subjective time in the center of philosophical investigations. Religions have cast objective time in the image of eternity. This they invested with attributes of subjective time, because religions are vitally interested in man's emotionality. They project their concepts of suffering, loneliness, and death onto the huge spatiotemporal panorama of unlimited space and eternal time.

Philosophy has overlooked the problem of death for a long time. Religions have always overemphasized the dread of death.

An interdisciplinary approach seeks to understand how we live into the objective and into the subjective future and tends to develop a theory of emotionality. It relies on logical analysis insofar as we live in objective time-space, causally structured, observing phenomena extrospectively, and compartmentalizing ourselves and the world about us. We gain intuitive insight insofar as we live in both forms of time-space, structured by both categories, follow both the extrospective and the introspective methods of observation, and combine compartmentalization with interpenetration. Interdisciplinary research accords emotionality the same rights as rationality, and it stresses the central importance in the human existence of suffering, loneliness, and the dread of dying.

Religions offer release from loneliness and they promise the conquest of death if one takes the leap of faith into unification, or "religation," with the godhead. The interdisciplinary way demands that we adopt the heroic life-style in order to overcome loneliness and the dread of death, because we have faith in man's creativity.

LITERATURE, THE ARTS, MUSIC, ARCHITECTURE, WORK

Tragedy deals intensively with the problems of suffering and death. One of the difficulties of tragedy in drama is that the death of one or more major characters in the play seems to be used as a device to bring the drama to a convincing ending. For instance, practically the whole cast of the *Hamlet* tragedy is dead by the end of the fifth act. Hamlet's father dies asleep under the influence of the poison, without the benefit of the last rites of his church. Polonius dies totally nonsensically by Hamlet's rapier, "like a rat," and out of sight. Ophelia sings herself to sleep during her suicidal journey down the river. The king and queen die in plain sight, punished for their wicked deeds. Only Hamlet dies as a hero, coming to grips with his imminent departure as he helps to prepare for the succession to the throne. His last words, "The rest is silence," convey his conviction that he has done his duty by avenging his father's murder. The strength, the nobility, and the heroism of Hamlet's character are underlined by Horatio's admiring, "Now cracks a noble heart." Hamlet conquered the dread of death through the ordeals of suffering that life had meted out to him.

Why is this bloody spectacle so elevating for us, the spectators or the readers of this play? Because, notwithstanding the weaknesses in his personality, Hamlet accepts his duty, he weighs killing his uncle within his own conscience, waging a ferocious battle within himself, accusing himself of indecision, cowardice, and emotional callousness. Shakespeare recreates Hamlet into a hero, who is caught between numerous dichotomies and fractures in his existence and who is going through constructive suffering as he synthesizes these contrasts. Hence, we are downcast as well as uplifted by this drama.

Henri Barbusse describes in *L'Enfer* a heartbreaking scene between a man who is dying from sarcoma of the jaw and a priest who was invited by the patient's wife to give him the last rites of the church. A struggle between the patient and the priest results. The priest goes to the limits of flexibility, while the patient stands firm in his agnostic conviction, that his life is going to end here and now with utter finality. The priest withdraws brokenhearted; the patient is exhausted. Barbusse takes no sides.[28]

Barbusse shows us an encounter between two human beings who are locked into what they have learned during a lifetime of sincere study, periods of doubt, and application in daily life of the results of their studies. Each is suffering deeply in the useless, mutual struggle to convince the other of his viewpoint. But what really separates them is their incompatible outlook about the future. The priest believes in eviternity: that in death we shed the temporary, spatial encapsulation of the soul in the body and that the soul will continue to exist in time eternal without spatial encumbrances. The patient believes in the finality of death, that he has fulfilled his goals in his terrestrial, spatiotemporal life, which has a future as long as body and mind are an undivided, interpenetrating whole.

Their encounter exemplifies destructive suffering because each man remains isolated in his particular island existence and therefore there is no interpersonal communication between them. But each man sticks heroically to his principles and, from this viewpoint, each suffers constructively.

Music and poetry together allow us to partake of most powerful forces to battle the dread of death. Gustav Mahler's *Resurrection* Symphony dramatizes in its fourth and fifth movements the ambivalence of human existence. First, man has to go through a phase of suffering, as the alto intones: "Man is lying in direst misery, man lies in greatest pain." Then, in a mysterious pianissimo, the choir sings: "Arise, yes, you will arise in an instant." And then the triumphant promise of the conquest of death: "You, death, who conquers all, now you have been conquered." The music enriches the words which proclaim that we have to go through a period of pain and suffering, in order to conquer the dread of death, so that we can enter a state of happiness.

However, music also can express the struggle of suffering without words. Beethoven's late quartets, especially the adagios, conjure up the fractures of abandonment and loneliness. In these exalted moments we are torn between feelings of deepest sorrow, expressed in most ethereal and beautiful tonal form. We suffer as we hear this music because we synthesize extreme aspects of our human existence. We feel at the same time the hurt of our human misery and also its conquest through exquisite beauty.

Music can fulfill this earthly gift of healing the dichotomies and the fractures in our existence because this form of art exists, mostly, in subjective time-space, while objective time-space stays in the background. Music expresses these two aspects of our being in the world most forcefully since it synthesizes time in a precious present. It is not fortuitous that most thinkers who have delved into the time problem, chose music as the realm of creativity which is rich in examples of temporal integration. Moreover, since music is the most abstract form of art, it is for many the chosen field to theorize about time. Saint Augustine has written a treatise on music; Bergson analyzes the temporal structure and the interpenetration of the elements of a musical phrase; Husserl often speaks of the temporal structure of a melody.

In contrast to music, architecture heals some of the dichotomies in human existence in three-dimensional space. It concretizes the idea and the feeling of being uplifted toward heaven, in contrast to being bound downward toward earth. Gothic architecture, which developed out of earlier Romanesque architecture, supplanted the semicircular arch by the pointed arch. The semicircular arch, the narrow windows, and the substantial walls give a self-enclosed, earthbound appearance to Romanesque buildings. The pointed gothic arch draws the eye upwards. Moreover, the walls of these churches became less massive with the invention of the externally supporting flying buttresses. The extensive use of stained glass windows gives a feeling of lightness and transparency which reinforces the upward striving power of the whole structure. The huge, tapering towers add from the outside to the antigravity image. The feeling of counteracting gravity is most successfully realized in the bell tower of the First Presbyterian Church in Stamford, Connecticut. It seems virtually to leap up into the air and because you know that you are looking at tons of steel and concrete, it can give you an eerie feeling of man's power to overcome gravity. For our analysis of the process of healing through suffering the important characteristic of architecture is its visually and physically uplifting quality. It counteracts gravity, and it defies this ever-present force in a metaphorical sense. It gives us a feeling of being raised up from the ground on which we stand. Downward has the meaning, in subjective space, of a depressing movement, of being dragged into the grave. Architecture counteracts the dread of death, in its silent way, by creating the illusion of upward-striving toward heaven, by integrating objective space with subjective space.

Architecture's power to counteract gravity is used in many religions to symbolize the conquest of death. The Egyptians were virtually obsessed by the dread of death and it is said that their entire religious metaphysical beliefs were designed to overcome and to overpower this dread. The massive pyramids point upward, and the gold-embellished top must have added a scintillating thrust toward heaven to this earthbound structure. The elaborate funeral rites and arrangements for the afterlife of the dead pharaoh are well known to cooperate toward the eternal preservation of the ka, or the soul, of the pharaoh. Tutankhamen's mummy was entombed not in one, but in three "nesting," interlocking sarcophagi.

Ramses III invented an ingenious arrangement to prove that he was an immortal god. His statue was placed at the end of a dark gallery in the tempel of Abu Simbel, a gallery lit by the rising sun in February and again in October. The manpower of the kingdom and the sophisticated knowledge of geometry and astronomy were put to work to ensure Ramses's immortality. Gerster reported on the dismantling, elevation, and restoration of the temple of Abu Simbel which otherwise would have been flooded by the man-made lake Nasser. Statues of the god Amun and of Ramses stand side by side in a dark recess in the sanctuary and it is so structured that the sun's rays shine upon them twice a year during the winter and the summer solstices.[28]

The dread of death urges on human beings to create amazing and admirable artifacts. The Aztecs built their temples on towering pyramids where they practiced their human sacrificial offerings to the gods. To us this seems excruciatingly cruel, but the Aztecs believed that the soul of the person whose heart was cut out of the living body had been brought as close as possible to heaven, where his soul was promised eternal happiness.

Sculpture too challenges gravity. The sculptor breathes life into a block of wood or stone and he creates something that seems to move, walk, dance, out of this immovable, unformed thing that has in our imagination become as light as a feather. Even Henry Moore's sculptures, heavy stubborn nonrepresentative structures that they are, create an illusion of movement, of lift, and they seem to defy gravity. A most mysterious contrast and resolution of movement while standing is Rodin's *John the Baptist*, whose feet are planted flatly and firmly on the ground. Yet Rodin has made the statue appear as if it were walking right off its pedestal by twisting the body, positioning the arms and shoulders, and extending the right hand in a forward-pointing gesture. The mystery which is grasped in this piece is that the positions of the legs and feet which in reality make a man stand fixed in space, are transformed into the illusion of forward movement, denying the law of body movement and of gravity. The positions of the two legs, the forward thrust of the right shoulder and arm, the sculpture as a whole, interpenetrates in subjective space and therefore it seems to interpenetrate in subjective time. Hence, it portrays projection of an act into the future fringe of a precious present, which was exactly the mission of John the Baptist.

Rodin revealed to Gsell that instantaneous photographs of a man walking would destroy the holism of this movement and would create the impression of someone who is congealed in space. It would interrupt registering the gesture because it would compartmentalize a purpose-striving movement. The sculpture transforms the banal activity of a man walking into a grandiose display of man's future-striving.[29]

Rodin's analysis of his sculpture recalls Bergson's differentiation between an unrolled and an unrolling motion. The photograph depicts the man walking as an unrolled movement, since it exists in objective time-space; the sculpture represents it as unrolling because the sculptor has synthesized objective with subjective time-space. We experience this synthesis when we view Rodin's sculpture as an overcoming of the heft of the massive bronze structure and hence as an overpowering of gravity.

Stance and movement interpenetrate in Rodin's piece; living and dying are copresent as depicted in Michelangelo's *Pieta*. The pathetic body of Jesus has finished its last hours of suffering. Mary, deep in thought, even now tries to console her son. The beauty of the sculpture unifies living with dying. Some consolation emanates from the sculpture which fills us with sadness, not unlike our becoming both downhearted and uplifted from seeing a performance of *Hamlet* or after hearing a late Beethoven quartet. The dichotomies and fractures which the playright, the composer, the sculptor places before us make us sad and we suffer. Mysteriously, within this experience of suffering we sense a modicum of bliss and ultimately peace of mind.

Work

We can forget our misery temporarily while we are working, but we return to the same state of mind after the day's work is done. Solzhenitsyn describes the healing power of work in *One Day in the Life of Ivan Denisovich*. The day begins with the ever-returning physical and mental misery of the prisoners, but once they are building a brick wall, every little detail, every little movement matters to them, even more than it would under normal working conditions outside the camp. The workers are totally lost in their work and the day goes by in a whiz. Ivan says: "When you think of a day, time seems to fly; when you think of the years it seems to stand stock still."[30] The integration of objective time-space with subjective time-space has been lost and therefore camp life seems unreal to the prisoners. How to keep one big toe a little warmer assumes outrageous importance. It is as if they try to forget the threat of death which overhangs the whole camp population, by getting lost in these minute details and in their work, which in itself has absolutely no meaning for them. The punishment of one of their comrades, who has tuberculosis and who is condemned to live in isolation in abysmal conditions, is only one of the incidents during the day which bores in the threat of death. Still, Ivan ponders before going to sleep, that this day has not been too happy, but neither has it been too bad.

Some heroic undercurrent keeps the prisoners alive, notwithstanding the fact that their distorted temporal structure makes it impossible for them to look into the objective future, since they do not know how many years they will be imprisoned. Therefore, time in the distant, objective future seems to stand still. These people who are being tortured slowly in subtle ways have been robbed of the conventional means of dealing with the dread of death. The amazing resilience of human mentation which leaps out of these pages shows that plain work can help in the fight against the concrete, here-and-now threat and dread of dying.

HUMOR

To talk in all seriousness about humor seems like a contradiction in terms. But the subject of humor belongs in a discussion about suffering because this exclusively human attribute is a powerful antidote against the miseries of suffering. It combines many hidden processes which energize this power. It is as if it sneaks on stage in the midst of tragedy and, lo and behold, it is a welcome guest, a ray of light in the somberness of the main theme of the play. It does not belong in the plot, it is a totally foreign intruder, and yet no one takes offense if humor is introduced tastefully. One of its techniques is to present a serious matter in comical language and dress, like the court jesters of old, who were allowed to criticize or even to insult the king. The modern cabaret is such a device which got away for a short time with poking fun at Hitler and Mussolini. Zijderveld says laughing at figures of power through the medium of satire puts limits on their awesome power for a while.[31]

The humoristic attitude requires that one possess the blessed ability to step aside from oneself and laugh at oneself, which is a sophisticated variation of the attitude of self-criticism. Take for example the strange incompatibilities of activities and emotional reactions in the following scene: Our father had to break up his family because, to rectify financial mistakes he had made, he had to go to the

274   The Meaning of Suffering

colonies, to earn enough that his children could have an education in keeping with the socioeconomic level of the family. His name had been freshly painted on one of the trunks which were standing around, and he rested himself for a moment against that trunk, so that his name was printed in big letters on his backside. The entire family laughed convulsively, including our harassed father. For a little while the tension was broken which gave everyone some measure of relief. It is as if the spirit of humor says to us: "I know it is all very bad right now, but look, if I can give you a bit of relief from your suffering, perhaps there is light at the end of the tunnel."

One of the wisest definitions of humor is *ein Lächeln unter Tränen* ("a smile covered over by tears"). Humor synthesizes two opposite, incompatible feelings from which we suffer, and it gives us brief sparks of relief. Humor holds out a promise of synthesis of greater dichotomies because it accomplishes a synthesis of minor fissures in its own inimitable manner. It kindles our hope that if through humor we can heal ever so briefly the totally opposite streams of mentation, then we may be able to heal the more serious dichotomies and even the most serious fractures in our existence.

Shakespeare often uses humor to interrupt or to relieve the mounting tension in his tragedies. Polonius who is usually portrayed as a somewhat unorganized, silly old man, reads Hamlet's love letter to Ophelia in his ostentatious, exteriorizing manner.

> Doubt thou the stars are fire,
> Doubt that the sun doeth move,
> Doubt truth to be a liar,
> But never doubt I love.[32]

A deeply felt and beautifully worded poem is pulled down to the level of the ridiculous by the stuffy old dignatary. Polonius's performance makes us smile, Hamlet's lofty words make us weep, and within the opposite pull of these ambivalent feelings we suffer. Shakespeare creates enormous tension in the spectator and listener or in the reader since he makes us partake of Hamlet's and Ophelia's suffering through humor.

In the opening scene of the fifth act of Hamlet Shakespeare inserts humor in a most grotesque vein in the characters of the down-to-earth and in-the-earth gravediggers, whom he introduces first as two clowns. They play on the themes of suicide and death with a riddle that is finally resolved by saying that the gravedigger builds the strongest house of all, because "The houses he makes last until doomsday."[33] This eerie scene hardly relieves the tension within us because it shows the ambivalent characteristics of humor under the most horrifying circumstances: the burial of the remains of a religious person in unsanctified ground.

Victor Frankl reveals the healing power of humor in his descriptions of the nightmarish experiences in a concentration camp. After the prisoners had learned to put their whole past lives behind them, they developed a sense of humor, poking fun at themselves and at each other. One of his coworkers was in real life a surgeon with a rather rigid outlook upon life. Frankl took it upon himself to literally teach this comrade a sense of humor. He says that humor as part of the art of living can be a solace even where suffering is omnipresent.[34] One might add that humor can be a softening factor even where dying and death are an ever present threat.

Compared to humor, laughter is quite another matter. It is rare indeed that people would laugh during a really serious situation and the one who does is severely criticized by looks and even verbally. Laughter serves to relieve tensions in a group of people who do not know each other very well and where the archaic primitive threat of the clan against the outsider hovers over a social gathering. Laughter brings people closer together because it dispels some of the initial anxiety and insecurity within the minority group and it diminishes the standoffishness of the majority. But once laughter takes hold of us it is primarily a hedonistic pleasure. The vigorous physical activities which occur during a hearty laughing spell, especially hyperventilation, cause a release of body tension and physical pleasure. Laughter does not have the synthesizing and the consoling effect of the tearful sense of humor. Bergson's essay on laughter is a wide-ranging analysis of this specifically human attribute. He uses the contrasting functions of extrospection and introspection to show that comical effects can be brought about when we don't or can't look into ourselves.

A child playing with puppets fantasizes that they are alive, knowing at the same time that they are not, thus creating a comic situation. When we treat a real person like a puppet we regard him as a thing whose strings we can pull. Because he does not know this, we laugh at this marionnettelike creature. A person acting as if he is a piece of machinery is being dehumanized, and laughter soothes the pain which this transformation causes in the onlooker. If the mechanized person can laugh together with us at his puppetlike behavior, then he can be restored to a whole, self-propelled human being who is no longer activated by the imaginary strings we are pulling.

> Laughter covers over the dehumanizing effect of man changing into a mechanical arrangement, and makes this condition less painful. Any arrangement of acts and events is comic which gives us, in a single combination, the illusion of life and the distinct impression of a mechanical arrangement. Laughter's function is to convert rigidity into plasticity, to readapt the person to the whole.[35]

Comedy means we laugh at others, while humor means we laugh at ourselves. Laughter is socially engendered, outward-turned, as it is brought about by group tensions and it is a socializing, communal release of tension. Humor flows from our ability to split ourselves into a watcher and a doer and it is inward-turned.

Comparing the first section of this chapter, one can see similarities and differences between healing through humor and healing *in abstracto*, which are the extremes within which the other modes of healing operate. We heal the contrasts between the two forms of time-space and the two structuring categories *in abstracto*, which satisfies us rationally over an extended period of time, but it gives us emotional satisfaction only in muted feeling tones. On the other hand, healing of suffering through humor gives us immediate, emotionally highly charged relief, but its effect is short-lived and ephemeral.

Between healing *in abstracto* and healing through humor lies a large region in which the heroic attitude is in the foreground. I refer to science, medicine, religion, culture, and to everyday living. Healing *in abstracto* is an emotionally restrained and wide-ranging attempt to heal dichotomies. Humor is a fleeting but an emotionally much deeper cutting approach to the relief of suffering within limited situations. Abstract healing and humor heal the core of our existence from the outside, as it were. Science, medicine, religion, culture, the heroic life-style, heal this core from within.

All modalities of healing through suffering are contrast-filled processes. They are characterized by ambivalent feelings, they are rational as well as emotional efforts, and they follow the extrospective as well as the introspective approaches. Since they are situated between abstraction and humor they confront suffering from a fundamental vantage point. Hence, these efforts lead toward a state of temporary peace of mind. However, we should not nurture the false hope of obtaining a permanent resting place in the restless flux of living. This hope rests on the conviction that once we have suffered, we have become immune to suffering. Alas, this is not so. Science is an ongoing process which comes up with temporary, partial solutions; medicine is seldom totally successful; religions depend for their effectiveness on the leap of faith which not everyone can take, nor does everyone see the necessity of taking this blind leap; philosophy is a field divided within itself and, like science, discovers true propositions which constantly require refinement and revision; partaking of the arts and of culture presupposes a fairly sophisticated state of personality development; as a result, many people, if not the great majority of mankind, are left with the toil and trouble of everyday work and with a reduced level of defenses of their existential core against the onslaughts of fissures, contrasts, dichotomies, and fractures. The red thread that holds the various healing processes together is the heroic life-style, on whatever gradation of emotional intensity and fleetingness, or rational depth and durability these processes go forward. This life-style is grounded on the insight that suffering is the core of our existence. We take countermeasures against the blemishes, discords, dichotomies, and fractures so that they do not destroy the meaning of our existence. We hold this meaning together through suffering.

Poets have always known intuitively that the spoken and the written word have great neutralizing power in softening the pain of suffering. Stuhlmueller thinks that Israel had to learn first to write down the wisdom gathered over long periods of time. Then, interpretation, the musical expression of these words, and the acting out in liturgy and ceremony of their beliefs gave a new voice to the suffering of Israel's people. At the same time, this power of language brought them closer to God and promoted the power of prayer. Stuhlmueller claims that without the written and spoken word Israel's beliefs would have been expressed only crudely in jumbled and crashing noises.[36]

Friedrich Rückert grasps similar thoughts more concisely and without contradictions by saying that we can carry the load of suffering more easily after we have expressed it in retroactive contemplation in the written word. "After all, it is easier to bear it as a written aftermath, / Than fermenting without form in the turmoil of our senses.[37]

THIRTEEN PERSONAL VIGNETTES

I have shown that suffering is grounded in our intellectual, emotional and spatio-temporal constitution. This interpretation may create the impression that I wish to follow a mostly rational, logical procedure to describe the healing power of suffering. But we can look at human behavior and experience from a more rational or a more emotionally slanted point of view, in which combined approach neither viewpoint needs to overshadow or even eliminate the other. The reason why my discussions of the healing function of suffering may strike the reader as a one-sided and excessively rational analysis of this emotion-laden problem, is due to the limitation of our method of observation. We describe and interpret the suffering of our fellow human beings by the methods of extrospection, identification, empathy, and projection. But introspection and intra-audition are left out of this approach because we can only introspect and intra-audit within our own personal selves. The foregoing descriptions and analyses of suffering and its healing power give us the insight that people indeed do suffer and that faults and contrasts are being healed through suffering. This insight dawns upon us when we study their behavior and their utterances by extrospection, identification, empathy, and projection. But this is a limited method of observation because introspection and intra-audition are not publicly available.

One is obliged to make use of one's own early, relatively distant past, and recent and present experiences of suffering in order to describe and to interpret the process of healing through suffering in greater depth. Thus we arrive at an interpretation of suffering's healing power which is more emotionally slanted as well as rationally grounded.

So far, two questions have been left dangling: (1) what do we actually experience when we suffer, or, what is the concrete physiognomy of this experience? and (2) in what manner do these experiences concretely heal contrapositions in our existence? If we can answer these questions, then we can state unequivocally that suffering heals blemishes, dichotomies, and fractures in our life and hence that suffering is meaningful. The following personal vignettes share concrete experiences and processes with the reader. I do not mean to project a maudlin image of my life as if it is a long road of miseries, since the vignettes are collected without any temporal continuity. My life, on the whole, has been happy, productive, and satisfactory—more and more so with the passage of time.

These brief glimpses add introspection and intra-audition to the channels of observation, and therefore they allow more firmly grounded insight into suffering's capacity to heal life's dichotomies. As we proceed from the first to the later vignettes we see a gradual widening of the horizons of objective and subjective time-space: the ways of dealing with life's vicissitudes become increasingly complex.

> In my middle and late teens I began to question the metaphysical tenets of the Dutch Reformed church more and more, especially when the time of confirmation was at hand. I was disturbed by the rationally unresolvable problems which flow from the concept of a personal God. The fact that miracles seem to challenge the mechanistic cosmological image of natural science troubled me. Thus far, I trod the same path of doubting and examining our religious training as did my fellow students in high school and later on in medical school. But most of them dropped their intense questioning and either embraced the church in which they had grown up or they simply forgot about their interest in religious questions altogether. Instead, I tried to find solutions for the basic questions about life, death, and suffering in philosophy and received some pointers from some older friends of the family. My sensitivity to the suffering of others was one reason I decided to study medicine. This concern about suffering reinforced my questioning why we should believe that it is the suffering of Jesus that will reconcile us and will build a close relationship with the personal God of the Judeo-Christian traditions. Suddenly and quite intuitively the thought emerged in my mind as if by "inspiration," that it is not Jesus' suffering which performs the miracle of our immersion in the godhead and our reconciliation with the personal God of Christianity, but that the personal suffering of each of us humans exerts this restitutive power. This proposition had for me the quality of a self-evident truth, and it cleared up much befuddled thinking.

> As I went along in studying philosophy—instead of going to church I devoted many hours on Sunday to this study—the metaphysical groundings of religion became intellectually less and less acceptable. However, this path did not lead to my denigrating or devaluing religion. I only tried to find different solutions for the same problems to which religion gives its own answers: why do we humans suffer? Therefore the title and the tenor of this book has come quite naturally to me.
> My finding a personally acceptable solution to the problem of suffering during a flash of intuitive insight decades ago in a precious present, may have contributed to my paying scant attention at present to the question *how* suffering heals the imperfections in our existence. I hope to correct this flaw in my exposition of this part of the problem by describing a few disconnected vignettes of my personal life which have mostly to do with loneliness and the dread of dying.

When I was three or four years old my mother, sister, and brothers were singing together, my mother at the piano. Instead of singing along with them I chose to whistle the same songs and mother told me either to sing along with the others or to keep quiet, which was a rather unexpected and abrupt command by a parent who was quite permissive and wrapped up in her children. I felt sad and left out of the group, alone, rejected, and I cried as I was sitting on the floor watching mother's foot working the pedal of the piano.

I interpret these feelings and behaviors as an inkling of the awareness of the fracture of loneliness and therefore of suffering. I see myself sitting "down there," separate from the others and alone, while they are singing "up there" and I sense retrospectively the dichotomy between my sadness and their cheer. This suggests that I experienced interpenetration of sadness-within-joy within a precious present. I was not torn apart by these self-inflicted contradictory aspects of my existence but held myself together and hence I suffered.

My earliest memory of the dread of death which I experienced at about the same age is associated with a story about a monkey who was a household pet. When the monkey hears a burglar entering the house through an attic window, he attacks the burglar who thrusts him from the roof and the monkey suffers a pain-ridden death. The last picture in the book shows a little boy shedding tears on the monkey's grave. I told my mother or whoever was reading the story to me, to stop at the point when we were getting to the final scene because "Now it is not pretty any more." How early do we learn to evade the dread of death!

While I am enjoying the humoristic and happy phases of the life of my pet, I know very well what is going to happen to him, that the "bogeyman," death, is going to snatch him away, but I do not want to hear about this final scene. I won't let go of the cheerfulness previous to the sad ending and I try to attenuate the anxiety about the approaching death which I keep in the background. I accomplish some measure of synthesis of the constructive and the destructive aspects of life and I suffer as a result of this process. I hold on, rather artificially, to the continuity of life by interrupting the story. The sadness of the ending is not eliminated, but it colors the pleasures of the lighter side. The foreknowledge of the monkey's death makes me suffer within the contrast living-versus-dying. This very suffering synthesizes this contrast and makes it possible for me to enjoy the cheerful part of the story over and over again. But I do not want to live the death of my pet over and over again. Previous sadness when they read the whole story to me and the memories of the pictures had strengthened me toward relieving the sting of death somewhat, and therefore I was able to enjoy the parts prior to the ultimate truncation. These memories show that childhood suffering is the "parent-suffering" of adult suffering.

I underwent a traumatic event a few years later: my father incised an abscess in my groin while my mother held me down; the three of us were quite tense, and mother told me that I might pull her hair, as if she wanted to share the pain with me. Father undoubtedly lanced the abscess within a second or so. I remember being frightened, overwhelmed, and of course feeling pain. But strangely, I do not remember this little drama in the mode of suffering, whereas I do remember the first example distinctly in this mode.

I explain this difference as follows and I trust that this interpretation does not force a childhood experience into the mold of adult rationalization: I did feel physical pain and strong negative affect during the little operation, but I did not feel a loss of either parent since in fact there was close togetherness between the three of us. When mother chastized me for my asocial behavior, it felt as if I had briefly "lost" my mother although she did not inflict any physical pain on me. On the other hand, the physical pain that my father made me undergo, did not make me feel as if I had momentarily "lost" him. In retrospect, the first instance was mental suffering due to loneliness and the third example was experiencing physical pain not intermingled with suffering because there was no inkling of separation from either parent. This memory reinforces the conviction that we must separate mental suffering from the sensation of physical pain.

I admired my father's power in many areas, for instance when I observed how he sutured a blood-spurting wound in the foot of a male patient. But I did not interpret the injury he inflicted on my body as death approaching me. Older children often think of death in terms of bodily injury (Kübler-Ross). I was frightened, but I can't think that I was frightened to death.

> Our family was broken up because our father decided to go to the colonies so that he could repay a heavy financial debt. At age eight I was taken by my middle-aged aunt into the little ménage of her and my grandmother. Here silence and quiet of a depressive quality prevailed. Father came to see me for a final goodby, and he tried to cheer me up as I sat on his lap crying. The parting at the railroad station drove in the finality of the truncation of our family. Later he wrote how sad this parting had been for him when he looked into my tear-filled eyes.

While I suffered through the separation from father there was also a supportive quality to this encounter since each suffered in his own manner and such socialization of suffering is an effective way of softening its impact. When we are separated from beloved ones, alone and lonely, we can make ourselves partake of the suffering of the others who may be alone and lonely also. Empathizing and identifying with them helps to overcome loneliness.

This memory is more complex than the previous ones since it was stretched out over a fairly long period of objective time-space; there had been many warning signs long before the final parting.

The lack of religious rituals such as prayer before meals and Bible reading after supper, were to me a painful loss in grandmother's home and I held on to these traditions fervently, going to Sunday school and to church without being told to do so. A picture of Jesus, lovingly touching the head of a child gave me a feeling of solace and I often went past the store where it was displayed. Also, studying the piano gradually became a highpoint during the day.

Here we see not just repression and forgetting of suffering as I did around age three and four, but concrete and rather effective countermeasures against aloneness and loneliness. It was as if I took some of the remnants of the split-up family with me into the new environment and tried to heal the contraposition between the old and the new circumstances. The fracture of loneliness made me suffer and this suffering healed this fracture.

I remember very distinctly having the following fantasy about age twelve: the only other fairly young thing present in this little household was a cute little poodle dog who slept during the greater part of the day. I was watching him while he slept during the twilight hour, when grandmother insisted that no light be turned on. During this hour I felt particularly bored and sad and I thought how nice it would be if I could go to sleep just like the dog. But then I rejected that idea, counterbalancing it with interesting things that did go on during the day and that soon I would be reading again when finally the lamp would be lit. I thought about going to school again next day which I enjoyed very much and visiting a family with four boys on Sundays. I had some dim awareness that although the dog was happy when he was awake and fell asleep when bored, our life as humans had some kind of worth which was missing in the life of the animal.

In later years I interpreted this fantasy as an affirmation of the intrinsic value of human life: we humans have to suffer and the dog has been spared this fate because he does not aspire to concretize values. I do not imply that I went that deeply into the differences between man and animal at age twelve.

I saw that there was no in-between phase from being actively awake to being fast asleep in the life of the animal while I, being bored, had to suffer in a depressive environment. Watching the sleeping dog I began to interrelate the happy animal existence without value and our suffering as a necessary accompaniment of human beings pursuing values. Later I learned to verbalize this insight as our striving to realize ethical deeds, true propositions, and beautiful things. With the development of a more extensive spatiotemporal perspective I became capable by age twelve to fourteen to think farther back into my subjective personal past and farther ahead into the objective distant future. As the temporal-spatial structure of the process of suffering becomes more complex, its healing power becomes more effective.

I heard the yearly performance of Bach's *Passion according to Saint Matthew* many times. The nadir of this composition are Jesus' last words which the baritone sings in Aramaic and the tenor repeats in German. This scene is a most emotionally disturbing and depressing portrayal of the fear of loneliness in conjunction with the dread of death. Bach's music helped me to internalize and to interpret Jesus' suffering and led to the flash of insight into the value of our individual suffering.

I came closer to understanding the difference between the unfractured existence of a being who does not suffer and our fractured existence with its core of suffering, which in turn heals the contrasts in human life.

I met a girl in my twenties. The encounter gave me the overwhelming conviction that we were meant for each other. Reliving this experience during the next few weeks, I began to believe that we could create a "hypostasis" (Charles Morgan, *The Fountain*). I wrote to her asking for further contacts and I went each morning to the post office, anxiously waiting for her reply. I was always uplifted in the main room of the post office with its ceiling a high, beautiful vault of elliptical shape made out of yellow glazed bricks. The girl's letter contained a bombshell: she was engaged. The whole atmosphere of the place changed in a split second; it took on an entirely different feeling tone, but the experience of its beauty was not destroyed. I went back to my rooms and played Beethoven's *Pathétique* sonata and Chopin's Polonaise in C Minor.

I experienced the collapse of a highly idealized and partially fantasized belief in the capture of holistic integration through completion by another human being. I struggled out of that apocalypse with the aid of music, literature, and especially the poetry of Rilke, writing a diary and sharing this experience with a friend who, unfortunately, was surd to my suffering. I continued my medical, musical, and philosophical studies and I did not become a recluse.

I did take an excessive risk on the basis of my belief-system and under the pressure of strong emotional drives. While reliving the brief encounter, I vacillated between being firmly convinced of its value and critically tearing at the foundations of this conviction. This internalized battle went on for a few weeks, until the experience at the graveside (described earlier) gave me the impetus and the courage to write the letter. Previous periods of suffering enabled me to climb out of this pit of suffering which makes manifest that suffering is a healing agent so that we can overcome life's miseries and challenges. The defenses which absorbed the brunt of suffering within the acute crisis were that the beauty and the power of the architecture was not lost and that my ability to empathize actively with the suffering of Beethoven and of Chopin continued. In fact, we grow through suffering.

My most meaningful encounter with death occurred when my future wife and I met for the first time. At that occasion I had to perform the duty of confirming her father's death. Identifying and empathizing with the grieving person I was deeply impressed by her dignified, serious, and heroic attitude, yet displaying dark sorrow in subdued undertones about her loss of both parents within two years. Later on when we had grown to know each other, I attempted to give her solace in her grief. I conjured up before her mind's eye the idea of music going on in two contrasting themes simultaneously, one representing in dark lower tones the destructiveness of death and the other melody in the higher register, embracing life and jubilating in its elevating powers. I tried to communicate to her through this metaphor that we have to live somehow through the heart-rending contrast between death within life and life within death.

Destruction of life and the promise of relief from loneliness was thrust upon her almost simultaneously. Our developing a real-ideal woman-to-man relationship did not eliminate her grieving over the loss of her parents, but it was a powerful healing agent to help my future wife out of an overwhelming depressive experience. Grief and happiness were companions in an uneasy balance. Harassed as we were by the future uncertainties in a deteriorating world situation, the transport of the hypostasis was an effective antidote. Previous sufferings had strengthened us toward undertaking many risky tasks that impeded a smooth path ahead.

I needed the extraction of an impacted and infected wisdom tooth under nitrous oxide anesthesia. It so happened that I had read in the *Connecticut Journal of Medicine* during the evening before this little operation, that a sizable number of patients die each year from anesthesia. Next morning I told the anesthetist about this statistic and urged him to be careful. I asked to see the extracted tooth and examined it with the dentist. The fascinating aspect of this very brief procedure was my state of mind for about thirty minutes while my wife was driving me home. There was a distinct feeling of uncertainty as to exactly what had been done to me, notwithstanding the fact that I had seen the extracted culprit. A slight feeling of unreality pervaded me for a short time, and it was a relief when I felt my normal self again. This type of anesthesia interrupts the stream of consciousness almost immediately and very briefly. Since it is induced by an outsider while one is a passive object oneself, the experience has a striking similarity to our powerlessness in the face of death. Sleep too interrupts this stream, but since it is gradual, semivoluntary, and, as it were, brought on by oneself this interruption is of an entirely different quality than the anesthesia. Incidentally, children at a certain age do seem to be afraid of going to sleep because they seem to feel this interruption might be the same process as dying.

The significant qualities of this rather unimportant interlude are the feelings of slight unreality, the slight lowering of my ego boundary, and a fleeting feeling of a change in my individuality. The relief which I felt when I was my normal self again suggests that some suffering under the dread of death in some mitigated way and in a subdued key was involved in this interruption of my mentation.

> I have listened to many dreams reported by patients in which they were *almost* killed in an accident or by being violently attacked. A young man who suffers from severe borderline schizophrenia and who had attempted several times to commit suicide, reported not only frequent violent dreams about battles in which he killed other people, but he insisted that he dreamed several times about being killed himself. Most people have anxiety dreams in which the brakes of the car do not work while going down hill, but they manage to slow the car down or otherwise avoid a fatal accident. In contrast I remember a nightmarish dream when an X-ray picture was taken of my chest; something went wrong with the apparatus and an atomic explosion destroyed my thorax. As I expired, in my last breath before dying, I said to my wife, "I love you," and awakened quite relieved. This dream comes close to an image of one's personal dying and death.

I interpret the meaning of the so-called last words in the dream as evidence of my conviction, even while dreaming, that death cannot destroy the fusion-communication relationship between woman and man. The value of this relationship continues to exist within the lifetime of the survivor, but it ends as a real-ideal existent with the death of this survivor. The memories of survivors who have witnessed the existence of this real-ideal relationship are real existents during the limited, objective time-space span of their lives.

> About fifteen years ago someone dear to us was afflicted by a serious illness which caused her to retreat into herself so that it was almost impossible for us to be in personal communication with her. At that time she and I attended a concert in which a singer performed Schumann's song about a widower who is looking at his deceased wife's picture. It seems to come to life for him and he exclaims, "I cannot believe it that I have lost you." As we were sitting next to each other in physically close proximity I felt at the same time that she was mentally miles removed from me. I empathized with the widower's feelings which are brought to life in agonizing despair by the music which reinforces the depressive imagery in the poem by a beautiful synergy between the two art forms. The fracture in real life is reinforced by the fractured existence of the widower depicted in the music and the poetry. This double layer of real fracture upon esthetically portrayed fracture engendered such a burst of deep suffering within me that I had to physically bite my tongue painfully to keep these upsurging emotions under control.

Fortunately this person was able, during the next six to eight years, to gradually climb out of the pit of misery into which she had sunk about the time of the concert. Recovering in a zigzag course of improved and regressed phases, much suffering was meted out to her and to us who love her. As for myself, the crucial experience occurred during the concert. This prepared me for later periods of suffering under the load of times of hope and despair, alternating periods of a closer personal relationship versus the chill of repeated distancing during the long period of her recovery. The nadir of suffering had given me strength to master the periods that followed over a long stretch of objective time-space.

COMMENTS

Insight into suffering flows from insight into the concept and process of interpenetration. Interpenetration is the grounding constituent of the power of suffering to heal life's imperfections. The mental triad and the precious present together are a paradigm of interpenetration in many concrete activities. Therefore we can use the mental triad functioning in a precious present as a model for the manner in which suffering heals the fissures in our existence. The act creates a new emergent by projecting condensed memories into the future-in-the-fringe. This new emergent has the character of an interpenetrating whole, since it has attributes both of the old and of the new. This emergent created in a precious present and which is now in its future fringe, initiates our bringing about changes in the present state of affairs which will go on in the objective near and distant future.

Similarly, suffering creates a new emergent out of the contrasting elements in life by forcing them to interpenetrate. While this is going on the person who is creating this interpenetrating whole is undergoing the state of suffering. We are being changed by this creation and suffering for it, into a more valuable, stronger, and wiser individual, and we achieve a higher level of personal integration. Stronger, because we are now better prepared for future sufferings. Wiser, because our temporal-spatial horizons have been enlarged.

When I suffered during the twilight hour and compared my existence with that of the sleeping dog, I began to see that the dog does not interrelate contrasts, that he is either awake or asleep. Man is able to, and animals cannot, make discordant, contrasting states of affairs and states of mind interpenetrate. A nonsuffering animal learns from experience only insofar as certain activities have concrete survival value for it which assure his homeostasis. The animal remains more or less the same organism before and after each new experience or after it has overcome a dangerous physical situation. Man seeks to maintain homeostasis also, but beyond this causally structured process in objective time-space, man also strives to realize ethical, esthetic, and logical values in subjective time-space, which values are purposively structured. Being between the worlds of causality and intentionality, man suffers. After he has been successful in making these two worlds interpenetrate, man is a different person. Through suffering he has created a new emergent. He has changed a small sector of reality by means of activating the mental triad in a precious present, and, since he has suffered, he has changed himself also. Suffering is the necessary condition for our mental and our personal growth and for our becoming (G. W. Allport). And so is interpenetration a necessary constituent of becoming.

The funeral march in the *Eroica* symphony expresses Beethoven's suffering, and it can make us listeners suffer, especially during the last few measures when the major theme is fractured. But the fourth movement is full of light and joy and power which celebrate survival after going through the pit of suffering and dying in a metaphorical sense in the funeral march. We can only experience this release from misery in the fourth movement if we have first lived through (*erlebt*) the drama of the sad movement. We are uplifted by the final movement because we make the two movements interpenetrate. The interpenetrating essence of music can express suffering within joy and joy within suffering. Suffering invests cultural objects with a deeper meaning and majestic power.

A religious person who visits a Gothic cathedral will admire the *grandeur*, its beauty, and its antigravity power. But beyond this esthetic experience he associates the meaning of this building with the suffering of Jesus, even if he is not aware that the ground plan represents the cross. In this environment he experiences very strongly the *mysterium tremendum* of his closeness to God. He immerses and fuses his personal suffering in Jesus' suffering, because he is able to make the blind leap of faith into the belief in a personal God who will bestow the state of grace upon him if he opens himself up to this gift.

A nonbeliever experiences the beauty and the upward-striving power of the cathedral in a mode which is similar to what I experienced in the elliptical vault of the post office in Utrecht. He finds strength in the beauty of the cathedral as a healing factor of his personal suffering. His private internal suffering interpenetrates with the beauty of the building.

A beautiful vase can uplift us and we admire the artist-craftsman who created this object. Did he suffer while he made it? This is questionable. But if the artist is confronted with a state of affairs that makes him suffer, then this object may take on a different meaning for him, and it may blossom forth with a new worth and meaning which it did not possess for him before the crisis. Looking back on his creative power over matter may give him strength and courage during this sad period, and he may tap this source of power, with which he created the beautiful object, to battle heroically through this period of suffering. The beautiful object interpenetrates with suffering and with healing. Even as it heals blemishes and contrasts, suffering transforms truth, beauty, and ethics. It imbues these values with a new seriousness and with deeper and more powerful emotional impacts than they had when the artist first looked at them before he went through a process of suffering and healing. They are the helpmates of the suffering which heals faults and fissures but truth, beauty, and ethics do not of themselves heal dichotomies and fractures.

These thirteen vignettes make the physiognomy of suffering manifest. Interpreting these personal experiences shows how suffering heals life's miseries.

NOTES

1. Carroll Stuhlmueller, "Biblical Voices of Suffering and Prayer" (Paper read at the First Congress on Human Suffering, Notre Dame University, Notre Dame, Ind., April 1979), pp. 22-23, by kind permission of Stauros International, West New York, N.J.
2. *Webster's New International Dictionary*, 2d ed., s.v. "Fates."
3. Ibid., s.v. "Norns."
4. Jean Piaget, *The Child's Conception of Time* (New York: Basic Books, 1969), p. 202.
5. Paul Fraisse, *The Psychology of Time*, trans. Jennifer Leith (New York: Harper & Row, 1963), p. 271.
6. R. Merleau-Ponty, *Phénoménologie de la Perception* (Paris: Edition Galimard, 1945), p. 333 (Author's translation).
7. Piaget, *Child's Conception of Time*, pp. 7-10.
8. John Bowker, "Suffering as a Problem for Religion" (Paper read at the First Congress on Human Suffering, Notre Dame University, Notre Dame, Ind., April 1979), pp. 1-3, by kind permission of Stauros International, West New York, N.J.
9. *New Columbia Encyclopedia*, 1975, s.v. "Zoroastrianism."
10. *Webster's New International Dictionary*, 2d ed. s.v. "Hesiod."
11. Ernst Cassirer, *An Essay on Man: An Introduction to a Philosophy of Human Culture* (New Haven: Yale University Press, 1944), p. 130.
12. Bowker, "Suffering as a Problem for Religion," pp. 11-12.
13. Stuhlmueller, "Biblical Voices of Suffering and Prayer," p. 5.
14. Johan Huizinga, *The Waning of the Middle Ages* (Garden City, N.Y.: Doubleday & Co., 1954), p. 242.
15. Aeschylus, *Prometheus Bound*, trans. David Grene (Chicago: University of Chicago Press, 1942), p. 589.
16. Job 15:2-6; 12-14.
17. Hans Bethge, "Die chinesische Flöte (Leipzig: Am Insel Verlag, 1907). (Author's version in English).
18. Viktor Frankl, *Man's Search for Meaning* (New York: Washington Square Press, 1963), p. 54.
19. Peter Kunstadter, "Living with the Gentle Lua," *National Geographic* 130, no. 1 (1966): 126.
20. Cicero, *Tusculan Disputations*, bk. 1, sec. 9, p. 71. *Apuleius Golden Ass*, bk. 9, p. 222. *Basic Works of Cicero*, ed. and with an introduction and notes by Moses Hadas (New York: Modern Library, 1951).
21. Søren Kierkegaard, "A Concluding Unscientific Postscript," in *Masterpieces of World Philosophy in Summary Form*, ed. Frank N. Magill and Ian P. McGreal (New York: Salem Press, 1961), pp. 629-30.
22. Martin Heidegger, *Sein und Zeit*, (Halle a.d.S.: Max Niemeyer Verlag, 1929), pp. 263, 255.
23. Eugène Minkowski, "Lived Time" in: *The Human Experience of Time: The Development of Its Philosophic Meaning*, ed. Charles M. Sherover (New York: New York University Press, 1975), pp. 457, 453.
24. Charles Sherover, *The Human Experience of Time* (New York: New York University Press, 1975), pp. 4, 15, 272.

25. Robert Kastenbaum, "Memories of Tomorrow: On the Interpenetrations of Time in Later Life," in *The Personal Experience of Time*, ed. Bernard S. Gorman and Alden Wessman (New York; London: Plenum Press, 1977), p. 211.

26. J. T. Fraser, *Of Time, Passion and Knowledge: Reflections on the Strategy of Existence* (New York: George Braziller, 1975), p. 205.

27. Henri Barbusse, *L'Enfer* (Paris: Albin Michel, n.d.), p. 250.

28. George Gerster, "Saving the Ancient Temples at Abu Simbel," in *National Geographic* 129, no. 5 (May 1966): 717, 718.

29. Auguste Rodin, *L'Art* (Paris: Bernard Grasset, 1924), pp. 102-13.

30. Alekandr Solzhenitsyn, *One Day in the Life of Ivan Denisovich*, trans. Bela von Bloch (New York: Lancer Book, 1963), p. 64.

31. Anton Zijderveld, "Jokes and Their Relation to Social Reality," in *Social Research* 35, no. 2, (Summer 1968): 288.

32. William Shakespeare, *The Tragedy of Hamlet, Prince of Denmark. The Works of Shakespeare*, ed. John Dover Wilson (Cambridge: At the University Press, 1977), Act 2, 2, 1414, p. 42.

33. Ibid., Act 5, 1, 115, p. 136.

34. Viktor Frankl, *Man's Search for Meaning*, p. 68.

35. Henri Bergson, *Laughter: An Essay on the Meaning of the Comic*, trans. Brereton and Rothwell (New York: Macmillan Co., 1912), pp. 69, 177.

36. Stuhlmueller, "Biblical Voices of Suffering and Prayer," p. 1.

37. Friedrich Rückert, *Ausgewählte Werke in einem Band, Herausgegeben und eingeleitet von Julius Kuhn* (Leipzig: Verlag von Philip Reclam Junior, n.d.), p. 174 (Author's translation).

11
The Meaning of Suffering

MEANING AND HEROISM
We can synthesize the imperfections in our existence if we commit ourselves to the heroic life-style. Heroism has many faces. Usually we think of it in connection with battles and war, especially when someone sacrifices his own life to save the lives of his fellow combatants, or when someone pulls a comrade to safety during an industrial accident under great danger to himself. In ancient times heroes went out to destroy a monster singlehandedly or to overcome dangerous situations by wit and presence of mind as did Odysseus. Max Scheler points out that ancient heroes took on a dangerous situation as a risky adventure, seeking fame and the adulation of others. This is not suffering for a profound purpose but for the sake of creating a potent self image to be admired by others."The ancient heroic attitude breaks down before the more profound suffering of the soul, i.e., remains completely dependent on the image which, in reflection, it creates of the other and, at least, of himself."[1]

The heroic attitude which is the correlate of suffering, carries an inner physiognomy which lacks the theatricality of the hero-warrior of ancient times. His battle and suffering are predominantly a physical, bodily affair. In modern times we usually think of the great creative minds when we honor someone for his heroism. Socrates assumed the heroic attitude when he took the poison. The late self-portraits of Rembrandt inspire us with admiration for this man who has suffered intensely and whose deeply carved features and penetrating eyes glow with heroic power. When Beethoven became deaf, he seriously considered committing suicide, but the urgent drive to express his musical ideas that were boiling in his mind, made him decide to accept his horrible fate heroically. He said, "Ich werde das Schicksal bei den Rachen greifen!" ("I shall grab Fate by its throat!") These are the exceptionally talented people of genius who put their own life in the service of an ideal.

But the heroic life-style is inserted also in everyday living, in the making of difficult decisions, or in the daring pursuits of inventors who find new means of manipulating and controlling things. In this progression from pedestrian business to the loftiest endeavors, the heroic life-style becomes imbued with more and more emotional charge, anxiety as to success or failure, hope or despair, and deeper modes of suffering. The heroic life-style ranges from average man caring for objects on hand, to the work of the most ideal-driven creators. In these activities which one pursues on different levels of rational and emotional intensity, the heroic life-style plays a constitutive role. The heroic warrior of the past, who attacked danger, destruction, and death, projected his suffering into the outside world; people who live heroically today may suffer privately and often their suffering is hardly noticed by others.

Usually, one would not classify routine tasks of doctors and nurses as heroic deeds, but if one is a patient oneself, these activities take on an unusual feeling tone. As a locum tenens, I saw many patients a day during a grippe epidemic. I went through the routine of pulse-counting, temperature-taking, and listening to chests, which soon became quite dull activities. But when I got a case of grippe myself, and a colleague came to see me, I felt that there was something awe-inspiring about his ministrations.

If we have to plunge into the depths of discontinuity, disharmony, rents and tears, dichotomies and factures, searching to forge synthesis and harmony, we have to be prepared that sometimes we will get the results we are looking for, but more often we are not so lucky. No matter what the outcome may be, we cling stubbornly to the belief that a solution is possible. But since we cannot predict the future-in-the-fringe of a precious present, and since we can calculate the future in objective time only with approximate, statistical accuracy, we have to accept that we do suffer in our attempts to heal the imperfections and the fractures in our existence. Adopting the heroic attitude of living into the future we accept courageously that suffering is the core of the human condition. Whether we suffer on the physical level due to outer circumstances or due to physical illness, or on the idealistic level in the concretization of values, the heroic attitude accompanies suffering and transforms it into a constructive, future-attuned, meaningful process.

The urge to heal imperfections in our existence presupposes that we accept the fact that these conflicts and fractures do exist. If we overlook or deny these facts of life, we should live only a half human life by convincing ourselves that we always exist totally harmoniously and happily. "Leibnitz was also an optimist about the essential goodness of man and the possibility of his perfections, and it is probably this view of the nature of man which more than any single factor led Leibnitz to his 'best of all possible worlds' doctrine."[2]

This optimistic view of ourselves and our surrounding world makes no sense to us modern children of the twentieth century with its global upheavals. The confidence of our forefathers who lived in the nineteenth century and who believed in a world evolving toward ever more perfect and happier stages of human development and in continued biological evolution, has been thoroughly shaken after two world wars and the threat of atomic disaster overhanging us.

Does it make any sense to return to the pessimistic view of the world of the late Middle Ages? Then, many people were convinced their purpose on earth was to pay outstanding debts on the sins committed by their forefathers, going back to Adam and Eve. The sooner out of this life, the better, since for the believers in the Christian religion there was the promise of a better life in the hereafter. This past-tending life-style was imbued with the dread of death, according to Huizinga. During the "Waning of the Middle Ages" people were as obsessed with death as the ancient Egyptians were.

We modern people cannot share the optimism of the nineteenth century either, since the ever-present threat of worldwide destruction by an atomic war hangs over us. Should we therefore give up striving for the healing of fissures and fractures in our existence and should we adapt an utterly pessimistic outlook upon our existence in a cowardly, deterministic style? Is there no meaning to life in the modern world? This style mocks our innate zest for life. Some young people give in, and they do give up; they live unorganized lives, wrecked by drug and alcohol abuse, they move recklessly from mate to mate and restlessly from place to place. They seem to lack goal-setting and limit-setting ethical ideals, and they no longer strive to become integrated, whole human beings. They find tranquility and peace of mind only on rare occasions; these resting points slip away from them, because they have not gone through the internal battle of suffering. Fortunately, many give up this hedonistic life-style after a number of years of experimenting with one compartmentalized pleasure after another.

Most young and old people do go on making something out of their lives regardless of the times of upheavals in which they are living. They continue to live into the future in objective and in subjective time-space, notwithstanding internal and external threats. They search for some measure of happiness, their confidence tempered by the knowledge of their eventual personal death from natural causes or by destruction by the calamities of modern warfare. In the midst of life we have to accept the possibility of dying. As Rilke says, we have to live with this fundamental and inescapable existential dread. "Death is formidable. / We are his, / With smiling lips; / When we fancy ourselves in the midst of life, / He dares to weep, / Right inside of us."[3] These dark foreboding words pictorialize in a tersely beautiful form our awareness of the dread of death and the zest for life interpenetrating mutually.

When and if we have learned to come to grips with the dread of our possible personal death at any time, we are aided and prepared to adopt the heroic life-style; then we can take the same countermeasures against the threat of external disasters, such as atomic war.

Then we will also be better equipped to overcome the internalized emptiness of loneliness, against which we can conjure up several of the defenses against the dread of death. The strongest and the most final of these defenses is submerging ourselves into a real-ideal marriage. These psychological defenses set us free, to a certain extent, to pursue goals, to attempt to realize what is valuable to us, to raise children and to build our marriages and other interpersonal relationships, because we are convinced that suffering for these ideals has meaning.

It seems to us as if the imperfections and the contradictoriness in our human existence are more forcibly grounded in our perception of ourselves and of our surrounding world than ever before in human history. This illusion flows from our unrealistic evaluation of the past, especially about the lost paradise of our childhood and then, by extension, about the glories of the distant, historical past of nations. Hence, we are apt to optimistically overvalue these personal and communal past times. The so-called good old times had their blemishes and fractures too. The people who were living in those times were willing to suffer for what they deemed to be the true, the good, and the beautiful and, if they adopted the heroic life-style, they tried to do something about the shortcomings of their own lives and of their contemporaries.

Suffering Integrates Pessimism with Optimism

If one designates suffering as the core of the human existence, does this mean that one has to adopt an excessively pessimistic life-style? Are we imprisoned in suffering? Is suffering meaningless? Is there no escape? This is hardly the thrust of this book. I hold, to the contrary, that suffering sets us free from a hopeless outlook toward the future and that we create meaning in the present. Also, suffering releases us to some extent from an all-encompassing deterministic view of the world which surrounds us. Suffering overcomes one-sided determinism and excessive pessimism, and it clears the road toward building an indeterministic life-style over and above and transcending the deterministic life-style. We should be resigned to the fact that the world which surrounds us is causally structured, which is true of the structure and function of our body also. We cannot and we should not even try to shed that area of our existence to which we are bound down, insofar as we cannot change very much in these cause-effect sequences in objective time-space, which lack intrinsic meaning. But the heroic attitude drives us onward to continued purpose-striving. Notwithstanding the roadblocks which the causally structured world and our body functions put in our way, we try to implant our purpose-striving tendencies into the deterministic world of causality. In the deterministic-indeterministic life-style we do have a measure of freedom of the will, we do create new emergents out of what seems unalterably fixed, and we do insert meaningful goal-striving activities into cause-effect sequences.

In the world of thought we dare to combine ideas which on first approach seem incompatible; the artist confronts us with images of the world which no one has seen before; the composer challenges our hearing with music which on first listening sounds cacophonous; philosophers come up with visions of ourselves and of the world which cause shock waves through the learned societies. New thoughts are condemned as false, new unconventional paintings and sculptures or buildings are ridiculed; new ethical and philosophical doctrines face fierce criticism. But all these innovations, new outlooks, reevaluation of values, never spring forth out of a vacuum. The original thinker, artist, scientist, deviates from the traditions of his contemporaries, but he always starts out from well-trodden pathways and he knows thoroughly what his predecessors have done. Noguchi has said that you have to renege on your teachers if you want to do something original, which is your very own. It is really quite astonishing that some of us can release ourselves from the shackles of the past, without, however, denying our indebtedness to our teachers. We look at renovators with astonishment because they perform the mysterious creation of a new emergent by making past and future interpenetrate in the present.

Isaac Newton said: "If I have seen further it is by standing on ye Shoulders of Giants."[4] Roy Schafer tells ". . . a very short story about Picasso. When told at first that his now celebrated portrait of Gertrude Stein did not resemble her at all, Picasso replied simply and with enviable confidence: 'It will.' "[5]

Creative activity is not solely projection into the future of subjective time and calculation of the future in objective time. In his early efforts, the creator of future, truly new emergents, is bound down and determined by the tradition in which he has grown up and from which he gradually releases himself. Thus far, he leaves a deterministic world behind and charges forth into an as yet, even to him, unknown indeterministic world. The tension and the uncertainty and the unpredictablity of where he is going, flares up within him as existential dread— he suffers under the pain of fusing aspects of his world which as yet have not taken their final form. When Richard Wagner was composing the scene of Tristan's bleeding, fatal wound, he experienced this existential dread because he had no idea where the music was going to take him.

The battle of being torn between conflicting aspects of our existence goes on in the minds of the great creators on a grandiose scale. This same battle is being fought in miniature and in more subdued tones in the daily lives of average people. None of us is free from problems, conflicting wishes, either/or decisions, fallow periods of boredom, versus uplifting times of accomplishment. We live between optimistic outlooks and pessimistic frustration, confident moods and diffident views of ourselves and of our surroundings, meaningful and meaningless phases of our lives. Pessimism characterizes the deterministic, static, and cowardly life-style which evades creative endeavors or commitment to the pursuit of values. It blocks entrance into new and untried roads of research and discovery which beckon our imagination. Pessimism shuts us off from moving forward into objective time-space, and it paralyzes the mental triad from projecting into the future in subjective time-space. It puts brakes on the mental triad, and it ruins the precious present. As a result, pessimism is colored with a melancholy, drab, unchanging feeling tone which saps our energy. These are the chronically discouraged people who think it is no use to try something new or different, because everything will remain the same anyhow. They are fatalists and they think that we have no freedom of the will whatever, and that all this fuss and bother about creating values is useless and meaningless.

Does the pessimist in his depressive mood suffer? Being in a depressive mood is not the same as suffering. Depression does accompany suffering, but it does not constitute suffering. The pessimist remains locked into the depressive state of mind, but suffering goes beyond this feeling tone. The pessimist is unwilling or unable to give free rein to the mental triad so that it can perform its act of projecting into the future-in-the-fringe of a precious present because he evades the threatening, anxiety-ridden state of uncertainty which is inherent in the act of projecting into the subjective future. Therefore he remains bogged down in his dull state of mind with its heavily loaded feeling tone, since he does not partake of a modicum of freedom of action. The mental triad holds out the promise that we are able to attain such freedom of decision. The misery of the pessimist is nonconstructive and therefore it lacks meaning, since he shies away from going through the crucible of suffering.

The optimist plunges courageously into the awesome, unknown future-in-the-fringe of a precious present and into the future in objective time. He is willing to endure the emotional turmoil of suffering as the inevitable stage through which he has to go, since he is determined to realize values. He suffers constructively and his suffering has meaning. Suffering gives us the needed self-confidence and awareness of selfhood, which keep the pessimistic, deterministic outlook upon the world and inlook into ourselves within bounds.

SUFFERING INTEGRATES INDETERMINISM WITH DETERMINISM

The dualistic life-style is pessimistic insofar as it concedes that our existence is in part determined by causally structured processes which go on in objective time-space. But over and above being resigned to determinism, we cherish the indeterministic possibilities within ourselves and in others, in view of the mental triad's options to operate in a precious present. In consequence we are weighted down with existential tension and we suffer in the back-and-forth flux between these two contrasting aspects of our world, within ourselves, and the tensions in our mind-body relationship. Only if optimism overlays pessimism, or even takes over from it, will we feel the confidence and the drive to opt for a synthesis of the deterministic and the indeterministic worlds through our creative activities. Then we conquer blissfully happy periods if we are lucky enough to successfully heal some of the dichotomies which separate these two worlds. If we should heal the fractures in our existence, then we will be blessed by a state of bliss. But in the process of bringing such a synthesis about, we do suffer.

The mind-body relationship is an area in which existential tension comes to the fore in a realistic context. The experience of joggers is an instructive example of this tension. A well-known cellist told me that he does not feel any bodily aches or pains nor any acute mental tension or elevation of mood while he is jogging, as so many other joggers report. But one or two hours after this exercise he feels a change of mind and body coming over him in the sense of relaxation, a mental and emotional uplift, and a feeling of bodily power, which allow him to do many more things over a longer period of time than he would have been able to accomplish without the previous exercise. Other joggers may tell you that they feel "just great" after they have gone through the initial ten to twenty minutes warming-up phase and they all sense increased power of mind over body. They find it hard to put this shift from determinism toward indeterminism into words. Once they have accomplished overpowering determinism with indeterminism, they leave the phase of suffering behind, and they partake of a state of euphoria.

Most of these examples are instances where people suffer because they set out to do something which they have planned consciously. Suffering is the price they have to pay for a value they want to concretize. However, on the whole we have very little control over the slings and arrows of our destiny. Many people feel that they suffer due to life's circumstances in which they are enmeshed, which they have not sought and which are none of their own doing. Some of us certainly get a raw deal from life. No one asks to be put in a concentration camp or in a slave labor camp. The Russian dissidents don't ask to be punished, but, although they know that they will be chastised for their straying from the party line, they hope that their criticism of the regime and their suffering will be a small factor that will bring about some changes for the better in Russia. They feel that their suffering has value and that it is meaningful.

Some people find ways to take countermeasures against their suffering even in the brutal environment of the Nazi prisons. A Norwegian physician who was incarcerated in such a prison in Norway, heard and saw his fellow inmates being tortured and killed outside his window, never knowing if he might be the next victim. He kept himself from going to pieces mentally and emotionally by playing an imaginary violin, and he listened by intra-audition to symphonies and violin concertos that he knew very well. He was a serious amateur violinist, and he was able to keep his musical value-system alive. These elevating and exceptional instances of the heroic life-style are powerful proof that human beings can wrest victory over determinism of a most destructive kind and that they can still partake of an indeterministic, value-laden life-style under incredible hardship. Such heroism seems to be latent and it can become manifest in almost all human beings during periods of stress and crises.

The battle against determinism is not fought invariably on the levels of the highest artistic and intellectual activities, as Frankl and the Norwegian doctor did. One can roll determinism back with more concrete measures. When the Squalus sank off the coast of Massachusetts in 1939, the survivors who were rescued by means of the Mommsen lung, told the story of their sitting silently in a watertight compartment, while they knew that other crew members had perished in the adjoining flooded sections of the submarine. One of the young sailors asked, "I wonder if we'll ever get out of here alive?" to which the captain answered, "We won't talk about that now." Silence again. Then, one of the men said, "Who do you think is winning the baseball game in Fenway park this afternoon?" From then on the tension was broken and a lively banter back and forth about baseball filled up the rest of the time they had to spend under the threat of death.[6] This story shows that there are many different ways and levels on which people cope with the dread of dying and of death.

We hear this same note of the consoling power of fascination with sport during the terminal phase of the life of a great man, G. H. Hardy. C. P. Snow describes these last days in a melancholy mood but he lets us see how deeply Hardy remained interested in the outcome of important cricket games. He said in effect, that if he knew the hour of his death he still would want to know the cricket scores. "He managed something very similar. Each evening before his sister left him, she read a chapter of a history of Cambridge University cricket. One such chapter contained the last words he heard, for he died suddenly in the early morning."[7]

Let us compare two life-styles, one quite simple, the other more complex. A boy of sixteen quits high school and goes to work as a punch-machine operator. At fifty-six he is still doing the same boring job, and he can hardly wait until he will retire at sixty-two. At sixteen he did not care to keep options open for choices, for flexibility, and for possibilities of various jobs. In that period of his life he could see only the immediate freedom from school chores and of making money right away. He had no interest in looking into the distant, objective future. Now, at fifty-six, he has very few options left, and he feels caught between his job and its repetitive manipulations which have no value for him, except financially, and in which he lost all interest years ago, and the other side of his life, involving his various obligations toward his family. His hierarchy of values is simple and fixed, and he sees no escape from this deterministic style, except in retirement. Here is someone fallen away into average man, who hardly hunts after values and who lives a concrete life, mostly, in objective time-space, which is causally structured and compartmentalized during eight or ten boring daily hours, six days a week. However, there is another aspect to his life which represents values to him, since he tries to keep his second marriage going and he supports his wife and children adequately. Even in this nearly deterministic life-style there is a facet of indeterminism and realization of values.

In contrast to this simple life-style, we get a different feel when we look at the life of a sophisticated, well-educated, professional person, who is successful in a challenging profession which fulfills him, who partakes of a satisfactory, durable marriage and whose children live up to the parents' expectations. He has built a complex hierarchy of numerous values over a long period of time, so that there is a certain flexibility in his life's circumstances which gives him a certain degree of freedom to make decisions and changes if he wishes.

It is as if these two people live in two different worlds: one, in an almost deterministic life-style, causally structured in objective time-space and caged in by compartmentalization; the other, almost in an indeterministic life-style, purposefully structured in subjective time-space, in which the different phases of his activities interpenetrate. One may feel that the punch-machine operator's life is almost meaningless, while the professional's cup runneth over with meaning. However, one should be aware that these two life-styles, which one can contrapose artificially, run in to each other by gradations and similarities. The first life-style is not exclusively deterministic, and the second life is not entirely indeterministic. Therefore, these two people have in common that they suffer from existential tension as they have to shift back and forth between these two extremes. The laborer has realized some values in his lifetime, and the sophisticated person cannot deny that he lives in a causally structured, deterministic world. Both live a mixed life-style and their attempts to keep the two aspects of their lives balanced out over against each other testifies to their heroism. Both lives, on different levels, are meaningful.

THE HIERARCHY OF VALUES

The pursuit to realize values gives meaning to our lives. We think of values as the carriers of continuity, steadfastness, and constancy over long periods of objective time, even into eternity, as the ancient Greek philosophers taught. They seem to be the kingpins which give us guidance and security throughout our life and beyond the grave. Reality, however, tells a different story. Values depend to a large extent on within what culture and in what period of history one happens to live. What is highly valuable in one culture may be scoffed at in another culture. An eloquent example of this discrepancy is the fact that opposite values are assigned to a bride in some countries north of the Mediterranean, compared to how families behave toward a bride in North Africa. North of the Mediterranean the bridegroom expects, or he may even demand, a dowry from the parents of the bride; in some North African tribes, on the other hand, the future husband has to transfer some of his livestock to the stable of the father of the bride. In the first instance, an unmarried daughter is somewhat of a financial burden to the family; in the other situation she is an economic asset to her family, since she is part of their work force.

Aside from cultural customs, one assigns changing and different weights and importance to various values from one period of one's life to another. Speaking for myself, I find that I have relegated playing the piano, which has meant a great deal to me throughout my life, and still does, to a subordinate plane, since writing this book requires that I put as much concentration, energy, and time as I can muster into this activity. The value of writing outweighs the immediate emotional gratification which I get from playing the piano. I do put about twenty minutes aside, several days during the week, into exercises which are necessary to keep neuromuscular mechanisms in shape, but this hardly compares with the uplift in a precious present which I experience when I "really" play the instrument.

Values shift their roles not only within the history of a society, but also within the lifespan of an individual. In particular, the values of interpersonal relationships change throughout our lifetime. First, the mother is the most important person in the baby's world. Then, the father and others are added to its cast of characters. At age three or four the young child begins to devaluate its parents and he may act this rejection out by escaping from home for short distances and brief time periods. He also gives vent to his aggressiveness against them by indulging in fantasies of not being really theirs, but only a foundling or of having been adopted.

In the following scene, closeness to parents, attempts at distancing from them, and recouping of the family bond, are acted out before one's eyes. We picked up our son who was thirteen, at the airport when he returned from summer camp. The first out of the plane were ten- to twelve-year-old boys, who scampered down the ramp, ran to their fathers and mothers and hugged and kissed them. Next the thirteen- to fifteen-year-old boys came sauntering down, much more restrained, acknowledged their parents with a "Hi there" and stuck closely to a pal in lively conversation. Finally the sixteen- to eighteen-year-old boys got out, and they hugged or kissed their mothers and gave their fathers either a warm handshake or a hug around the shoulders.

Teenage rebellion, though, is real and it can get a whole family in turmoil. Rapprochement with his parents and with the rest of the family later on, when the adolescent's inner turmoil has subsided, insures a normal development in the late teens and early twenties. This temporary, separating maneuver and devaluation of the parents helps the young teenager establish his self-evaluation and self-awareness. Then he can build a healthy self-image as he becomes convinced there is something worthwhile within himself. Starting out on his own initiative and geographically separated from his parents, he finds new values in employment, in caring for himself, or in the rigors and rewards of college life. He may end up establishing his own family, when he must develop a whole new hierarchy of values.

In creating values we vibrate between reality and ideality. We experience existential dread, anxiety, when we work toward bringing these two worlds together. Our striving to concretize values makes us suffer. But we suffer not only under our preoccupation with particular values; we have to build and rebuild, organize and reorganize the several values into different hierarchies as our life flows along. This never-ending, multifaceted, restless activity of building and rebuilding value systems imbues our suffering with meaning. We go through the reevaluating of all values in order to assign a meaningful place and sequence of values on the ladder of the hierarchy. Correlating values among each other in consonance with our personality structure, the stage of our development, life's circumstances, and the activities which are going on in the society of which we are an integral part, is a restless process of making contrasting elements interpenetrate into a harmonious whole. Since we never quite succeed in reaching this goal, we suffer under the burden of building hierarchies of values.

The highest value in this hierarchy which two people can strive for in our Western cultures, is a real-ideal marriage. Two separate, whole, mature individuals enter into the mystery of fusion-communication between woman and man, in which they create a new emergent which stands above them, which is a different entity compared to the sum total of the twosome and of which both partake so as to be fulfilled as woman and as man. Two real people partake of this ideal, supra-individual emergent. There are numerous pitfalls in this undertaking because neither of the two gives up personal identity, assets, and deficits, and each brings his own history into the union. Each has built a hierarchy of values in which each was fulfilled prior to entering into the real-ideal marriage.

If they cherish their marriage as the *summum bonum*, the supreme good, then how do they correlate this mysterious hypostasis which is exclusively theirs, with their concretely living together, each doing his specific tasks, each having his own separate and different interests? How do they fit the mothering and the fathering roles into a harmoniously functioning unit if children should add to the completion of the marriage? The woman has to split herself into wife and mother, the man into husband and father. And now we have the difference between mothering and fathering. The mother is always there with the child, she deals continuously with details, her day is truncated into short periods of five, ten, fifteen minutes. The father usually falls into the role of intervening during crises, when unusual, occasional problems come up. Besides, he not being there all day, the child is more apt to listen to his praises or warnings, and therefore he gets results with minimal effort, while the mother has put her whole self and her whole day into the child. The differences between the female and the male roles are apt to stir up resentment against the father-husband.

Some differences between the two personalities may throw a shadow on the relationship, but other differences may be constructive, as is the case between my wife and myself. She being more visually oriented, has taught me the beauty of Egyptian sculpture and of van Gogh; I am more auditorily inclined, and I have opened her ears to the worlds of Bach and Beethoven.

How best can one fit these various roles together into a viable hierarchy? Two doctors who were married found out, to their dismay, that they talked together almost exclusively about each others patients so that they knew a great deal about these patients' lives, but very little about how the other was living outside his professional life. They made the simple rule not to talk medicine after office hours. Without devaluating either their professional or their marital lives, they kept the hierarchy of their values intact. Unfortunately, the solution is not always that simple.

Suffering Furthers Holism

We are whole individuals if we are capable of synthesizing objective time-space with subjective time-space, determinism with indeterminism, the real world with the world of ideals. A whole person moves in many directions and this diversity is held together by unifying forces within our personality. These forces are the heroic life-style, the acceptance of suffering, and an overriding belief that life is meaningful. These faculties make it possible for us to be flexible in our pursuit of values, such as our alternating between an island existence when we are preoccupied with our own goals, problems, and ideals, and the shift toward a continent existence when we submerge ourselves in companionship with our fellow humans, communicating to them what we deem as valuable to ourselves and to them. But during such communication we also have to adjust some of our own values to those of the others. Our belief that life does have meaning encourages us to do something about the dichotomies and the fractures in our own lives, and we try to facilitate a similar healing process in our fellow humans' existence. Many of us can contribute something toward a solution of society's problems or the personal problems of a few of our contemporaries. Some of us can do a great deal to help others, such as Free Dommisse's psychiatrist, who had the insight that in her he had met a very sick patient who would be capable of reintegrating herself. Visiting her regularly and patiently in her room in solitary seclusion and immobilized in body-restraining envelopes, he suggested to her that gradually she would be able to release one arm, and, if she would be able to control that limited amount of freedom to move, then another part of her body would be released from restraints, until she could move freely again among the other patients. He worked with her toward discharge from the institution at some risk to his own reputation, because family and friends raised eyebrows at the prospect of having a potentially assaultive patient back in noncontrolling society. He acted heroically upon his belief in the hidden faculties of this young girl and he helped her to build a meaningful life for herself.

We gain deeper insight into the meaning of suffering when we compare the normal life-style with the life-style of abnormal people. Let us consider the hedonist, the depressed, the manic, and the schizophrenic person. Because their temporal-spatial structure is deficient or inadequate, they cannot partake of holism as long as the deviations of their afflictions plague them. In some pathological states this lack of temporal-spatial and holistic organization is extreme, and we, the observers, are overwhelmed by the painful doubt that these lives seem meaningless to us.

The hedonist is not a whole person. His life-style is put together out of incoherent experiences that do not belong together, as he flits from one pleasurable stimulation to another, like compartmented frames in a moving picture film. He lacks that which makes human life into a coherent entity, because he lacks the capacity to integrate objective and subjective time-space. He strings his life together by grasping one euphoric state after another and if he does not get what he is after, his life feels empty and arid to him when he does not live in a—to him—precious present. He lives in the here-and-now in which he does not interweave his past with his future in his present and as a result, rationality and emotionality are not cooperating. Only emotions count in the life of a hedonist, his future horizons are very limited, and he does not opt for the realization of values. His life-style is bent on the exclusion of suffering. The compartmented frames on his life's film are not held together by suffering. Neither are his neutral and empty and euphoric states of mind in contact with each other. His life-style lacks meaning. The so-called fleeting states of happiness of the hedonist are fleeting cheerful moods, which are only the accoutrements of happiness, in contrast to the secure bliss we gain through suffering.

Hedonistic pleasures are engendered within us mostly by physical stimuli which cause bodily sensations with a pleasant feeling tone, but without intrinsic value. There is nothing wrong with enjoying bodily sensory stimuli, as long as we are aware that they lack deeper value, on account of their truncated temporal-spatial structure occurring mostly in objective time-space, which passes over the integration of body with mentation. Metaphorically speaking, we are temporarily a happy body, in which state of consciousness mentation is in the background and plays a subordinate role.

Kierkegaard interprets the hedonistic life-style in similar terms, but he sees despair as the ultimate outcome of this compartmentalized type of living. He feels that the hedonist is not capable of committing himself to the decision between either/or, which underlies our becoming a whole human being. Ideally, this style should be transformed and completed by the ethical life-style in which true decisions are made. He is convinced that the ethical style should be elevated by religion into a truly valuable mode of existence.[8]

I should like to simplify Kierkegaard's exposition of the hedonistic life-style. It lacks meaning because it lacks the synthesizing leavening of suffering. Because the hedonist does not synthesize objective with subjective time-space, his conscience cannot operate in this defective temporal milieu. Our ethical conscience evaluates what we are doing now, it approves or disapproves of what we have done before, and it cautions us about what we are going to do. But since the hedonist forgets about his past and hardly thinks about the consequences of his act in the future, an ethical conscience cannot function within this temporally unstructured, distorted, arid mental climate. He is a piecemeal, noncohering, incomplete human being. Suffering does not bind his disjointed life-style together. It therefore lacks meaning.

Some neurotic patients and some psychotic persons are poorly integrated individuals also. But in contrast to the hedonist, they suffer very painfully.

Some neurotics live in a fractured life-style as they harbor disturbing and frightening fantasies, which they keep a deep secret from others, while, in contrast, in their professional and their social lives they may feel and behave quite normally and effectively. Neurotics suffer as they shift back and forth between the two extremes of getting gratification from their work and their friendships, but then they are dragged back into their neurotic life-style of utter loneliness, unyielding anxiety, concrete fantasies of dying and depression. They are aware of living between these opposite life-styles which many cannot bridge over and form into a stabilized whole. Their suffering is unconstructive. The meaning of their professional and social life is marred and encumbered by the meaninglessness of the neurotic aspects of their existence.

Depressed patients feel that they have been thrown out of their normal mode of living into a depressed phase, in which they do not recognize themselves, and they torture themselves with guilt feelings about wrongdoings which are usually distortions of minor infractions against their personal rules of conduct. Such minor misdeeds hardly bothered them during the period of living healthily, but which they remember now as if they had committed heinous sins. What hurts them most, is that their depression prevents them from experiencing emotional gratification from activities which they did enjoy before they became depressed.

They are frightened to death because they have lost all hope of ever again enjoying reading, being uplifted by music, or of ever feeling close once more to the people whom they love. Now, they feel quite isolated from them. They may be so obsessed with the thought that they will never find their normal selves again, that they contemplate taking their own lives and they may go on to executing this act. These are periods of nonconstructive, meaningless suffering.

Manic patients may appear cheerful and euphoric on the surface, but if they are able and willing to describe how they really feel, they will tell us reluctantly that they do feel depressed and frightened underneath their overactive behavior and their happy exterior. Their manic drive often leads to reckless actions which are totally incompatible with their usual life-style. In the midst of fantasizing grandiose schemes, they may suddenly burst out crying and cling to the people who are with them, as if to grasp for protection against their inner turmoil and their loss of selfhood.

One can back up the opinion that periods of endogenous depression and manic excitement are indeed meaningless to the patient, since they usually devaluate their sick periods after they have recovered. Most of them do not want to reconstruct and to relive these decompensated periods, to which they attach neither value nor meaning. They are well aware that there has been a break in their life-style, but they despise it as weakness of character and now they suffer under painful self-devaluation. If they should decide to partake of psychoanalysis or of intensive long-term psychotherapy after recovery from the acute episode, then we can often help them to develop insight in the background of these periods and they may become better-integrated individuals. Successfully treated patients may, as an aftermath, reevaluate their sick periods as meaningful instigators of the subsequent personality growth.

Many schizophrenic patients have insight into the unreality of their hallucinations, and they may test whether these experiences are real or imaginary. Usually this insight does not help them to curtail these inner experiences, and they are frightened when they find out that, in fact, they are hallucinating. These people too may enjoy periods of relative normality and freedom from psychotic symptoms, but they suffer gravely when they are aware of a recurrence of symptoms of their underlying chronic illness. This anticipation is more than some patients can tolerate; therefore the risk of suicide is very great at these times. The horror of losing touch again with reality strikes those patients who have insight into their being ill, with the meaninglessness of their psychotic symptoms. Their hallucinations, paranoid ideas, and feelings of unreality may be structured similar to everyone's dreams, but unlike dreams, they do interrupt the feelings of being an integrated self, and this disintegration causes the patient to suffer destructively and without meaning. They often feel helpless against the onslaughts from within and they have to live through the horror of the loss of feeling of selfhood. In their battle against these fissures and splits in their personality they need help from companions, psychiatrist, and medications. They do not feel like whole human beings.

It takes courage for the recovered manic-depressive patient and the reintegrated schizophrenic person to enter into a psychotherapeutic relationship. In fact, it is a sign of their latent heroism if they so decide and go through a long and painful road of treatment. Successfully treated neurotic or psychotic patients can often attain a large degree of holistic reintegration. But they have to accept that this process of self-evaluation and reconstruction will cause them to suffer. Suffering is the very process which forges our holism. This mode of suffering is constructive and meaningful.

The person with osteoarthritis of both hip joints was frightened to death of becoming a wheelchair case, but instead of being split asunder by this fear, he tapped his reservoir of courage and he kept the holism of his body and mind intact. Through the depth of suffering he grew into a stronger, more self-confident and more admirable, whole person, and he is living a meaningful existence.

We grow into more valuable, better-integrated human begins as we live from crisis to crisis; we learn from each foregoing unsettling experience which makes us suffer, to do things differently in the future. This learning and growing process, painful as it is, causes some elements of a previous crisis to be incorporated into the period of relative stability which follows an upsetting experience, and we eliminate undesirable elements from our future activities. Thus, we are better prepared to tackle the next problem that comes our way.

This life-style is based upon the integration of the two forms of time-space and on the interpenetration of the deterministic with the indeterministic life-style. Integration through interpenetration is the hallmark of holism, which suffering brings about.

THE MEANING OF SUFFERING

Suffering is meaningful because it interweaves with our efforts to seek values. The synthesis of reality with value systems causes us to suffer and, beyond the realization of individual values, our building them into a hierarchy is a painful process.

To the extent that man is an island unto himself, loneliness is one of the deepest sources of suffering. We seek to soothe this state of mind by various means: by interpersonal contact; by enjoying works of art or music together; scientists by working in a group with fellow scientists; by attempting, as a therapist to set patients free from destructive suffering. Relations between parents and children, or close friendships, often ameliorate and soften the pain of loneliness, but since these relationships are fragile, they are a healing factor only from time to time. We overcome loneliness in a real-ideal marriage, but the ever-present threat of death of one's mate hangs over the couple: This close bond does not eliminate the dread of death unless one believes with Charles Morgan that the supraindividual, emergent hypostasis which the couple have created has an everlasting existence of its own. In that case, one is convinced that a real-ideal marriage does overcome the finality of death. The harsh truth of life is that most people never get over the loss of a truly beloved mate.

The thinking religious person who believes that his religious faith overcomes the pain of loneliness as well as the threat of dying, veers between his emotionally driven faith and his rational critique of the metaphysical underpinnings of religion. Rational doubt over against faith makes him suffer.

Suffering heals the imperfections in the human existence, but it does not accomplish a permanent synthesis, because suffering is also one of the discontinuous constituents of our discontinuous, nonpermanent sojourn on earth. We have to resign ourselves to the fact that healing happens only during temporally limited and separated periods. Suffering has meaning because it connects disconnected phases of life which stand in contraposition to each other. As a result we obtain periods of secure, inalienable happiness. Although suffering unifies that which is disjointed, which synthesis engenders within us a state of bliss, the specific meaning of suffering is not to make us happy forever. Suffering subdues, mollifies, and balances out moods of diffidence, impotence, and despair, compared with feelings of confidence, omnipotence, and jubilation. But the uncertainty of the future warns us against unrealistic and unwarranted overconfidence in ourselves and in our future.

SELF AND SOUL

We can get a clearer view of the meaning of suffering by placing it in a context with the concepts of Self and of Soul. The core concept of metaphysical interpretations of man is the concept of the immaterial Soul, which exists in eviternity. Metaphysical systems regard suffering as one of the encumbrances of our terrestrial existence from which we can be liberated after our personal death, when the Soul continues to exist. The concept of Soul tends to slough suffering off as undesirable and as a sign of our imperfections. Self, to the contrary, cooperates closely with suffering and accepts it as a valuable and meaningful constituent of human existence. Self and suffering work in consort with the mental triad in a precious present and each component of this threesome has a share in the time-binding process which the mental triad accomplishes in a precious present. But suffering goes beyond the narrow confines of the future-in-the-fringe, because our plans and our decisions stretch out into the distant future in objective time-space. We suffer when we contemplate calamitous eventualities which may happen in the distant future of days, weeks, months, years, decades ahead, beyond the immediate future-in-the-fringe.

Suffering is the background for and the agency which brings about the unity of the Self. We suffer in respect to Self, because we go through periods of self-doubt, suspension of the awareness of Self, discontinuity of self-awareness, or even the loss of selfhood. The fact that self-awareness is interspersed and interwoven with death awareness, drives home the fragility of the Self. Through suffering we create and attain a modicum of continuity of self-awareness. The fact that self-awareness and death-awareness are discontinuous in actual experience challenges the unity and the continuity of the Self. The threat of our personal death is the chief destabilizing factor which our suffering tries to stabilize. The Self, locked into objective and subjective time-space, *sub specie temporis*, has to live heroically through the emotional turmoil which we are constrained to pass through in our existence this side of death. As one of the results of suffering, an individual develops a Self, and therefore he "has" a Self which he experiences as self-awareness and death-awareness. Unlike Eccles's "Self Conscious Mind," which is a hidden homunculus, Self is one of the constituents of the whole individual but Self does not stand for this whole human being, as if Self were a homunculus.

Self is not a homunculus. Although it is tempting to elevate Self to that dignity, we should not use Self as the stand-in for the whole, psychophysical human being. In contrast to Self, Soul does carry whatever is valuable within the human individual. The concepts of Soul and of Self have some attributes in common, but they are also markedly contrasting concepts. Self exists since childhood and up to our personal death, when Self is dissolved together with the dissolution of the holistic mind-body relationship. The suffering Self comes to grips with its own dissolution; hence our dread of dying.

The concept of Soul, created by religious metaphysical systems, evaluates the meaning of suffering and of dying very differently. Soul is a captive of mind and body holism during its sojourn on earth, and the death of the body releases Soul from the limitations of objective and subjective time-space, since it continues to exist *sub specie eternitatis*, on the further side of death. Soul denies the most excruciating source of suffering, which is the termination of our existence in death; it eliminates the dread of dying. The metaphysics which underlies the concept of Soul, sublimates the dread of dying with elaborate rational constructs and emotionally supported rituals, which lie beyond verification by experience. One bases the concept of Soul on faith. Most religions teach that brief periods of terrestrial happiness are only a weak foretaste of celestial bliss which awaits us as a reward for a life free from sin, insofar as this lies within the power of our ethical conscience. These truncated periods of earthly bliss are not regarded as an accomplishment of suffering. The religious person suffers because he can grasp the ideal for which he strives, which is the unification with the godhead, only in particular precious presents, which he interprets as a *nunc stans*, in which he gets a brief glimpse of eternity. In between these blissful periods he suffers from doubting whether his Self has been at fault for partaking of and losing the *summum bonum* of immersion in the godhead, and he may question whether he has lost his belief in his religious metaphysical system. He suffers heroically during these periods of emotional, intellectual, and Self doubt.

Both the religious and the humanistic interpretation of human existence are deeply concerned with the fact that we do suffer. Religions ascribe a different meaning to suffering *sub specie eternitatis*, compared to the humanist who searches for the meaning of suffering *sub specie temporis*. In a time-oriented system of thought, suffering wrests happiness from life before death, if we courageously pursue the heroic life-style. Then we become aware of the intrinsic meaning of suffering.

Religious metaphysics regards mind-body holism as being devoid of intrinsic value. In fact, mind-body derives its ethical and other values from being the temporary indwelling place of Soul's earthly sojourn, and Soul suffers under this encapsulation. Soul welcomes death since, having been stripped of earthbound body and mind, Soul lives in eviternity in unmitigated bliss.

From the viewpoint of time and space, man finds his roots in suffering as a future-striving, dynamic, creative center. Suffering has intrinsic and constructive meaning in a time-oriented interpretation of human existence. The whole human being, including his Self, values suffering as a time-binding, healing agent which overcomes during discrete periods the vicissitudes of life, if man is willing to follow the heroic route. Constructive suffering fires and energizes the heroic life-style.

FROM SUFFERING TOWARD BLISS

Suffering does not eliminate suffering. Or, more clearly worded, once we have gone through a period of suffering and attain a state of bliss, this sequence of mental states does not guarantee that from now on our life will be continuously happy. We do achieve periods of inalienable, firmly grounded happiness as a result of suffering, but we have to become resigned to these phases in our life coming to an end. The discontinuity of human life exists precisely in these alternating arid and colorful phases; fallow, boring periods and exciting, creative times. In conjunction with these phases, we traverse a varied spectrum of dull feeling tones, uplifted moods, and exhilaration and omnipotence. The thread which holds these discontinuous states and moods together is our awareness of selfhood. But the Self, holding present, past, and future together, suffers under these fissures, dichotomies, and fractures. The meaning of suffering lies hidden in the binding and healing acts of the Self, which reconciles discontinuity with continuity, determinism and indeterminism, failures with successes. While establishing harmony and unity between these aspects of human existence, which stand in contraposition, the Self suffers, and becomes a true Self which has both self-awareness and death-awareness. The hedonist, who does not suffer, or evades suffering at all cost, does not develop a true Self in his disjointed, distorted life-style, nor does he attain periods of true happiness.

The meaning of suffering comes to light in the heroic life-style, when we dare to go on living into the threatening subjective and objective future. The act of the mental triad, relying on our past experiences and supported by tradition, synthesizes past-in-the-fringe with future-in-the-fringe within a durational present, which is an act of courage because we suffer tension, anxiety, fright, and sometimes the overwhelming dread of death during these acts. Viktor Frankl shows that suffering can be a healing agent if the prisoners in a concentration camp can safeguard the functioning of their mental triad within the heroic life-style. Those who cannot embrace this style exist in a living death; they have died already in a metaphorical sense. Destructive suffering has destroyed their humanity. The heroic life-style accepts that suffering is the core of the human condition. Even with death constantly at hand suffering has value and meaning.

Viktor Frankl urged his fellow prisoners to adopt this life-style from a viewpoint of suffering, dying, and heroism when he talked to them in the utter darkness of the nighttime barracks. There is a Heideggerian quality to this deeply felt admonition to his fellow sufferers when he reveals his conviction to them that there can be dignity and value in dying heroically. When he falls silent several men press his hand in a gesture of heartfelt gratefulness for this revelation. "Suffering is an ineradicable part of life, even as fate and death.... He will have to acknowledge the fact that even in suffering he is unique and alone in the universe. No one can relieve him of his suffering in his place.... For us, the meaning of life embraces a wider cycle of life and death, of suffering and dying."[9]

Frankl was able to liberate himself from the narrow present of this hellish environment by titanic effort. He kept his temporal-spatial structure intact by giving a fantasized lecture, reconstructing his manuscript, in an imaginary, uplifting talk with his wife. Here we admire with fear and trembling the heroic life-style of superhuman proportions.

In what is for us Western people normal daily life, we can become aware of similar struggles, sufferings, heroism, and coming to grips with the dread of dying in times of personal crises, loss of important peers, disappointments, creative activities, in the flux of happy and depressive stages of our life. We suffer for our partaking of the world of values, which rewards us with feelings of gratification, emotionally bright feelings, of happiness, and sometimes of bliss. We wish the world of values to be stable, unchangeable, and absolutely reliable, but discontinuity, uncertainty, and fluctuation are inserted in this world too. Dichotomies and fractures plague us also in the world of values, which cause us to suffer. We continue to build, change, and rebuild the hierarchy of values, with hope toward the future in which we predict with confidence that what we are accomplishing here and now in a precious present will come to pass in the objective future. We strive heroically to reconcile the disparate aspects of our existence, we suffer while we strive toward the realization of ideals and values. Then, we reach periods of bliss. Our belief that suffering gives meaning to our lives leads to the victory of blissful states of mind.

CONCLUDING THOUGHTS

I have evaluated suffering from several perspectives. It will be worthwhile to define the meaning of suffering along lines laid out by Ogden and Richards.

> When we define *words*, we take another set of words which may be used with the same referent as the first, i.e., we substitute a symbol which will be better understood in a given situation. With *things*, on the other hand, no such substitution is involved. A so-called definition of a horse as op-

posed to the definition of the word, "horse," is a statement about it, enumerating properties by means of which it may be compared with and distinguished from other things. All definitions are essentially *ad hoc*. They are relevant to some purpose or situation, and consequently, are applicable only over a restricted field of "universe of discourse." Thus, if we wish to indicate what we are referring to when we use the word "beauty," we should proceed by picking out certain starting-points, such as *nature, pleasure, emotion,* or *truth*, and then saying "beauty is lying in a certain relation (*imitating* nature, *causing* pleasure or emotion, *revealing* truth) to those points."[10]

When we include a concept like beauty in our "universe of discourse," we have to gather certain relevant points within this concept together and then we have to state how they actively focus upon or point toward the direction of these differentiating points. Thus beauty becomes a dynamic concept, which shows that beauty does something to us and for us, in particular that it arouses valuable emotions within us.

We might say of the meaning of suffering that it relates to the imperfections and the fractures in our existence, that its function is to heal these imperfections and fractures, that its relevance is found in the transformation of apprehension, anxiety, or dread, into more secure states of mind, and that its physiognomy is represented by destructive suffering and constructive suffering. This list of attributes demands that we take a closer look at the function of healing. Suffering and healing intertwine because suffering derives meaning from healing, since suffering heals blemishes, fissures, dichotomies, conflicts, and fractures in our existence.

Historical and cultural convention will have it that healing is the domain of the physician. Within this context, healing has the two meanings of cooperating with the "healing forces" of nature and the opposite meaning of counteracting causally constituted factors in nature which are set on the loose.

An orthopedist puts a broken limb in a cast to prevent a bend in the bone after the fracture has healed. He interferes with the forces of muscular contraction impinging on the bone fragments, because he foresees that the limb will be bent and foreshortened if he does not prevent the effects of these causally determined forces of nature, which are embodied in the nervous system and in the musculoskeletal apparatus. On the other hand, the doctor regards new bone formation between the two fragments as a "healing force" of nature.

A psychiatrist may prevent a depressed patient from committing suicide by promptly administering electro-shock treatment if he judges the risk too severe to wait for the beneficial effect of an antidepressant medication. He predicts that the distorted mentation of a deeply depressed patient is likely to destroy this person as a whole mind and body. His healing activities interfere with the natural processes of the depression and the powers that ravage the patient in the mental realm.

Both doctors interfere with causally and emotionally determined processes because they have the foreknowledge of the future course of these processes in objective time-space. Nature has none of this foresight. But most doctors think that they guide and cajole the "healing forces of nature" in a direction which is beneficial to their patients. This is a metaphorical use of the concept of "healing," and it rests on a mystical, uncritically accepted belief that such a teleologically structured entity actually exists in nature. Most of us harbor the ingrained, archaic conviction that nature and natural processes are purposeful, which shows that we project our own capacity of foresight, foreknowledge, and purpose-striving onto natural processes. Projecting our own meaningful activities upon nature, we confound meaningfulness with purposiveness, we equate meaning with purpose.

One can use the term *healing* in a metaphorical sense in the following concrete situations. These situations can be observed introspectively and without relying on the highly problematical concept of the pervasive teleological organization of natural phenomena. When an athlete pushes his body to the limits of its capacity, he may die in a metaphorical sense, as he comes to grips with his dread of dying. He comes out of this painful experience a more integrated, more mature, more self-confident, more self-aware and more death-aware individual. His suffering on the physical and on the mental levels not only heals the fracture of living-versus-dying in a metaphorical sense, but it also heals the conflict between his recalcitrant body and his purpose-striving mentation. Therefore this experience has meaning for him. Dr. Casey described his conquest of the dread of death during a horror-stricken period of paralysis due to anectine.[11] His suffering derives lasting meaning for him because he knows that he will face the actual fact of his future death stoically and heroically, since by facing death he has healed the dread of dying in a metaphorical sense.

In this "universe of discourse" one can define the concept of meaning in relation to suffering more effectively by negation and affirmation. Let me use two words from everyday language with opposite intent as starting points: *empty* and *full*. Narrowed down to what is intended by the term *the meaning of suffering*, *empty* denotes *meaninglessness*, and *full* denotes *meaningful*. That is, the opposites, an empty life versus a full life, are similar to the opposites, a meaningless life and a meaningful life.

A life filled with activities is not necessarily a meaningful life. A hedonist, or a hypomanic person, or a workaholic, who is immersed in hectic activities, does not interconnect these activities, which remain compartmentalized, occurring one after the other, mostly in objective time-space. Because these crowded-in activities do not interpenetrate mutually, these seemingly full lives are, in the eyes of a critical observer, meaningless. However, if this type of person should allow us a glimpse into his inner feelings and if by chance he might be capable of introspection and self-evaluation, we would probably feel that he does suffer from time to time. The hedonist and the hypomanic person live, mostly, in objective time-space, but we should hold the possibility open that they may live on occasion in subjective time-space. A workaholic may suddenly be startled to realize he really has accomplished very little of what he set out to do.

In contrast, the life of a quiet, sedate person whose sparse, occasional activities are temporally separated from each other by fairly empty time slots, is not necessarily an empty, meaningless life. If such a person makes his infrequent functions interpenetrate with the arid periods, then such a life is a full and meaningful life. This person lives in objective as well as in subjective time-space and he synthesizes causality with intentionality and compartmentalization with interpenetration, which engenders existential tension, or anxiety, or dread. The sedate person may feel emotionally quite bland and grey during the empty time slots when he is not doing anything of importance. But when he liberates the mental triad in a precious present, this person liberates his activities into an interpenetrating whole and he experiences a lively feeling tone. The hedonist, the hypomanic person, and the workaholic usually experience vivid feelings during their multiple, mutually unorganized activities which therefore do not bring about existential tension. Therefore, meaningful does not equal filled with emotionality, nor does meaningless signify lack of emotionality. The meaningful life is characterized by existential vibrations and dread, which are lacking in a meaningless life. A meaningful life is tension-filled because we try to balance conflicting goals and constituents, fissures and fractures, against each other. Life is meaningful in consequence of our integrating our activities with several of the constituents of our personality, of our body, and our surrounding world. We integrate Self with the mental triad in a precious present, objective time-space with subjective time-space, causality with intentionality, rationality with emotionality, compartmentalization with interpenetration. This multifaceted synthesis and effort at integration makes us suffer. Suffering, emotionality, and meaning ought to integrate and interrelate, since they are the energizing powers which underlie the effectiveness of healing.

Analogies and metaphors can enlighten us about the healing function of suffering. As Ogden and Richards say, "A normal mind, however, still requires the aid of instances, analogies and metaphors."[12] We need these aids in order to realize that the significance of suffering occurs in the two modes of destructive and of constructive suffering. Suffering and healing are sometimes disjointed, but most of the time they are coordinated. This separation of suffering from healing may cast doubts on the thesis that suffering is meaningful. We can resolve the disjunction of destructive versus constructive suffering if we bring the heroic life-style into the discussion. Within that context, suffering can have meaning in the destructive mode also. Arthur Miller's *The Price* can enlighten this obscure point.

One can look upon this play as a concatenation of four people who suffer sometimes destructively and at other times constructively.

The theme of death runs throughout the play: the chair in which the father of the two brothers died many years ago stands center stage.

Victor, a policeman, is trying to sell old furniture to a dealer, Solomon, and they haggle over the price.

His wife, Esther, worries that the stream of their lives has frozen over. She wants to move ahead into a better future. She believes in her simplistic way that the money which they will make from the sale of the furniture will open up possibilities for her husband so that he can retire from the police force.

Victor is afraid of making any change at all. His fear of the future feeds into his dread of dying which, at age fifty, seems to be edging closer and closer. He looks back on his life as if it was not real, as if it is not his own life.

Walter, his brother, is a successful surgeon, and he appears in the middle of Victor's wheeling and dealing with Solomon. Walter tries to heal the rift which has kept the brothers apart since late adolescence, but these attempts bounce off on Victor's resentment against his brother. Walter reveals that he used to be driven by sheer terror in his surgical practice so that he had to undertake compulsively to operate on hopeless cases. Now he is doing medicine without acting out his drive toward omnipotence and he is happy in his work.

Solomon, the eighty-year-old dealer, plays on the possibility of his dying in the near future right from his first entrance. He makes a tragicomical pun, indicating that climbing up the stairs to the apartment might be too much for his unstable heart condition. Solomon's suffering is particularly elevating because one sees here-and-now how he has come to grips with his dread of dying. Miller shows the old man's time-binding capacity by giving him the humorous opening remarks, over against the closing scene: here he listens to the laughter on an old record of a comic performance in which he took part as a young man, some sixty years ago. He joins the fun on the record and falls into a fit of laughter with tears in his eyes.

The healing power of humor and laughter comes out forcefully against the macabre stage setting of symbolized death and the utterances of the characters who let the others know that they dread dying. When Solomon says something funny, Walter bursts out laughing and when Victor and Esther also fall into this spell they suddenly feel closer to each other. An insightful concretization of the healing power of humor, right in the middle of their back and forth skirmishing and battling.

Victor and Esther suffer destructively as long as they attack and criticize each other, she in her thirst for change, he held back from changing by his dread of dying. They distance from each other, come closer together, fall away, and they finally find some modicum of healing the fissures and the dichotomies which tarnish their union. By synthesizing their past and their future within the present suffering they achieve the dignity of constructive suffering.

Walter seems to suffer destructively when he storms out of his brother's apartment in a rage, over the failure to rebuild the former warm relationship. But we know that he has also suffered constructively when he reevaluated and reconstructed his professional life. His heroic attitude leaps to the fore since he does accept the fact that he has to suffer in his fruitless attempt to heal some of the dichotomies in his life. Miller's "Production Note" underscores the conclusion that Walter's heroic attitude lifts his destructive suffering onto the level of constructive suffering:

> Walter is attempting to put into action what he has learned about himself and sympathy will be evoked for him in proportion to the openness, the

> depth of need, the intimation of suffering, with which the role is played. Each has proved to the other what the other has known but dared not face; each is left with touching the structure of his life.[13]

This final sentence can be interpreted in view of the temporal structure of human existence. These four people have learned a good deal about the temporal structure of their lives since they have faced their past and they have dared to take a look at the dread of their future dying and death. Miller's play looks deeply into the human condition. It deals with loneliness and the dread of dying and of death; it makes the heroic attitude audible and visible; it shows us how painful the building of hierarchies of values can be; these four people suffer as they reveal the vicissitudes of their past to each other, and in so doing they synthesize their past, with their future, within their present. This time-binding process makes it possible for each "to be in touch with the structure of his life." One hears that suffering in both modalities is a determining feature of interpersonal communication. They suffer privately and secretly in isolation, when they become introspectively aware of their failures, the fears and the blemishes of their personal relations. They also suffer openly and communally in a continent existence within the context of their confrontation.

They are set free to suffer constructively after their revelations of past sufferings have become mutually known, and this, in turn, sets them free to synthesize their past with their future in their present.

The deeper we dig into the foundations of suffering, the more numerous become the layers of meaning which we uncover and the qualities of meaning multiply in proportion to the depth of the inquiry. Life has meaning if through suffering we synthesize the two time-space forms, the two categories, the two methods of observation, rationality with intentionality and emotionality, compartmentalization with interpenetration, and living both as an individual sufficient onto ourselves and immersing ourselves in society as we partake of interpersonal communication. These synthesizing activities lay the groundwork for the loftiest cultural and ethical endeavors in our lives, which actualize ideality within reality. In their turn, these endeavors are a grave source of suffering since they overarch the chasms, constrasts, and fractures of our existence, which we try to overcome and heal. Man suffers when he strives to create new emergents and urges and compels reality to interpenetrate with ideality.

Loneliness and dying are the major fractures in our existence and they are the roots of the direst modes of suffering.

We challenge the blight of loneliness by entering into a real-ideal marriage, which is the most valuable kind of interpersonal communication. But we have to pay a debt of suffering in order to create the synthesis of two human personalities into a suprapersonal emergent.

We can heal the fracture living-versus-dying in a metaphorical sense and in reality when we face the fact of our personal death, if we assume and defend the heroic attitude toward death. Many people heal this severest fracture in our existence by going through this process of constructive suffering.

A humanistic interpretation of man's terrestrial sojourn from the viewpoint of time and space rests on the conviction that man, while he suffers, creates a meaningful life, since he overcomes loneliness through interpersonal communication and he integrates living with dying through the medium of the heroic life-style.

NOTES

1. Max Scheler, "The Meaning of Suffering," in *Centennial Essays*, ed. Manfred Fringe (The Hague: Martinus Nijoff, 1974), p. 141.

2. Gottfried Wilhelm von Leibnitz, "Theodicy," in *Masterpieces of World Philosophy in Summary Form*, vol. 1, ed. Frank Magill and Ian McGreal (New York: Salem Press, 1961), p. 455.

3. Rainer Maria Rilke, *Schlusstück* in *Buch Der Lieder*, vol. 2 (Leipzig: Ernest Finn, Insel Verlag, 1955), p. 477 (Author's translation).

4. Robert K. Merton, *On the Shoulders of Giants. A Shandean Postscript* (New York: Harcourt, Brace & World, 1965), p. 9.

5. Roy Schafer, "The Psychoanalyst's Empathy," Lecture given for the Westchester Division of the New York Hospital, White Plains, N.Y., May 4, 1981, p. 33.

6. Peter Maas, *The Rescuer: The Rescue of the Squalus* (New York: Harper & Brothers, 1967).

7. G. H. Hardy, *A Mathematician's Apology* (Cambridge: At the University Press, 1967), p. 56.

8. Søren Kierkegaard, "Either/Or," in *Masterpieces of World Philosophy*, vol. 2, ed. Frank Magill and Ian McGreal (New York: Salem Press, 1961), pp. 614, 619.

9. Viktor Frankl, *Man's Search for Meaning* (New York: Washington Square Press, 1963), p. 107.

10. Charles K. Ogden and I. A. Richards, *The Meaning of Meaning* (London: Kegan, Trench, Trubner & Co.; New York: Harcourt Brace & Co., 1936), pp. 110, 114.

11. Jesse P. Casey, "A Psychiatrist's Experience of Terror," *Psychiatric Annals* 8, no. 5 (May 1978): 76-80.

12. Ibid., p. 114.

13. Arthur Miller, *The Price: A Play* (New York: Viking Press, 1968), p. 114.

SUFFERING HEALS IMPERFECTIONS AND FRACTURES IN HUMAN EXISTENCE.
SUFFERING MAKES US INTO WHOLE HUMAN BEINGS.
UNIFYING REALITY WITH IDEALITY, WE SUFFER.
LIFE IS MEANINGFUL BECAUSE WE SUFFER.
SUFFERING FILLS LIFE WITH MEANING.

Glossary

1. *Cause-effect* sequences — occur in objective time-space.
2. *Compartmentalization* — separates elements within an entity from each other.
3. *Constructive suffering* — keeps the future open; it enables growth and becoming.
4. *Destructive suffering* — forecloses the future; it stifles development and change.
5. *Emergent* — a new entity is created by making past and future interpenetrate within the present.
6. *Future fringe* — a constituent of the precious present. Unlike the objective future, the future fringe interpenetrates with the durational present, and it protrudes for varying, short distances into the objective future.
7. *Future-in-the-fringe* — accommodates one of the activities of the mental triad which is projection into our private, personal future.
8. *Holism, Gestalt* — are results of the process of interpenetration.
9. *Interpenetration* — makes elements of an entity or of an activity influence and change each other.
10. *Intra-audition*, or *introspection* — we listen to an internal dialogue going on within ourselves.
11. *Janusian* (A. Rothenberg) — looking at the same time back into the past and ahead into the future, from the vantage point of the present.
12. *The mental triad* — a time-binding agent; it makes the past and the future interpenetrate within the present.
13. *nunc fluens* — the moving, durationless now which stands for the flow of objective time from past to future.
14. *nunc stans* — the stillstanding, durationless now. Mystics describe experiencing eternity in a *nunc stans*.
15. *Objective time-space* — the form in which we experience our surrounding world; this form is homogeneous and it can be measured.
16. *Pain* — is not synonymous with suffering; it does not cover the whole gamut of suffering.

Glossary

17. *Past fringe* — a constituent of the precious present. Unlike the continuous, objective past, it refers to discontinuous periods in our private past and in the communal past.
18. *Past-in-the-fringe* — accommodates one of the activities of the mental triad which is condensation of our memories in the near, the distant, or the far away past.
19. *The precious present* — consists of a durational present, a long past fringe, and a short future fringe.
20. *The punctiform instant* — is durationless; we represent it with a geometric point without spatial dimension.
21. *Subjective time-space* — the form in which we know about our inner states; this form is heterogeneous, and it can not be measured.
22. *Time-synthesizing mentation* — generates emotional tension, anxiety, and existential dread.
23. *Thrown-into-the-world* — Heidegger's term points up aspects of our existence of a deterministic nature which we did not initiate and over which we have little control.

Bibliography

BOOKS

Adler, Mortimer. *Aristotle for Everybody: Difficult Thought Made Easy*. New York: Macmillan Co., 1978.
Aeschylus. *Complete Tragedies*. Translated by David Grene. Chicago: University of Chicago Press, 1942. Vol. 1, *Prometheus Bound*.
Albert, Stuart, and Jones, William. "The Significance and Structure of Children's Bedtime Stories." In *The Personal Experience of Time*, edited by Bernard S. Gorman and Alden E. Wessman, pp 109-32. New York; London: Plenum Press, 1977.
Allport, Gordon W. *Becoming: Basic Considerations for a Psychology of Personality*. New Haven: Yale University Press, 1955.
Amundsen, Roald. *My Life as an Explorer*. Garden City, N.Y.: Doubleday, Page, 1927.
Aquinas, Saint Thomas. *Summa Theologica*. Translated by A. J. Pomerans. New York: Basic Books, 1969.
Aristotle. *Physics*, bk. 2, chaps. 3, 8. *The Basic Works of Aristotle*. Edited by Richard McKeon. New York: Random House, 1966.
———. *Physics*. With an English translation by Phillip Wicksteed and Francis W. Conford. Vol. 2. Cambridge, Mass.: Harvard University Press, 1935.
Augustine, Saint. *Confessions*. With an English translation by William Watts. Bk. 11 London: William Heinemann; New York: G. P. Putnam's Sons, 1931.
Ayers, Alfred J. *Language, Truth and Logic*. New York: Dover Publications, 1935.
Bannister, Roger G. *The Four Minute Mile*. New York: Dodd, Mead & Co., 1960.
Barbusse, Henri. *L'Enfer*. Paris: Albin Michel, n.d.
Bell, Alan P., and Weisberg, Marvin S. *Homosexualities*. New York: Simon and Schuster, 1978.
Bergsma, Jurrit. *Somatopsychologie: Op Zoek naar een Psychosociale Dimensies van de Geneeskunde* [Somatopsychology: Searching for psychosocial dimensions in medicine]. Lochem, Kaatsheuvel, The Netherlands: Uitgeversmaatschappij De Tijdstroom, 1975.
Bergson, Henri. *Durée et Simultanéité*. 7th ed. Paris: Presses Universitaires de France, 1968.
———. *Laughter: An Essay on the Meaning of the Comic*. Translated by Brereton and Rothwell. New York: Macmillan Company, 1912.

———. *Matière et Mémoire: Essai sur la Relation Du Corps à L'Esprit*. Paris: Librairie Félix Alcan, 1925.
Berrill, N. J. *Charade of the Honeybee*. Vol. 1. Illustrated Library of the Natural Sciences, edited by Edward M. Weyer, Jr. New York: Simon and Schuster, 1958.
Bethge, Hans. *Die chinesisehe Flöte*. Leipzig: Im Inselverlag, 1907.
Brown, Margaret W. *The Dead Bird: Four Children Prepare a Woodland Funeral*. New York: Dell Publishing Co., 1979.
Bunge, Mario. *Causality: The Place of the Causal Principle in Modern Science*. Cleveland; New York: World Publishing Co., Meridian Books, 1963.
Burckhardt, Jacob. *Die Kultur der Renaissance in Italien*. 14th ed. Leipzig: Alfred Kroner Verlag, 1925.
Cassirer, Ernst. *An Essay on Man: An Introduction to a Philosophy of Human Culture*. New Haven: Yale University Press, 1944.
Childers, Perry, and Wimmer, Mary. "The Concept of Death in Early Childhood." In *Child Development*. Milwaukee: University of Wisconsin, 1971.
Cicero. *Tusculan Disputations*, bk. 1, sec. 9, p. 71; *Apuleius Golden Ass*, bk. 9, p. 222. *Basic Works of Cicero*. Edited by and with an Introduction and notes by Moses Hadas. New York: Modern Library, 1951.
Cottle, Thomas J. "The Time of Youth." In *The Personal Experience of Time*, edited by Bernard S. Gorman and Alden E. Wessman. New York; London: Plenum Press, 1977.
Descartes, René. *Meditations on First Philosophy*. Translated by E. S. Haldane and G.R.T. Ross. Cambridge: At the University Press, 1968.
———. *Philosophical Works*. Rendered into English by E. S. Haldane and G.R.T. Ross. Vol. 2. New York: Dover Publications, 1955.
Diesterweg, A. *Diesterweg's Populäre Himmelskunde und Mathematische Geographie*. Leipzig: Akademische Verlagsgesellschaft, Becker und Erler Kam, 1941.
Doob, Leonard W. *Patternings of Time*. New Haven: Yale University Press, 1971.
Drake, Stillman. *Galileo Galilei: Two New Sciences*. Madison, Wis.: University of Wisconsin Press, 1974.
———. *Galileo Studies: Personality, Tradition and Revolution*. Ann Arbor: University of Michigan Press, 1970.
Eckhart, Johannes [Meister Eckhart]. *Sermons* ("Nothing above the Soul," Sermon 14; "God beyond Time," Sermon 25). A modern translation by Raymond B. Blakney. New York: Harper & Brothers, 1941.
Engelhardt, H. Tristram, Jr. *Mind-Body: A Categorial Relation*. The Hague: Martinus Nijhoff, 1973.
Fearing, Franklin. *Reflex Action: A Study in the History of Physiological Psychology*. Baltimore: Williams & Wilkins, 1930.
Feibleman, James K. *Ontology*. Baltimore: Johns Hopkins University Press, 1951.
Fraisse, Paul. *The Psychology of Time*. Translated by Jennifer Leith. New York, Harper & Row, 1963.
Frankl, Vicktor. *Man's Search for Meaning*. New York: Washington Square Press, 1963.
Fraser, J. T. *Of Time, Passion and Knowledge: Reflections on the Strategy of Existence*. New York: George Braziller, 1975.
Freedman, Alfred, and Kaplan, Harold L. *Comprehensive Textbook of Psychiatry*. Baltimore: Williams and Wilkins Co., 1967.

Freud, Sigmund. *Das Ich und Das Es*. In *Collected Works*. Vol. 13. London: Imago Publishing Co., 1940.
———. *Die Traumdeutung*. 3d. enlarged ed. Leipzig; Vienna: Franz Deuticke, 1911.
———. *Jenseits Des Lustprinzips*. London, Imago Publishing Co., 1940.
Freudenthal, Hans. "Johannes Kepler." *Dictionary of Scientific Biography*. New York: Charles Scribner's Sons, 1970, vol. 7, pp. 289-313.
Gay, John. *Red Dust on the Green Leaves: A Kpelle Twin's Childhood*. Thompson, Conn. 1973.
———and Cole, Michael. *The New Mathematics and an Old Culture: A Study of Learning among the Kpelle of Liberia*. New York: Holt, Rinehart and Winston, 1967.
Goethe, Johann Wolfgang von. *Faust I, Faust II*. Leipzig; Utrecht: Pfeil Verlag, G.m.C.H., 1924.
Goldstein, Kurt. *Der Aufbau Des Organismus: Einleitung in die Biologie unter besonderer Berücksichtigung am kranken Menschen*. The Hague: Martinus Nijhoff, 1934.
Gorman, Bernard S., and Wessman, Alden R., ed. *The Personal Experience of Time*. New York; London: Plenum Press, 1977.
Grinker, Roy R., and Bucy, Paul C. *Neurology*. 4th ed. Springfield, Ill.: Charles C. Thomas, 1949.
Hardy, G. H. *A Mathematician's Apology*. Cambridge: At the University Press, 1967.
Hartmann, Heinz. "Comments on the Formation of Psychic Structures." In *The Psychoanalytic Study Of The Child*, edited by Ruth Eisler et. al. New York: International Universities Press, 1948.
———. *Ego Psychology and the Problem of Adaptation*. Translated by David Rapaport. New York: International Universities Press, 1958.
Heidegger, Martin. *Sein und Zeit*. Halle a.d.S: Max Niemeyer Verlag, 1929.
Höffding, Harold. "René Descartes." In *Histoire De La Philosophie Moderne*. Translated by F. Bordier. Paris: Librairie Félix Alcan, 1924.
Huizinga, Johan. *Homo Ludens: A Study of the Play Element in Culture*. Boston: Beacon Press, 1966.
———. *The Waning of the Middle Ages*. Garden City, N.Y.: Doubleday & Co., 1954.
Hume, David. *A Treatise on Human Nature*. Edited by L. A. Selby-Brigge. Oxford: At the Clarendon Press, 1965.
Husserl, Edmund. *The Phenomenology of Internal Time Consciousness*. Translated by James S. Churchill. Bloomington, Ind.: Indiana University Press, 1973.
Jacobs, William T. *Hannibal*. New York: McGraw-Hill, 1973.
James, William. *The Principles Of Psychology*. Vol. 1. New York: Henry Holt & Co., 1870.
Jones, Ernest. *The Life and Work of Sigmund Freud*. New York: Basic Books, 1961.
Kant, Immanuel. *Critique of Pure Reason*. Translated by J. M. Meiklejohn. London: Dell & Daldy, 1870.
Kastenbaum, Robert, and Aisenberg, Ruth. *The Psychology of Death*. Concise ed. New York: Spring Publishing Co., 1976.
Kent, Rockwell. *Rockwelliana*. New York: Harcourt, Brace & Co., 1933.
Kerferd, G. B. "Aristotle." In *Encyclopedia of Philosophy*. New York: Macmillan Publishing Co., 1936, Vol. 1, p. 157.
Kierkegaard, Søren. "A Concluding Nonscientific Postscript." In *Masterpieces of World Philosophy in Summary Form*, edited by Frank Magill and Ian McGreal. New York: Salem Press, 1961.

———. "Either/Or." In *Masterpieces of World Philosphy in Summary Form*, edited by Frank N. Magill and Ian McGreal. New York: Salem Press, 1961.
Kirwan, B.E.R., and Gore, P. A. *Elementary Luganda*. Kampala, Uganda: Bookshop, 1951.
Kübler-Ross, Elizabeth. *On Death and Dying*. London, Macmillan & Co., 1969.
Lawick, Baronesse Jane van. *My Friends: The Wild Chimpanzees*. Washington D. C.: National Geographic Society, 1967.
Leibnitz, Gottfried Wilhelm von. "Theodicy." In *Masterpieces of World Philosophy in Summary Form*, edited by Frank Magill and Ian McGreal. New York: Salem Press, 1961.
Lewis, Sinclair. *Babbitt*. New York: Harcourt, Brace & World, 1922.
Lidz, Theodore. *The Person: His Development throughout the Life Cycle*. New York: Basic Books, 1968.
Loewald, Hans. "The Experience of Time." In *The Psychoanalytic Study of the Child*, edited by Ruth Eisler et. al. New York: International Universities Press, 1972.
Lorenz, Konrad. *On Aggression*. Translated by M. K. Wilson. New York: Harcourt, Brace & World, 1966.
Maas, Peter. *The Rescuer: The Rescue of the Squalus*. New York: Harper & Brothers, 1967.
McDougall, William, ed. *Body and Mind: A History and a Defense of Animism*. London: Methuen and Co., 1911.
Maeterlinck, Maurice. *Les Sentiers Dans La Montagne*. pt. 8, *Monde Des Insectes*. Paris: Bibliothèque Charpentier, 1919.
Mead, George Herbert. "Mind, Self and Society." In *Masterpieces of World Philosophy in Summary Form*, edited by Frank Magill and Ian McGreal. New York: Salem Press, 1961.
Merleau-Ponty, R. *Phénoménologie de la Perception*. Paris: Edition Galimard, 1945.
Merton, Robert K. *On the Shoulders of Giants: A Shandean Postscript*. New York: Harcourt, Brace & World, 1965.
Miller, Arthur. *The Price: A Play*. New York: Viking Press, 1968.
Minkowski, Eugène. "Lived Time." In *The Human Experience of Time: The Development of Its Philosophical Meaning*, by Charles M. Sherover. New York: New York University Press, 1975, pp. 457-458.
———. "Das Zeit und das Raumproblem in der Psychopathologie." Edited by Dr. Kronfeld, In *Weiner Klinische Wochenschrift*, 1931, 12, p. 348d. (Author's translation).
Morgan, Charles. *The Fountain*. New York: Alfred A. Knopf, 1932.
Moulyn, Adrian C. *Structure, Function and Purpose: An Inquiry into the Concepts and Methods of Biology from the Viewpoint of Time*. New York: Liberal Arts Press, 1957.
Nordentoft, Kresten. *Kierkegaard's Psychology*. Translated by Bruce H. Krimmse. Pittsburgh: Duquesne University Press, 1972.
Nozick, Robert. *Philosophical Explanations*. Cambridge, Mass.: Harvard University Press, Belknap Press, 1981.
Ogden, Charles K. and Richards, I. A. *The Meaning of Meaning: A Study of the Influence of Language upon Thought and of the Science of Symbolism*. London: Kegan, Trench, Trubner & Co.; New York: Harcourt, Brace, & Co., 1936.
Piaget, Jean. *The Child's Conception of Physical Causality*, translated by Marjorie Gabain. London: Kegan, Trench, Trubner & Co.; New York: Harcourt, Brace & Co., 1930.

———. *The Child's Conception of Time*. Translated by A. J. Pomerans. New York: Basic Books, 1969.
Plato. "Phaedrus." In *Works of Plato*, selected and edited by Irwin Edman. New York: Modern Library, 1928.
Poortman, J. J. *Indeterminism or Determinism*. Assen, The Netherlands: van Gorcum & Co., 1949.
Popper, Karl R., and Eccles, John C. *The Self and Its Brain*. Berlin: Springer Verlag, 1977.
Rickert, Heinrich. *Der Gegenstand Der Erkenntnis: Einführung In Die Transcendentalphilosophie*. 5th ed. Tübingen: Verlag von J.C.B. Mohr, 1921.
Rilke, Rainer Maria. "Schlusstück." In *Buch Der Lieder*. Vol. 2. Leipzig: Ernst-Finn, Insel Verlag, 1955.
Rodin, Auguste. *L'Art: Entretiens Avec Paul Gsell*. Paris: Bernard Grasset, 1924.
Rothenberg, Albert. *The Emerging Goddess: The Creative Process in Art, Science and Other Fields*. Chicago: University of Chicago Press, 1979.
Rückert, Friedrich. *Ausgewählte Werke in einem Band. Herausgegeben und eingeleitet von Julius Kuhn*. Leipzig: Verlag von Philip Reclam Junior, n.d.
Rue, Leonard Lee. *The World of the Beaver*. Philadelphia; New York: J. B. Lippincott Co., 1964.
Ryle, Gilbert. *The Concept of Mind*. London: Hutchinson of London, 1949.
Santillana, Giorgio de, and Dechend, Hertha von. *Hamlet's Mill: An Essay on Myth and the Frame of Time*. Boston: Gambit, 1969.
Scheler, Max. "The Meaning of Suffering." In *Centennial Essays*, edited by Manfred Fringe. The Hague: Martinus Nijhoff, 1974.
Scott, John P. "Teaching Animals to Talk." In *Animal Behavior*. Garden City, N.Y.: Doubleday & Co., Anchor Books, 1963.
Sherover, Charles M. *The Human Experience of Time: The Development of Its Philosophical Meaning*. New York: New York University Press, 1975.
Solzhenitsyn, Aleksandr. *One Day in the Life of Ivan Denisovich*. New York: Lancer Book, 1963.
Sophocles. *Oedipus Rex, Oedipus at Colonus, Antigone: The Oedipus Cycle*. An English version, Dudley Fitz and Robert Fitzgerald. New York: Harcourt, Brace & Co., 1939.
Spicker, Stuart F. *The Philosophy of the Body: Rejection of Cartesian Dualism*. Edited and with an introduction by Stuart F. Spicker. Chicago: Quadrangle Books, 1970.
Spitz, René. *A Genetic Field Theory of Ego Formation: Its Implications for Psychopathology*. The Freud Anniversary Lecture Series, New York Psychoanalytic Institute. New York: International Universities Press, 1959.
Stone, Joseph, and Church, Joseph. *Childhood and Adolescence: A Psychology of the Growing Person*. New York: Random House, 1968. Chapter: Fears.
Strauss, Anselm. "Awareness of Dying." in *Death and Dying: Current Issues in the Treatment of the Dying Person*, edited by Leonard Pearson. Cleveland: The Press of Case Western Reserve University, 1969.
Swift, Jonathan. *Gulliver's Travels*. New York: Dodd, Mead & Co., 1950.
Syz, Hans. *Vom Sein und Vom Sinn: Ein Bericht ueber ein fruehes erlebnis*. Zurich: Editio Academica, 1972.
Teilhard de Chardin, Pierre. *The Phenomenon of Man*. Translated by Bernard Wall. New York: Harper & Row, 1961.

Terrace, Herbert S. *Nim*. New York: Alfred A. Knopf, 1979.
Weigand, Hermann J. *Thomas Mann's Novel "Der Zauberberg"*. New York: Appleton-Century, 1933.
Weiss, Robert S. *Loneliness: The Experience of Emotional and Social Isolation*. Cambridge, Mass.: MIT Press, 1973.
Whitehead, Alfred North. *Adventures of Ideas*. New York: Macmillan Co., 1935.
Wittgenstein, Ludwig. *Tractatus Logico-Philosophicus*. Translated by D. P. Pears and D. F. McGuinness. London: Rutledge & Kegan Paul, 1961.
Wright, Georg Henrik von. *Causality and Determinism*. New York: Columbia University Press, 1974.
Wulf, Maurice de. *An Introduction to Scholastic Philosophy*. Translated by P. Coffey. New York: Dover Publications, 1956.
Wundt, Wilhelm. *Grundlagen der Physiologischen Psychologie*. 4th rev. ed. Leipzig: Verlag von Wilhelm Engelmann, 1893.
Zimmer, John T. "Army Ants in Ceaseless Circle." In *Illustrated Library of Natural Science*. Vol. 1. New York: Simon and Schuster, 1958.
Zorach, William. *Zorach Explains Sculpture: What It Means and How It is Made*. New York: American Artists Group, 1947.

ARTICLES

Benoit, Emile. "Human Survival and the Fear of Death." *Harvard Magazine* (March-April 1980): 21-25.
Bonaparte, Marie. "Time and the Unconscious." *International Journal of Psychoanalysis* 21 (October 1940): 429-68.
Brines, Michael L. "Bees Have Rules." *Science* 206, no. 4418 (November 2, 1979): 571-73.
Burrow, Trigant and Syz, Hans. "Two Modes of Social Adaptation and Their Concomitant Ocular Movements." *Journal of Social Psychology* 44, no. 2 (April 1949).
Campbell, Thomas W. "Death Anxiety on a Coronary Unit." *Psychosomatics* 21, no. 2 (February 1980): 131, 135.
Carlisle, Olga. "Reviving the Myths of Holy Russia," *New York Times Magazine*, September 16, 1979.
Casey, Jesse F. "A Psychiatrist's Experience of Terror." *Psychiatric Annals* 8, no. 5 (May 1978): 76-80.
Dwyer, John M. "Understanding Modern Immunology." *Connecticut Medicine* 39 (March 1975): 170-73; (April 1975): 216-18.
Fraser, J. T. "Temporal Levels and Reality Testing." *International Journal of Psychoanalysis* 62 (1981):3.
Galdikas, Birute. "Living with the Great Orange Apes—Indonesia's Orangutans." *National Geographic* 157, no. 6 (June 1980).
Gardner, Allen R., and Gardner, Beatric T. "A Standardized System of Gestures Proven a Means of Two-way Communication with a Chimpanzee." *Science* 165 (August 15, 1969).
Gerster, George. "Saving the Ancient Temples at Abu Simbel." *National Geographic* 129, no. 5 (May 1966).
Goodall, Jane. "Life and Death at Gombe." *National Geographic* 155, no. 5 (May 1979).
Gorney, James E. "The Negative Therapeutic Interaction." *Contemporary Psychoanalysis* 15, no. 2 (1979).

Hankoff, L. D., and Munver, Ultan Chandram L. "Prenatal Experience in Hindu Mythology." *New York State Journal of Medicine: History of Medicine*, December 1980.

Harrison, Saul I., and Guiora, Alexander Z. "The Intelligent Use of Human Nature." *Journal of Operational Psychiatry* 2, no. 1 (1980).

Heyerdahl, Thor. "The Voyage of Ra II." *National Geographic* 139, no. 1 (January 1971).

Hoffman, Paul. "Pope Paul Compares Defecting Priests to Judas." *New York Times*, April 9, 1971.

Hugh-Hellmut, Hermione von. "The Child's Concept of Death." Translated by G. Kris. *Psychoanalytic Quarterly* 34 (October 1965).

Kripke, Saul. "New Frontiers in American Philosophy." *New York Times Magazine*, August 14, 1977.

Kuhn, Clifford C., and Bradman, William A. "Pain as a Substitute for the Fear of Death." *Psychosomatics* 20, no. 7 (July 1979).

Kunstadter, Peter. "Living with the Gentle Lua." *National Geographic* 130, no. 1 (1966).

Leaky, L.S.B. "Finding Earliest Man." *National Geographic 118*, no. 3 (September 1960).

Leen, Nina. "Conversations with a Chimp." *Life*, February 1972.

Loewald, Hans. "The Superego and the Ego-Ideal: pt. 2, Superego and Time." *International Journal of Psychoanalysis*, vol. 43, 1962.

Moulyn, Adrian C. "The Functions of Point and Line in Time Measuring Operations." *Philosophy of Science* 19, no. 2 (April 1952).

———. "The Limitations of Mechanistic Methods in the Biological Sciences." *Scientific Monthly* 71 (July 1950).

———. "Mechanisms and Mental Phenomena." *Philosophy of Science* 14, no. 3 (July 1947).

Nemiroff, Martin J. "Reprieve from Drowning." *Scientific American* 237, no. 2 (August 1977).

Pannekoek, A. *A History of Astronomy*. London: Allen & Unwin, 1961; New York: Barnes & Noble, 1969.

Patterson, Francine. "Conversations with a Gorilla." *National Geographic* 154, no. 4 (October 1978).

Pepper, Curtis Gill. "Moshe Dayan: Reflections on War and Peace." *New York Times Magazine*, Sunday May 4, 1980, p. 39.

Premack, David. "The Education of Sarah: A Chimp Learns Sign Language." *Psychology Today* 4, no. 4 (1970).

———. "Language in a Chimpanzee?" *Science*, vol. 172, 1971.

Robin, Eugene. "Claude Bernard: Pioneer of Regulatory Biology." *Journal of the American Medical Association* 242, no. 12 (September 21, 1979).

Sabon, Michael. "The Near Death Experience." *Journal of the American Medical Association* 244, no. 1 (July 4, 1980): 29-30.

Schafer, Roy. "The Psychoanalyst's Empathy," manuscript, n.d.

Seiden, Anne. "Overview: Research on the Psychology of Women; 2: Women in Families, Work and Psychotherapy." *American Journal of Psychiatry* 133, no. 10 (October 1976).

Sheehan, George. "Athlete and Death: Easier to Live and Easier to Die." *New York Times*, November 23, 1975, Sports Section, p.2.

Solla Price, Derek J. de. "The Tower of the Four Winds: Piecing Together an Ancient Puzzle." *National Geographic* 149, no. 4 (April 1, 1976).
Sperry, R. W. "Hemisphere Deconnection and Unity in Conscious Awareness." *American Psychologist*, vol. 23, 1968.
Stent, Gunther S. "Limits to the Scientific Understanding of Man: Human Sciences Face an Impasse since Their Central Concept of Self Is Transcendental." *Science* 187 (March 21, 1975).
Stuart, George E., and Imboden, Otis. "The Maya Riddle of the Glyphs." *National Geographic* 148, no. 6 (December 1975).
Suomi, Stephen J., and Harlow, Harry F. "Social Rehabilitation of Induced Depressive Disorders in Monkeys." *American Journal of Psychiatry* 133, no. 11 (November 1976).
Syz, Hans. "Two Modes of Social Adaptation and Their Concomitant Ocular Movements." *Journal of Social Psychology* 44, no. 2 (April 1949).
Whittaker, Linton. "Talking about Death with My Son." *American Medical News*, May 23, 1980.
Yerkes, Robert M. "The Intelligence of Earthworms." *Journal of Animal Behavior*, vol 2, 1912.
Zijderveld, Anton C. "Jokes and Their Relation to Social Reality." *Social Research* 35, no. 2 (Summer 1968).

REFERENCE BOOKS

Bible, RSV. New York: Thomas Nelson & Sons, 1952.
Dictionary of Philosophy, edited by Dagobert Runes. New York: Philosophical Library, n.d.
Encyclopaedia Britannica. 13th ed. London: Encyclopaedia Britannica Co., 1926.
Encyclopedia of Philosophy. Vol. 1. New York: Macmillan Co., 1967.
New Catholic Encyclopedia. New York: McGraw-Hill Co., 1967.
New Columbia Encyclopedia. New York: Columbia University Press, 1975.
Universal Standard Encyclopedia. New York: Standard Reference Works Publishing Co., 1958.
Webster's New International Dictionary, Unabridged. 2d ed. Springfield, Mass.: G. & C. Merriam Co., 1949.
Winkler Prins Encyclopedia, 6th ed., Amsterdam; Brussels, 1954.

MISCELLANEOUS

Papers

Permission has been given by Stauros International, West New York, N.J., for the use of quotations from the following papers presented at the First Congress on Human Suffering at Notre Dame University, Notre Dame, Ind., April 1979.
Bowker, John. "Suffering as a Problem of Religion."
Breton, Stanislaw. "Human Suffering and Transcendence."
Gajardo, Joel. "Suffering Coming from the Struggle against Suffering."
McGill, Arthur C. "Our Suffering and the Suffering of the Christ."
Oates, Wayne, E. "Forms of Grief: Diagnosis, Meaning, and Treatment."
Stuhlmueller, Carroll. "Biblical Voices of Suffering and Prayer."

LETTERS

Orr, Harry. Personal communication to the author, May 20, 1980.

MUSIC

Haggin, B. H. *Toscanini*. New York: Atheneum, 1975.
Marek, George R. *Beethoven: Biography of a Genius*. New York: Funk and Wagnalls, 1969.
Mahler, Gustav. Symphony no. 2, pts. 4 and 5.
Murdoch, William. *Chopin: His Life*. New York, Macmillan Co., 1935.
Rubinstein, Arthur. "In Paris Hall," *Life*, March 2, 1959.

Index

abstraction from movement, 16-17
Adam and Eve, 261-62
Adler, Mortimer, 3
Aeschylus, 262-63
Albert, Stuart, 212
Allport, Gordon, 96-97, 240-41, 286
Amundsen, Roald, 43-45, 212
Andrea Doria, 97-98, 257
animism, child and causality, 66-67, 87, 253-54
anthropomorphism, and Nature, 79
Antigone, 237
approach, interdisciplinary, 3, 12,
Aristotle, 3, 24, 64-65, 70, 146
ASL (American Sign Language), 118-20
astrology, 57-58
astronomy, 58-59; and teleology, 80
Augustine, Saint, 24, 109, 270
auxins, and growth of plants, 112
Average man: in *das Man*, 29, 54; and falling away, 197, 258; and objective time-space, 54
Ayers, A. J., 66

Bach, Johann Sebastan, 235, 282
Bannister, Roger, 45
Barbusse, Henri, 269-70
beaver, deterministic and indeterministic behavior of, 116, 208
because, disclosure of cause and of purpose, 64

bedtime stories, 212
Beethoven, Ludwig van, 270, 282, 286, 289
being-there (*Dasein*), 4
Bergsma, Jurrit, 172-73
Bergson, Henri, 16, 24, 28, 73, 74, 151-52, 252, 270, 272, 275
Bernard, Claude (*milieu intérieur*), 111-12
Berrill, N. J., 207-8
billiards: neurophysiologist's approach to, 77-78; physicist's predictive conception of, 74-77; player's emotional involvement in, 84-86
bliss, and suffering; 10, 12, 246
Bonaparte, Marie: and awareness of death, 211; and time, 130-32
Bowker, John, 260-62
Bradnam, William, 236
Breton, Stanislas, 192, 235-36
Brines, Michael, 115, 183-84
Buddhism: and Stepping out of oneself, 222-23; and suffering constituent of being, 222-23; and suffering punishment for sin, 222-23
Bunge, Mario, 83-84
Burckhardt, Jacob, 50
burial rites: in animals, 209; and Antigone, 237; and Aztecs, 271-72; and Egyptians, 8, 271; in *Forsyte Saga*, 237; of Oedipus, 91

calendar, 6, 15
Calmette, Léon, 259
Campbell, Thomas, 215
Casey, Jesse, 214, 309
Cassirer, Ernst, 224, 262
Caterpillar, Processionary, 114
causality: and causal equation, 64, 75; and compartmentalization, 54, 64, 74; and contemporaneousness, 81, 83; and derivation from mental triad, 252-56; and efficient, final, formal, material cause, 65; and human action, 68-71; and integration with purposiveness, 64; *of necessity*, 65-66; and objective time-space, 252
Child: and animism, 66-67; and childhood paradise, 5; and concept of causality, 253-54; and differentiation of time from space, 29-30; and dread of death, 8; and infinite time, 130
Childers, Perry, 210-11
chimpanzees: ASL, 118-19; beyond determinism, 121; chimp to chimp communication, 119; and no emergents, 122; and imitation, 120; and killers, no memory of, 208-9; and nest building, 118; Project Nim, 120-23; no suffering, 123; time words, 121; tools, 118
Chopin, Frédéric, 236, 282
Christianity, 19, 222-24, 261, 278
Church, Joseph, 211
Clay, E. N., 27, 33
Clock, 6, 15-16, 19-20
communication, interpersonal: and speaking and listening, 53, 56; and eye contact, 55
compartmentalization: and causality, 67; and interpenetration, 39, 40-41; and measurement, 15; and objective time-space, 23, 24
conditions: necessary, 80-81; sufficient, 80-81
conscience: and inner conviction, 9, 240; and our deeds, 239-40; and religious revelation, 238; and suffering, 240; and superego, 232, 239
Copernicus, 18

creativity, 8, *See also* emergents; emotionality

Dante. *See* Burckhardt
das Man. See Heidegger
data of consciousness, immediately given: and interpenetration, 23-24; and precious present, 23-24
Dayan, Moshe, 212
death: and denial, postponement, sublimation, 7; and discontinuity, 29; and dread of, 7, 188, 311; and foreknowledge of, 7; and meaning of, 210; *memento mori*, 210; and metaphorical, 210, 217; and parent-suffering, 219; and phases of, 211
decision making: constructive, 97; destructive, 97-98; by Hannibal, 98; limited freedom of, 93; pathological, 97; process of, 96; of a profession, 98-100; and objective and subjective time-space, 89-90; and selfhood, 100
decompartmentalization, 39
defenses against suffering: denying, 7; forgetting, 7; postponing, 7; sublimation, 7
derivation: of causation from mental triad, 12, 252-53; of objective space from subjective space, 251-52; of objective time from precious present, 12, 247-48
de Santillana, Giorgio, 92, 111
Descartes, René, 17-18, 109, 146-48, 167, 252, 257
de Solla Price, Derek J., 16
Després, Jacqueline, 171
determinism: coexistent with indeterminism, 11; and decision making, 96, 100; and fate, 90; and individual's history, 89; and memory, 95; and mind, 70, 95; and mythology, 90; and objective time-space, 89; and suffering, 257, 294-95
Deutero-Isaiah, 193-94
devaluation of suffering, 6
de Wulf, Maurice, 109
dialogue, internal: and intra-audition, 52-53; and introspection, 52; and second self, 53

dichotomies: and life in opposite time forms, 6; and values, 10, 13
diving reflex, 78
dog, and time-space, 30, 116-17; and injured dolphin, 117; and kamikaze dolphin, 208
Domisse, Free, 234-35
Donne, John, 7, 196-97
Doob, Leonard W, 12, 23, 36
'down below' toward 'up above', 86
dualism, 3, 7, 13
Dwyer, John, 78-79

earthworm, in T-tube, 113-14
Eccles, John, 153-56, 165, 305
Eckhardt, Meister, 26
EEG (electroencephalogram), 56
Einfühlung. See empathy
Einstein, Albert, 11, 31, 229
EKG, electrocardiogram, 56
emergents: and creativity, 35, 86; versus determinism, 89; and interpenetration, 203-4; and mental triad, 35; and subjective and objective time-space, 35, 85; and suffering, 35, 203-4, 285
emotionality: and dread, 7, 188; and dynamism, 84, 224, 256; and fright, 188-89; and mental triad, 35; and rationality, 18, 72-73, 254; and tension, 76, 224; and shift from objective to subjective time-space, 39, 189; and suffering, 72-73, 225
empathy: auditory and visual, 53-54; and feeling to others, 4, 53; and second self, 53-54
Engelhardt, H. Tristram, 95, 160-63, 169, 174, 185
Epimetheus, and Pandora, 261
epistemological subject, 107-8
eternity: without beginning, without end, 25; and religion, 263-65; and soul, 305
ethics. *See* conscience
eviternity: with beginning, without end, 25; and soul in, 305-6
evolution, theory of, 110; and spatial structural coherence, 110-11
existence: continent, 7; contrasts in, 6; eternal, 246; human, 13; island, 7;
subhuman, 93; temporo-spatial, 264-65; terrestrial, 12, 265; truncated, 197-98
extrospection, 11; and measurement, 12; and natural science, 47; and outward turned observation, 49

falling away: into average man, 197-98; 258; of memories, 28; and *das Verfallen in das Man*, 258
Fate: and determinism, 90-91; and Moirae, 249; and Norns, 249-50
feedback, 77-78; and cybernetics, 77-78; and sensory motor reflex, 78
feeling, oceanic, 59; of omnipotence, 70, 82, 86
Feibleman, James, 109
Fermat, Pierre, 17
filtering out, of sensory impressions, 50-51
fish, territorial behavior of, 115
foreknowledge, 29, 87, 207, 308-9
foresight, 6, 14, 149, 172
fractures, 5, 7, 10, 12
Fraisse, Paul, 9, 31, 250
Frankl, Victor, 194-95, 204, 263, 274-75, 295, 307
Fraser, J.T.: and chaos and articulation, 138-39; and dread of death, 267; and 'nesting' of five temporal levels, 139; and subconscious and prototemporal Umwelt, 140
freedom: of decision making, 11, 92-93; limited, 100; of will, 96
freight train starting: and cause effect sequence, 82-83; and compartmented movement, 82
Freud, Sigmund, 11, 33, 124-30, 140-41, 191, 200, 218, 239
Frisch, Karl von, 115, 207
future: anticipation of, 19; calculation of objective future, 79; dread of, 8, 189; projection into subjective future, 35, 84, 189-90; and suffering, 190, 225; truncation of in neurotic illness, 231

Gajardo, Joel, 223
Galdikas, Birute, 209

Index

Galileo, and mechanistic conception of cosmos, 64
Galsworthy, John, *Forsyte Saga*, 237
gap, between primate and man, 112
Gardner, Allen, 118
Gardner, Beatrice, 118
Gay, John, 21, 92, 210
geometry: abstraction from movement in geometric point, 17; analytical, 17; geometric point, 16; plane, 16; solid, 16, straight line, 17; time immobile in geometric point, 17; time and space coordinates, 17-18, 75
Gestalt; beyond constituent parts, 38; and decompartmentalization, 39; and emergent whole, 39; and feeling, 39; and interpenetrating whole, 39
Gezelle, Guido, 13
Giacometti, Alberto, 199
Goethe, 9-10, 13, 125, 157
Goldstein, Kurt, 86
Goodall, Jane van Lawick-, 181, 208, 209
Gore, P.A., 36
Guiora, Alexander, 61

Haggin, B.H., 42
Hamlet, 269, 274
Hannibal, 98
Hardy, G. H., 192-93, 198, 296
Harlow, Harry F., 177
Harrison, Saul, 61
Hartmann, Heinz, 132-35
Harvey, William, 147-48
Hayeses, the, 118
healing: *in abstracto*, 11-12, 247-57; *in concreto*, 12, 257-58; and culture, 260-77; and growth, 246-47; and interpenetration with fractures, 244-45, 246; by physician, 199-200, 258-60; and selfhood, 285-86, 306; and suffering, 153, 282; of wound, 245-46
healing physician: and conquest of death, 258-59; of nature, 308-9; and foreknowledge, 308-9; and postponement of death, 258-59
heart attack, 78
hedonist, 300-301, 306, 310

Heidegger, Martin, 4, 29, 54, 93, 94, 197, 215, 217, 258, 266, 307
Heisenberg, Werner, and indetermancy principle, 89, 95-96
Heraclitus, 249
Herder. *See* Copernicus; Kepler
heroic lifestyle: and acceptance of suffering, 9; toward challenge of dichotomies and fractures, 291; and internal process, 289; and realization of values, 289; beyond violent deeds, 289-90
Hesiod, 261
heterogeneous discontinuum. *See* time-space subject
Heyerdahl, Thor, 19
Hinduism, 260, 261, 263
historian: and emotionality, 60; and identification, 60; and introspection, 60; and objectivity, 60
Hitler, Adolph, 194, 273
holism, and constructive suffering, 300-301. *See also* Gestalt
Homeostasis, 77, 285
Homer, 108
homogeneous continuum. *See* time-space, objective
homo ludens. *See* man
homunculi, 79, 156, 157
honey dance, 115
Huizinga, Johan, 12, 70, 210, 262, 291
Humanism: and intrinsic meaning of terrestrial existence, 38; and intrinsic meaning of values, 38; and man's ability of creation of values, 265; and rational intuitive method, 264; and systematic doubt, 257
humanistic interpretation: and irrationality within rationality, 134; and man an emotional and rational being, 18, 56, 59; man's terrestrial existence, 13, 38; and minimum of metaphysics, 38; and suffering *sub specie temporis*, 305-6
Hume, David, 65-66, 80, 149-50, 256
humor: a defence against suffering, 273; and emotional safety value, 274; and laughter, 311; and laughter and

hedonism, 275; and promise of hope, 274
Husserl, Edmund, 28-29, 270

identification: awareness of same drives in self and others, 4, 54; with living organisms, 77
immortality, metaphorical: and beautiful things, 10; not dependent on immortal soul, 38; and ethical deeds, 10; through mental triad and precious present, 38; and synthesis of two forms of time-space, 38; and true propositions, 10
immunology, causal teleological approach, 78-80
indetermancy principle, Heisenberg's, 89, 95-96
indeterminism: beyond causality, 89; and limitations of free will, 96; not as negation of determinism, 92-93; synthesis of determinism with indeterminism, 93, 95
information processing, 51
integration: mechanisms with purposiveness, 85; of two forms of time-space, 89
interpenetration: and change of elements of an entity, 40; and Gestalt, 39, 40; and holism, 84, 85, 86; and subjective time-space, 85, 86; walls between elements made porous, transparent, 42. *See also* "nesting"
intra-audition: and auditory dominance over visual, 53; and conscience, 240; and dialogue, internal, 52-53; and interpersonal communication, 53; and introspection, 52; unlike auditory sensory impressions, 52
introspection: and ancillaries and interpersonal communication, 52-53, 60; and late development of, 49-50; and intra-audition, 53; and interpenetrating whole, 48; unlike visual sensory impressions, 52
Iocaste, 90-91

Jacobs, William J., 98
James, William, 24, 27-28, 52

Janusian, 37, 69, 130, 169, 218, 222
Jehovah, 261-62
Jennings, Herbert, 113
Jesus, 19, 235, 261, 263, 272, 278, 281, 286
Job, 263
jogger, and metaphorical death, 170, 228-29
Jung, Carl, 107

Kant, Immanuel, 26, 109
kappa and tau, 23
Kastenbaum, Robert, 267
Kent, Rockwell, 9, 13
Kepler, Johannes, 18-19, 58
Kerferd, G.B., 65
Kierkegaard, Søren, 104-5, 124, 198, 224, 266, 268, 301
Kirwan, B.E., 36
knowledge: deductive, 264; beyond doubt, 264; faith, 264; inspirational, 264; intuitive, 264; propositional, 264
Kripke, Saul, 59
Kübler-Ross, Elizabeth, 211, 235, 236, 280
Kuhn, Clifford, 236
Kunstadter, Peter, 67-68, 210, 263

Laputians, and isolation, 54-55
Lawick, Baron Hugo van, 118, 181-82
Leibnitz, Gottfried Wilhelm von, 51, 290
Le Verrier, Jean Joseph, and Neptune, 58, 59
Lévy, Jaques, 58
Lewis, Sinclair, 86-87
Lidz, Theodore, 106-7, 216
light year, 18
lingua Adamica, 262
Li Tai Po, 263
locomotion, three modalities of, 73
Loewald, Hans: and interpenetration of future, past and present, 136; and superego and future, 137
Loneliness: versus aloneness, 7, 196; and connectedness and separateness, 187, 196; and marriage, 200-205; and dread of death, 4, 196; and lack of relationships, 201; and man vis-à-vis woman, 200; and religion, 206

Lorenz, Konrad, 115-16, 183, 208
Lua. *See* Kunstadter, Peter
Luganda: future tenses, far, near, 36; past tenses, far, near, 36

MacDougall, William, 150, 173
McGill, Authur, 192, 223-24
Maeterlinck, Maurice, 114-15
Mahler, Gustav, 270
man: and average man, 29; and communication, 54; and completion in religion, 263-64; and continent existence, 7; and *das Man*, 29; emotional and rational, 192; *homo faber*, 65; *homo ludens*, 70; *homo ridens*, 70; *homo sapiens*, 70. *See also* humanism; suffering
man, as an object and as a subject: anaesthetist, 180-81; physician and patient, 56-57, 260
Mann, Thomas, 238
Marek, George, 193
Maya calendar, 57
meaning, and suffering: constructive and destructive, 308, 310; from existential dread toward bliss, 307; and imperfections and fractures, 5, meaningful and meaningless, 93, 309-10; *sub specie eternitatis*, 305; *sub specie temporis*, 305
measurement, of objective time-space, 6, 19, 28, 36-37
mechanistic view of man, limits to, 112
medicine: and advent of death, 258; and constructive and destructive suffering, 12, 259; and holistic medicine, 172-73
mental triad: and act in the present, 3, 33-35; and apprehension toward the future, 34-35; and condensation of memories, 33; and determinism and indeterminism, 11, 95; and emergents, 35; and mentation, 33; and precious present, 34-35; and projection into the future fringe, 35-36; and superposition on causality, 85, 95; and synthesis of objective with subjective time-space, 39; and values, 35

Merleau-Ponty, Maurice, 20-21, 29, 48, 251
metaphysics, religious, 12-13, 278
metapsychology, Freud's: and causality, 127; and ego passive, 128; and deterministic influences, 125; and natural science, 126; and Newtonian time, 11, 126; and past determinacy, 125, 128; and reflex model of mentation, 127; and spatial schema of mentation, 126, 129; and timeless subconscious, 128, 132
Michelangelo, 235, 272
Miller, Arthur, *The Price*, 310-12
mind body relationship: and animal spirits and pineal gland, 148; and antedating, 154; body conscious matter, 162; and body divisible, 148; body and interidentification with mind, 161; body, poorer category, 160; and body in space, 145; and cerebral modules, 153-54; communication and mind, and body, 175; and descriptions, two kinds of, 159; and energy, conservation of, 173; and ghost in the machine, 159, 164; and hemisphere deconnection, 156-57; and holism and mineness, 168-69; and homunculus, 156; and interpenetration of mind body and existential tension, 166; memory not spatial; 151; mind, body incarnate, 162; mind over body 172; mind, richer category, 160; mind in time, 145; and movement and neural mechanisms, 151; and self-conscious mind, 153, 156; and synthesis of two categories, compartmentalization with interpenetration, two methods of observation, two forms of time-space, 163, 185; temporal atomicity, 146; three worlds, 153
Minkowski, Eugène, 22-23, 24, 266
miracle, and natural science, 174, 278
Moore, Henry, 272
Morgan, Charles, 106, 203, 282, 303
Muslims, 19, 260, 263
movement, and evolution: of higher animal species, purposeful, 115-24; of

inanimate matter, mechanical, 74; of lower animal species less purposeful, 113-15; of plants, mechanical, 112-13
movement, mechanical: causally structured, 71; compartmentalized, 71; emotionally neutral, 71, 76; measurable, 64; and unrolled, 74
movement, purpose-striving: emotionally charged, 71; and interpenetrating, 74; not measurable, 64; not separable from mechanical movement, 15, 63; unrolling, 74
music: auditory gestalt, 42; beauty in sadness, 211, 270; and leading tone into the future, 42; and musician and bar lines, 42; and suffering in bliss, bliss in suffering, 270; and time synthesis, 41, 42, 270; and vehicle of communication, 199
mutation, 110-11
multiple sclerosis, 171
mystery: and creation of emergents, 85, 86; and superposition of purposiveness on causality, 86, 167, 173-74; and synthesis of objective with subjective time-space, 86, 167
mythology, 90-91
myths, 8

NASA (National Space Administration), 76
Natonson, Maurice, 80
natural scientist, 14, 30-31, 47, 76, 184
Nature, animism, 86-87, 79
Nemiroff, Martin, 78
"nesting", and interpenetration of elements of an entity, 11, 95, 139
neural events: and axon, 77; and chemical processes, 77, 114; and dendrite, 77; and electrical impulses, 77, 115; and feedback, 77; and homeostasis, 77; and interneuronal gap, 77, 114; motor end plate on muscle fibers, 77, 127; and neural wiring, 115; and receptor organ, 127; and sensory-motor reflex, 127
neurophysiologist: and billiards, 77, 80; and tennis ball kicked, 68-70

Newton, Isaac: Newtonian time, 26, 126; and his statement, 293
Noguchi, on teachers, 293
Nozick, Robert, tension between determinism and free will, 100
nunc fluens, 25
nunc stans, 25-26

oarsman and watcher. *See* Allport, Gordon
Oates, Wayne, on parent-suffering, 219
Oedipus, 90-91, 222
Ogden, Charles, 307-8, 310
optimism: and pessimism, 12, 292-94; and suffering, 12, 292-94
Orr, Harry, 211-12
osteoarthritis, 171-72, 227, 303

pain: and suffering, 191; and as a warning signal, 191
Pandora, 261
Pannekoek, A., 58
pathology, of decision making, 97
Patterson, Francine, 120
Paul VI (pope), 241-42
perception, sensory: exteroceptive, 50; interoceptive, 50; proprioceptive, 50; and not of time and space, 14
philosophy, and death, 265-67, 268
physician: and compartmentalization, 56; and empathy, 260; and extrospection, 56; and identification, 260; and introspection, 56-57; and measurement, 56
physicist, and billiards, 74-77
Piaget, Jean, 12, 29-30, 66-67
Picasso, Pablo, 72, 293
picture drawing: artist and interpenetration, 72; student and compartmentalization, 72; and van Meegeren, 72
Plato, 108, 265
Poortman, J.J., 95-96
Popper, Karl, 105, 153, 154, 156
Portielje, and behavior of monkey, 118
precious present: askew structure, 35; and durational present, 24; and emotionality, 11; ephemeral, 34; future fringe, 7-8, 27, 85; interpenetration, 11, 34;

and not measurable, 35; and past fringe, 27
Premack, David, 119-20
projection: of own attributes on others, 54; of mentalistic attributes on organisms, 77, 79-80
Prometheus, 91, 262-63
psychiatrist: and empathy, identification, introspection, projection, 53-54; to meaningful from meaningless suffering, 4, 12, 125; and psychotherapy, 60-61; and transference, 60, 124-25
punctiform instant: and interval of objective time, 16; and linear time, 26, 28
purposiveness: and causality, 11, 63-64; and interpenetration, 77; not separable from causality, 63; and subjective time-space, 63, 77

realization of values, 9
reification of time and space, 27
relativity, theory of, 30-31
religion: and bliss, 12; and brevity of man's life, 261, 263; and conquest of dread of death, 260; and cosmologies of huge durations, 260; and immortal soul, 304; and relief from loneliness, 206; and sin and suffering, 5, 261-62; and suffering constitutive of being, 265; and suffering *sub specie eternitatis*, 305
Rembrandt, 289
Richards, I.A., 307-8, 310
Rickert, Heinrich, 107-8
Rilke, Rainer Maria, 282, 291
robot, and robotic cue, 76-77
Rodin, Auguste, 10, 13, 237, 272
Rückert, Friedrich, 277
Rue, Leonard Lee, 116, 208
Ryle, Gilbert, 159-60

Sabon, Michael, 214-15
saddle back, and precious present, 27
Schafer, Roy, 53-54, 293
Scheler, Max, 209-10, 222-23, 289
Schroedinger, 95-96
Schumann, Robert, 284

Seiden, Anne, 198
self: and awareness of, 102; and body image, 102; and choice, 104-5; and continuity in discontinuity, 101; death awareness and, 105; not a homunculus, 305; and memories, 101; and mental triad and precious present, 106; second, 53; and series from psychophysical person, concrete, to soul immaterial, abstract, to subject, epistemological, 102-3, 107-8
sensory motor reflex: and causality and teleology, 77-78; and cybernetics, 77
Shakespeare, William, 269, 274
Sheehan, George; 228-29
Sherover, Charles, 12, 37-38, 266-67
Sherrington, Charles, 50, 117
sin, and suffering, 5, 261-63
Snow, C. P. 193 296
Socrates, 108-9, 265, 289
solipsism: and empathy, 53; and indentification, 54; and interpersonal communication, 52, 175; and introspection, 53; and projection, 54
Solzhenitsyn, Alexandr, 272-73
Sophocles, 90-91, 222, 237
soul: and body's existence, 13; ethereal, 108; in eviternity, 25-26, 29, 103; and gravity, 108; and immaterial substance, 304; indivisible, 104; *nous*, 109; not in objective time-space, 305; and relief from dread of death, 103; *res cogitans*, 146; *res incorporea*, 109; *res spiritualis*, 109; not in subjective time-space, 305
South Sea islanders, 262
space, celestial, 18; cosmology Ptolemaic, 18; terrestrial, 18
Spitz, René, 177
space-time. *See* time-space
specious present: *See* Clay, E.N.; James, William
Sperry, R.W, 156-57
Spicker, Stuart, 148
Squalus, 295
statistical methods, 12, 83
Stent, Gunther, 50-51, 152
Stone, L. Joseph, 211

Strauss, Anselm, 216
Stuhlmueller, Carroll, 90, 193-94, 223, 247-48, 262, 276-77
suffering: and Amundsen, 44-45; and conscience, 237; constructive, 4, 224, 310; and death, 235-36; destructive, 4, 302, 311; emotional tension and future, 198-99; and growth, 246-47, 303; and happiness, 5; and healing of fractures, 5; hedonism and exclusion of, 300-301; and imperfections, 5; and interpenetration of elements, 244, 247, 285; intrinsic value of, 222; meaningful and meaningless modes of, 4, 292, 294; and sin, 5; and synthesis of fractures, 285
Suomi, Stephen J., 177
superego: and conscience, 239-40; Freudian, past-oriented, punitive, 239; and sin, 239
Swift, Jonathan, 54-55
synthesis: ideality and reality, 9, 13; of objective time-space with subjective time-space, 7, 89; of opposite aspects of the person, 8-9, 97
Syz, Hans, 55, 213-14

Teilhard de Chardin, Pierre, 14
teleology: cosmology and all existents purposeful, 65, 67; and living organism's actualization of their particular purposes, 77, 79-80; and holism, 22-23
tempus, with beginning, with end, 25
tennis ball kicked: analysis of, 68-70; and causality and holism, 69
Terrace, Herbert, 120-23
Thomas Aquinas, Saint, 25, 33, 109
thrown-into-the-world, 4, 93
time-space, objective: and before-and-after, 33-34, 244, 248; and calendar, 6, 15; and clock, 6, 15, 16-17, 19-20; and compartmented, 15, 244, 248-49; and derivation from subjective time-space, 247-51; and failure of perception of, 14; and geometric point, 16; and homogeneous continuum, 14; inseparable from movement, 15; and interval, 15-16; and line, 16; and measurability of, 6; and measuring rod, 18; and number series, 20; and physical illness, 226-27; and predictability of, 58; and punctiform instant, 15, 248-49; and spatialized time, 16, 252
time-space, subjective: not compartmentalizable, 21, 31, 35; and discontinuous, 14; and inseparable from objective time-space and movement, 15; and meaning of, 21; and meaningful directions in, 21; and not measurable, 21, 34; and projection into the future fringe, 33; and unpredictability of future in fringe, 35
Toscanini, Arturo, 42
tragedy, 6, 269, 274
trip, outward versus homeward bound: first from sign to sign, 21-22; later, through the sign, 21-22, 40
tropisms, and protozoa, 113
truth-value tests, and introspectivae data, 51-52
turgor, and nastic plant movements, 112-13
typists' errors, 39-40

values: hierarchy of, 297-99; and realization of, 9, 35, 38; and reevaluation of values, 188, 298; and suffering, 193-94
vignettes, personal: and how suffering heals, 277-85; and intra-audition and introspection, 277
Virchow, Rudolph, 111, 126

Wagner, Richard, 293
wasp, solitary ritual, 114
Weigand, Hermann, 238
Weiss, Robert, 195-96
Whitehead, Alfred North, and foresight, 6
Whittaker, Linton, 216-17
whole, mental and physical, 22
why: and inquiry into cause, 69; and inquiry into intent, 69
will, free, 93
Wimmer, Mary, 210-11

Wittgenstein, Ludwig, 85-86
workers on production line, and compartmentalization in objective time-space, 94
wound healing: and causality, 254; and compartmentalization, 245; and holism, 245; and restoration to previous condition, 245
Wright, Georg Henrik von, 66
Wundt, Wilhelm, 51, 150-51

Yerkes, Robert, 113-14, 118
yin and yang, 8, 204
yuga, and cyclical time and destruction and resurrection, 91

Zeus, 261, 262
Zorach, William, from compartmentalization to interpenetration, 40-41
Zoroastrianism, 260

About the Author

ADRIAN C. MOULYN is a psychiatrist on the Emeritus staff of Stamford Hospital in Stamford, Connecticut. His previous works include *Structure, Function, and Purpose: An Inquiry into the Concepts and Methods of Biology from the Viewpoint of Time* and numerous articles on mental health issues that have appeared in medical and scholarly journals.